Other Titles in This Series

86 **Sorin Popa,** Classification of subfactors and their endomorphisms, 1995
85 **Michio Jimbo and Tetsuji Miwa,** Algebraic analysis of solvable lattice models, 1994
84 **Hugh L. Montgomery,** Ten lectures on the interface between analytic number theory and harmonic analysis, 1994
83 **Carlos E. Kenig,** Harmonic analysis techniques for second order elliptic boundary value problems, 1994
82 **Susan Montgomery,** Hopf algebras and their actions on rings, 1993
81 **Steven G. Krantz,** Geometric analysis and function spaces, 1993
80 **Vaughan F. R. Jones,** Subfactors and knots, 1991
79 **Michael Frazier, Björn Jawerth, and Guido Weiss,** Littlewood-Paley theory and the study of function spaces, 1991
78 **Edward Formanek,** The polynomial identities and variants of $n \times n$ matrices, 1991
77 **Michael Christ,** Lectures on singular integral operators, 1990
76 **Klaus Schmidt,** Algebraic ideas in ergodic theory, 1990
75 **F. Thomas Farrell and L. Edwin Jones,** Classical aspherical manifolds, 1990
74 **Lawrence C. Evans,** Weak convergence methods for nonlinear partial differential equations, 1990
73 **Walter A. Strauss,** Nonlinear wave equations, 1989
72 **Peter Orlik,** Introduction to arrangements, 1989
71 **Harry Dym,** J contractive matrix functions, reproducing kernel Hilbert spaces and interpolation, 1989
70 **Richard F. Gundy,** Some topics in probability and analysis, 1989
69 **Frank D. Grosshans, Gian-Carlo Rota, and Joel A. Stein,** Invariant theory and superalgebras, 1987
68 **J. William Helton, Joseph A. Ball, Charles R. Johnson, and John N. Palmer,** Operator theory, analytic functions, matrices, and electrical engineering, 1987
67 **Harald Upmeier,** Jordan algebras in analysis, operator theory, and quantum mechanics, 1987
66 **G. Andrews,** q-Series: Their development and application in analysis, number theory, combinatorics, physics and computer algebra, 1986
65 **Paul H. Rabinowitz,** Minimax methods in critical point theory with applications to differential equations, 1986
64 **Donald S. Passman,** Group rings, crossed products and Galois theory, 1986
63 **Walter Rudin,** New constructions of functions holomorphic in the unit ball of C^n, 1986
62 **Béla Bollobás,** Extremal graph theory with emphasis on probabilistic methods, 1986
61 **Mogens Flensted-Jensen,** Analysis on non-Riemannian symmetric spaces, 1986
60 **Gilles Pisier,** Factorization of linear operators and geometry of Banach spaces, 1986
59 **Roger Howe and Allen Moy,** Harish-Chandra homomorphisms for $\mathfrak{p}$-adic groups, 1985
58 **H. Blaine Lawson, Jr.,** The theory of gauge fields in four dimensions, 1985
57 **Jerry L. Kazdan,** Prescribing the curvature of a Riemannian manifold, 1985
56 **Hari Bercovici, Ciprian Foiaş, and Carl Pearcy,** Dual algebras with applications to invariant subspaces and dilation theory, 1985
55 **William Arveson,** Ten lectures on operator algebras, 1984
54 **William Fulton,** Introduction to intersection theory in algebraic geometry, 1984
53 **Wilhelm Klingenberg,** Closed geodesics on Riemannian manifolds, 1983
52 **Tsit-Yuen Lam,** Orderings, valuations and quadratic forms, 1983
51 **Masamichi Takesaki,** Structure of factors and automorphism groups, 1983

(Continued in the back of this publication)

Hopf Algebras and Their Actions on Rings

Conference Board of the Mathematical Sciences

CBMS

Regional Conference Series in Mathematics

Number 82

Hopf Algebras and Their Actions on Rings

Susan Montgomery

Published for the
Conference Board of the Mathematical Sciences
by the
American Mathematical Society
Providence, Rhode Island
with support from the
National Science Foundation

Expository Lectures
from the NSF-CBMS Regional Conference
held at DePaul University, Chicago
August 10–14, 1992

Research partially supported by the
National Science Foundation Grants DMS 9114403 and DMS 9203375.

2010 *Mathematics Subject Classification.* Primary 16W30, 16S40;
Secondary 17B35, 20G99, 14L99.

Library of Congress Cataloging-in-Publication Data

Montgomery, Susan.
Hopf algebras and their actions on rings / Susan Montgomery.
p. cm. — (Regional conference series in mathematics; no. 82)
"Expanded version of ten lectures given at the CBMS Conference on Hopf Algebras and Their Actions on Rings, which took place at DePaul University in Chicago, August 10–14, 1992"–Pref.
Includes bibliographical references and index.
ISBN 0-8218-0738-2
1. Hopf algebras—Congresses. 2. Associative rings—Congresses. I. CBMS Conference on Hopf Algebras and Their Actions on Rings (1992: DePaul University) II. Title. III. Series.

QA613.8.M66 93-25786
CIP

∞ The paper used in this book is acid-free and falls within the guidelines
established to ensure permanence and durability.
This volume was printed directly from copy prepared by the author using AMS-LaTeX,
the American Mathematical Society's TeX macro system.
Visit the AMS home page at http://www.ams.org/

10 9 8 7 6 5 19 18 17 16 15 14

Contents

To my parents

Preface

These lecture notes are an expanded version of ten lectures given at the CBMS conference on Hopf Algebras and Their Actions on Rings, which took place at DePaul University in Chicago, August 10-14, 1992.

It was a very good time to have such a conference, for several reasons. The most obvious of these is the current great interest in quantum groups; these are Hopf algebras which arose in statistical mechanics and now have connections to many areas of mathematics. However there have been a number of significant recent developments within Hopf algebras themselves. Several old conjectures of Kaplansky have recently been solved, the most striking of which is a kind of Lagrange's theorem for Hopf algebras. In a different direction there has been a lot of work on actions of Hopf algebras, which unifies earlier results known for group actions, actions of Lie algebras, and graded algebras.

The object of the meeting, and of these notes, was to bring together many of these recent developments; in fact there is a great deal of interconnection between the various directions. The point of view throughout, however, is the algebraic structure of Hopf algebras and their actions and coactions. Quantum groups are treated as an important example rather than as an end in themselves; never-the-less the reader interested in quantum groups should find much basic material here.

Most of Chapters 1 and 2 is old, and in fact appears in the books on Hopf algebras by Sweedler [S] and Abe [A]; this is also true of parts of Chapters 5 and 9. I have included this material in order to be as self-contained as possible; moreover some of the arguments are new. The rest of these notes has not previously appeared in book form. Although many of the proofs are only sketched, and even occasionally omitted (with appropriate references to the literature), enough detail is given so that this book could be used for a graduate level course. In fact these notes grew out of courses I gave at USC in 1989 and in 1992. A standard first-year graduate algebra class should be a sufficient prerequisite.

There are many people I wish to thank. First of all is Jeff Bergen, who organized the conference and made all the arrangements, and second are the Supporting Lecturers: Miriam Cohen, Yukio Doi, Warren Nichols, Bodo Pareigis, Donald Passman, David Radford, Hans-Jürgen Schneider, Earl Taft, and Mitsuhiro Takeuchi. Many of their lectures are being collected and will appear in the volume [BeM 93].

In writing the notes, my deepest gratitude goes to Maria Lorenz, whose careful reading of the entire manuscript was invaluable, and to Hans Schneider, who provided many historical references as well as simplifying a number of proofs in the literature. I also want to thank Bill Chin, Davida Fischman, and Bodo Pareigis for their comments, Robert Blattner for making available to me his course notes from UCLA and for many conversations about Hopf algebras over the years, and the students in my two classes at USC whose questions on earlier versions of these notes were particlarly helpful: Ioana Boca, Paul Glezen, Michael Jochner, Horia Pop, and Yegan Satik. Finally, thanks to Lesley Newton for typing the manuscript.

Susan Montgomery
Los Angeles, June 1993

Chapter 1

Definitions and Examples

§1.1 Algebras and coalgebras

Throughout we let $\mathbf{k}$ be a field, although much of what we do is valid over any commutative ring. Tensor products are assumed to be over $\mathbf{k}$ unless stated otherwise.

We first express the associative and unit properties of an algebra via maps so that we may dualize them.

1.1.1 DEFINITION. A $\mathbf{k}$-*algebra* (with unit) is a $\mathbf{k}$-vector space A together with two $\mathbf{k}$-linear maps, multiplication $m : A \otimes A \to A$ and unit $u : \mathbf{k} \to A$, such that the following diagrams are commutative:

a) associativity

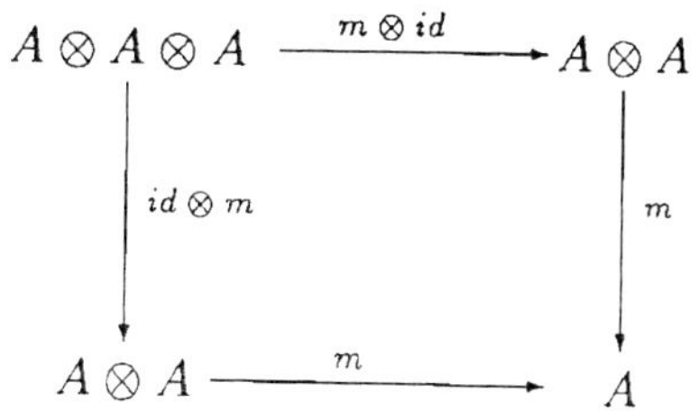

b) unit

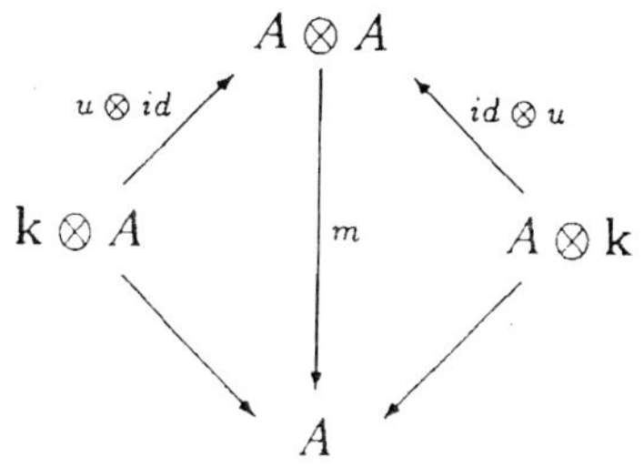

The two lower maps in b) are given by scalar multiplication. 1.1.1, b) gives the usual identity element in A by setting $1_A = u(1_{\mathbf{k}})$.

1.1.2 DEFINITION. For any $\mathbf{k}$-spaces V and W, the *twist map* $\tau : V \otimes W \to W \otimes V$ is given by $\tau(v \otimes w) = w \otimes v$.

Note that A is commutative $\Leftrightarrow m \circ \tau = m$ on $A \otimes A$.

We now dualize the notion of algebra.

1.1.3 DEFINITION. A $\mathbf{k}$-*coalgebra* (with *counit*) is a $\mathbf{k}$-vector space C together with two $\mathbf{k}$-linear maps, *comultiplication* $\Delta : C \to C \otimes C$ and *counit* $\varepsilon : C \to \mathbf{k}$, such that the following diagrams are commutative:

a) coassociativity

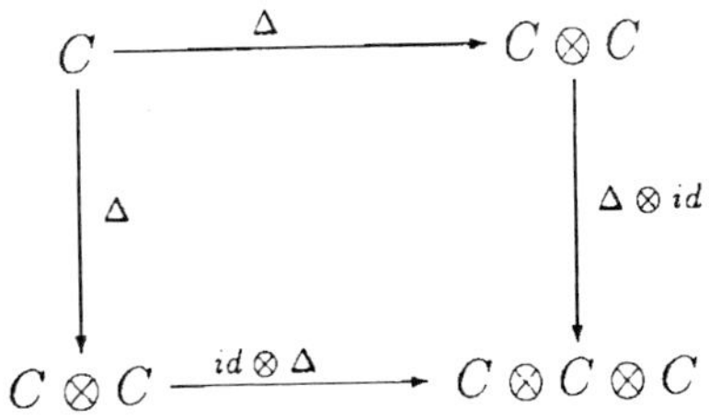

b) counit

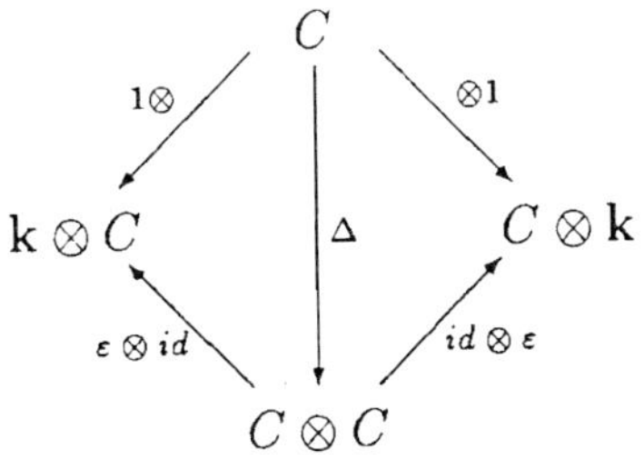

The two upper maps in 1.1.3.b) are given by $c \mapsto 1 \otimes c$ and $c \mapsto c \otimes 1$, for any $c \in C$. We say C is *cocommutative* if $\tau \circ \Delta = \Delta$.

Note that 1.1.3 b) gives that Δ is injective, just as 1.1.1 b) gives that m is surjective.

1.1.4 DEFINITION. Let C and D be coalgebras, with comultiplications Δ_C and Δ_D, and counits ε_C and ε_D, respectively.

a) A map $f : C \to D$ is a *coalgebra morphism* if $\Delta_D \circ f = (f \otimes f)\Delta_C$ and if $\varepsilon_C = \varepsilon_D \circ f$.

b) A subspace $I \subseteq C$ is a *coideal* if $\Delta I \subseteq I \otimes C + C \otimes I$ and if $\varepsilon(I) = 0$.

It is easy to check that if I is a coideal, then the $\mathbf{k}$-space C/I is a coalgebra with comultiplication induced from Δ, and conversely.

Finally, we may also use the twist map to dualize the notion of opposite algebra. For a given algebra A, recall that A^{op} is the algebra obtained by using A as a vector space, but with new multiplication $a^\circ \cdot b^\circ = (ba)^\circ$, for $a^\circ, b^\circ \in A^{op}$. In terms of maps this new multiplication is given by $m' : A \otimes A \to A$, where $m' = m \circ \tau$.

1.1.5 DEFINITION. Let C be a coalgebra. Then the *coopposite coalgebra* C^{cop} is given as follows: $C^{cop} = C$ as a vector space, with new comultiplication Δ' given by $\Delta' = \tau \circ \Delta$.

It is easy to see that C^{cop} is also a coalgebra.

§1.2. Duals of algebras and coalgebras

We shall now see that there is a very close relationship between algebras and coalgebras, by looking at their dual spaces.

For any $\mathbf{k}$-space V, let $V^* = \mathrm{Hom}_{\mathbf{k}}(V, \mathbf{k})$ denote the linear dual of V.

V and V^* determine a non-degenerate bilinear form $\langle\ ,\ \rangle : V^* \otimes V \to \mathrm{k}$ via $\langle f, v\rangle = f(v)$; we write it as a form since we frequently wish to think of V as acting on V^*. If $\phi : V \to W$ is k-linear, then the *transpose* of ϕ is $\phi^* : W^* \to V^*$, given by

$$\phi^*(f)(v) = f(\phi(v)), \tag{1.2.1}$$

for all $f \in W^*, v \in V$.

1.2.2 LEMMA. *If C is a coalgebra, then C^* is an algebra, with multiplication $m = \Delta^*$ and unit $u = \varepsilon^*$. If C is cocommutative, then C^* is commutative.*

The Lemma is proved simply by dualizing the diagrams; one needs only the additional observation that since $C^* \otimes C^* \subseteq (C \otimes C)^*$, we may restrict Δ^* to get a map $m : C^* \otimes C^* \to C^*$. Explicitly, m is given by $m(f \otimes g)(c) = \Delta^*(f \otimes g)(c) = (f \otimes g)\Delta c$, for all $f, g \in C^*, c \in C$.

If we begin with an algebra A, however, difficulties arise. For, if A is not finite-dimensional, $A^* \otimes A^*$ is a proper subspace of $(A \otimes A)^*$ and thus the image of $m^* : A^* \to (A \otimes A)^*$ may not lie in $A^* \otimes A^*$. Of course if A is finite-dimensional, all is well, and A^* is a coalgebra. For the general case, we require a definition.

1.2.3 DEFINITION. Let A be a k-algebra. The *finite dual* of A is $A^\circ = \{f \in A^* \mid f(I) = 0$, for some ideal I of A such that $\dim A/I < \infty\}$.

1.2.4 PROPOSITION. *If A is an algebra, then A° is a coalgebra, with comultiplication $\Delta = m^*$ and counit $\varepsilon = u^*$. If A is commutative, then A° is cocommutative.*

Explicity, $\Delta f(a \otimes b) = m^* f(a \otimes b) = f(ab)$, for all $f \in A^\circ, a, b \in A$.

We will prove 1.2.4 in Proposition 9.1.2. Some additional characterizations of A° will also be discussed in Chapter 9. In particular A° is the largest subspace V of A^* such that $m^*(V) \subseteq V \otimes V$.

§1.3 Bialgebras

Now we combine the notions of algebra and coalgebra.

1.3.1 DEFINITION. A k-space B is a *bialgebra* if (B, m, u) is an algebra, (B, Δ, ε) is a coalgebra, and either of the following (equivalent) conditions holds:

1) Δ and ε are algebra morphisms

2) m and u are coalgebra morphisms.

As expected, a map $f : B \to B'$ of bialgebras is called a *bialgebra morphism* if it is both an algebra and a coalgebra morphism, and a subspace $I \subseteq B$ is a *biideal* if it is both an ideal and a coideal. The quotient B/I is a bialgebra precisely when I is a biideal of B.

1.3.2 EXAMPLE. Let G be any group and let $B = \mathbf{k}G$ be its group algebra. Then B is a bialgebra via $\Delta g = g \otimes g$ and $\varepsilon(g) = 1$, for all $g \in G$.

1.3.3 EXAMPLE. Let $\mathbf{g}$ be any $\mathbf{k}$-Lie algebra and let $B = U(\mathbf{g})$ be its universal enveloping algebra. Then B becomes a bialgebra by defining $\Delta x = x \otimes 1 + 1 \otimes x$ and $\varepsilon(x) = 0$, for all $x \in \mathbf{g}$.

Note that examples 1.3.2 and 1.3.3 are cocommutative.

In any coalgebra, elements whose Δ is as in 1.3.2 or 1.3.3 are very important; thus we give them a name.

1.3.4 DEFINITION. Let C be any coalgebra, and let $c \in C$.

a) c is called *group-like* if $\Delta c = c \otimes c$ and if $\varepsilon(c) = 1$. The set of group-like elements in C is denoted by $G(C)$.

b) For $g, h \in G(C), c$ is called *g, h-primitive* if $\Delta c = c \otimes g + h \otimes c$. The set of all g, h-primitives is denoted by $P_{g,h}(C)$. If $C = B$ is a bialgebra and $g = h = 1$, then the elements of $P(B) = P_{1,1}(B)$ are simply called the *primitive* elements of B.

It is not difficult to prove that in any coalgebra, distinct group-like elements are $\mathbf{k}$-independent [S, 3.2.1], [A, 2.1.2]. As a consequence, if $B = \mathbf{k}G$, then $G(B) = G$, the original group.

If $B = U(\mathbf{g})$ and char $\mathbf{k} = 0$, then $P(B) = \mathbf{g}$, the original Lie algebra. However if char $\mathbf{k} = p \neq 0$, then $P(B)$ is the span of all $x^{p^k}, k \geq 0, x \in \mathbf{g}$; it is a restricted p-Lie algebra. See §5.5.

As another example of group-like elements, let A be any algebra and define

$$\text{Alg}(A, \mathbf{k}) = \{f \in A^* \mid f \text{ is an algebra map }\}. \tag{1.3.5}$$

In 9.1.4 we will see that $\mathrm{Alg}(A, \mathbf{k}) = G(A^\circ)$, the set of group-like elements in the coalgebra A°.

We continue with our examples of bialgebras.

1.3.6 EXAMPLE. If B is any bialgebra, then B° is a bialgebra; this is proved in §9.1. In particular, we consider the special case when $B = \mathbf{k}G$. In this case B° is called the set of *representative functions* $R_{\mathbf{k}}(G)$ on G. It can also be described as follows:

$$B^\circ = R_{\mathbf{k}}(G) = \{f \in (\mathbf{k}G)^* \,|\, \dim_{\mathbf{k}} \text{ span } \{G \cdot f\} < \infty\},$$

where G acts on $(\mathbf{k}G)^*$ via $(x \cdot f)(y) = f(yx)$, for all $x, y \in G, f \in (\mathbf{k}G)^*$. The algebra structure on B° (or on B^*) is given by

$$(fg)(x) = \Delta^*(f \otimes g)(x) = (f \otimes g)(x \otimes x) = f(x)g(x),$$

all $x \in G, f, g \in B^*$; that is, it is the usual pointwise multiplication. The coalgebra structure is given, as for any bialgebra B, by

$$\Delta f(x \otimes y) = m^* f(x \otimes y) = f(xy),$$

all $x, y \in B, f \in B^\circ$. However, this does not give an explicit formula for Δf as an element of $B^\circ \otimes B^\circ$. When B is finite-dimensional, that is $|G| < \infty$, we can give such a description, as follows:

Let $\{p_x \mid x \in G\}$ be a basis of $(\mathbf{k}G)^*$ dual to the basis of group elements in $\mathbf{k}G$; that is $p_x(y) = \delta_{x,y}$, all $x, y \in G$. Then

$$\Delta p_x = \sum_{uv=x} p_u \otimes p_v. \tag{1.3.7}$$

1.3.8 EXAMPLE. Let $B = \mathcal{O}(M_n(\mathbf{k})) = \mathbf{k}[X_{ij} \,|\, 1 \le i, j \le n]$, the polynomial functions on $n \times n$ matrices. As an algebra, B is simply the commutative polynomial ring in the n^2 indeterminates $\{X_{ij}\}$. For the coalgebra structure, think of X_{ij} as the coordinate function on the ij^{th} entry of the ring $M_n(\mathbf{k})$ of $n \times n$ matrices. Then Δ is the dual of matrix multiplication; that is $\Delta X_{ij} = \sum_{k=1}^n X_{ik} \otimes X_{kj}$. By setting $\varepsilon(X_{ij}) = \delta_{ij}$, B becomes a bialgebra.

If we let $X = [X_{ij}]$, the $n \times n$ matrix with ij^{th} entry X_{ij}, then one may check that $\det X \in G(B)$.

1.3.9 EXAMPLE. The "quantum plane". Choose $0 \neq q \in \mathbf{k}$ and let $B = \mathcal{O}(\mathbf{k}^2) = \mathbf{k}\langle x, y \mid xy = qyx \rangle$. B has a bialgebra structure given by setting $\Delta x = x \otimes x$, $\Delta y = y \otimes 1 + x \otimes y$, $\varepsilon(x) = 1$, $\varepsilon(y) = 0$. Note that $x \in G(B)$ and that $y \in P_{1,x}(B)$, the set of $1, x$ -primitive elements.

1.3.10 SOME QUANTUM GROUPS. The Appendix gives generators and relations for $U_q(\mathbf{g}), \mathbf{g}$ a finite-dimensional semi-simple Lie algebra, and for $\mathcal{O}_q(M_n(\mathbf{k}))$, and describes their coalgebra structures.

1.3.11 EXAMPLE. If B is any bialgebra, we can form a new bialgebra by taking the opposite of either the algebra or coalgebra structure. Thus B^{op} means B has the opposite multiplication but the same comultiplication, B^{cop} has the same multiplication but the opposite comultiplication, and $B^{op,cop}$ has both opposite structures.

§1.4 Convolution and summation notation

Before proceeding to the definition of Hopf algebra, we introduce another definition and a very useful notation.

1.4.1 DEFINITION. Let C be a coalgebra and A an algebra. Then $\mathrm{Hom}_\mathbf{k}(C, A)$ becomes an algebra under the *convolution product*

$$(f * g)(c) = m \circ (f \otimes g)(\Delta c)$$

for all $f, g \in \mathrm{Hom}_\mathbf{k}(C, A), c \in C$. The unit element in $\mathrm{Hom}_\mathbf{k}(C, A)$ is $u\varepsilon$.

A useful formula for $f * g$ is given in 1.4.4, below.

Note that we have already seen an example of convolution; namely, for any coalgebra C, the multiplication $m = \Delta^*$ in $C^* = \mathrm{Hom}(C, \mathbf{k})$ (see 1.2.2).

One can also define the *twist convolution* (or *anti-convolution*) product on $\mathrm{Hom}_\mathbf{k}(C, A)$ via

$$(f \times g)(c) = m \circ (f \otimes g)(\tau \circ \Delta(c)).$$

The following notation was introduced by Heyneman and Sweedler.

1.4.2 NOTATION. Let C be any coalgebra with comultiplication $\Delta : C \to C \otimes C$. The *sigma notation* for Δ is given as follows: for any $c \in C$, we write

$$\Delta c = \sum c_{(1)} \otimes c_{(2)}.$$

The subscripts (1) and (2) are symbolic, and do not indicate particular elements of C; this notation is analogous to notation used in physics (where even the $\sum$ may be omitted). In these notes we usually simplify the notation by omitting parentheses.

The power of the notation becomes apparent when Δ must be applied more than once. In particular, the coassociativity diagram 1.1.3 a) gives that $\sum c_{(1)} \otimes c_{(2)_{(1)}} \otimes c_{(2)_{(2)}} = \sum c_{(1)_{(1)}} \otimes c_{(1)_{(2)}} \otimes c_{(2)}$; this element is written as $\sum c_{(1)} \otimes c_{(2)} \otimes c_{(3)} = \Delta_2(c)$. Iterating this procedure gives

$$\Delta_{n-1}(c) = \sum c_{(1)} \otimes \ldots \otimes c_{(n)}$$

where $\Delta_{n-1}(c)$ is the (necessarily unique) element obtained by applying coassociativity $(n-1)$ times.

In this notation, the reader should check that the counit diagram 1.1.3 b) says that, for all $c \in C$,

$$c = \sum \varepsilon(c_{(1)})c_{(2)} = \sum \varepsilon(c_{(2)})c_{(1)} \tag{1.4.3}$$

and that the convolution product in 1.4.1 is given by

$$(f * g)(c) = \sum f(c_{(1)})g(c_{(2)}). \tag{1.4.4}$$

§1.5 Antipodes and Hopf algebras

1.5.1 DEFINITION. Let $(H, m, u, \Delta, \varepsilon)$ be a bialgebra. Then H is a *Hopf algebra* if there exists an element $S \in \operatorname{Hom}_{\mathbb{k}}(H, H)$ which is an inverse to id_H under convolution $*$. S is called an *antipode* for H.

Note that in sigma notation, S satisfies

$$\sum (Sh_1)h_2 = \varepsilon(h)1_H = \sum h_1(Sh_2) \tag{1.5.2}$$

for all $h \in H$.

We also have the obvious definitions of morphisms and ideals: a map $f : H \to K$ of Hopf algebras is a *Hopf morphism* if it is a bialgebra morphism and $f(S_H h) = S_K f(h)$, for all $h \in H$. A subspace I of H is a *Hopf ideal* if it is a biideal and if $SI \subseteq I$; in this situation H/I is a Hopf algebra with structure induced from H.

1.5.3 EXAMPLE. $H = \mathrm{k}G$, the group algebra, is a Hopf algebra by defining $Sg = g^{-1}$ for each $g \in G$. More generally, in any Hopf algebra H and for any $g \in G(H)$, equation 1.5.2 implies $Sg = g^{-1}$; in particular all group-like elements are invertible in H.

1.5.4 EXAMPLE. $H = U(\mathbf{g})$, the enveloping algebra, is a Hopf algebra by defining $Sx = -x$ for each $x \in \mathbf{g}$. More generally, in any Hopf algebra H and for any $x \in P(H)$, 1.5.2 implies $Sx = -x$.

The reader should check that if $x \in P_{g,h}(H)$, we must have $Sx = -h^{-1}xg^{-1}$.

1.5.5 EXAMPLE. If H is any Hopf algebra, then H° is also a Hopf algebra with antipode S^*. This is proved in 9.1.3.

1.5.6 EXAMPLE. The smallest non-commutative, non-cocommutative Hopf algebra has dimension 4, and is unique (and so self-dual) for a given $\mathbf{k}$ of characteristic $\neq 2$. It is described as follows:

$$H_4 = \mathrm{k}\langle 1, g, x, gx \mid g^2 = 1, x^2 = 0, xg = -gx\rangle$$

with coalgebra structure $\Delta g = g \otimes g$, $\Delta x = x \otimes 1 + g \otimes x$, $\varepsilon(g) = 1, \varepsilon(x) = 0, Sg = g = g^{-1}$, and $Sx = -gx$ (since $g \in G(H)$ and $x \in P_{1,g}(H)$). Note that S has order 4.

This example was first described by Sweedler. More generally, Taft has constructed an infinite family of finite-dimensional Hopf algebras with antipodes of order $2n$, for any $n > 1$ [T71].

1.5.7 EXAMPLE. The bialgebra $B = \mathcal{O}(M_n(\mathrm{k}))$ of 1.3.8 is not a Hopf algebra, since the group-like element $\det X$ is not invertible in B. However, there are several Hopf algebras closely related to B:

a) $\mathcal{O}(SL_n(\mathrm{k})) = \mathcal{O}(M_n(\mathrm{k}))/(\det X - 1)$

b) $\mathcal{O}(GL_n(\mathrm{k})) = \mathcal{O}(M_n(\mathrm{k}))[(\det X)^{-1}]$.

These bialgebras become Hopf algebras by defining $SX = X^{-1}$; that is, SX_{ij} is the ij^{th} entry of X^{-1}.

1.5.8 EXAMPLE. One can also localize the quantum plane $\mathcal{O}(\mathrm{k}^2)$ of Example 1.3.9 to obtain a Hopf algebra $H = \mathcal{O}(\mathrm{k}^2)[x^{-1}]$. The antipode is given by $Sx = x^{-1}$ and $Sy = -x^{-1}y$, as above. This example appears in [S, p.89]; note that S has infinite order.

1.5.9 EXAMPLE. The quantum groups $\mathcal{O}_q(SL_n(k))$ and $\mathcal{O}_q(GL_n(k))$ are obtained from $\mathcal{O}_q(M_n(\mathrm{k}))$ analogously to 1.5.7; see the Appendix. Their antipodes are also given there.

The following gives some important properties of the antipode [S, 4.0.1; A, p.62].

1.5.10 PROPOSITION. *Let H be a Hopf algebra with antipode S. Then*
1) S is an anti-algebra morphism; that is

$$S(hk) = S(k)S(h), \text{ all } h,k \in H, \text{ and } S(1) = 1$$

2) S is an anti-coalgebra morphism; that is

$$\Delta \circ S = \tau \circ (S \otimes S) \circ \Delta \text{ and } \varepsilon \circ S = \varepsilon.$$

In summation notation, 2) says $\sum (Sh)_1 \otimes (Sh)_2 = \sum S(h_2) \otimes S(h_1)$.

For a bialgebra B, recall that B^{cop} is the bialgebra with opposite comultiplication as in 1.3.11. It may happen that B^{cop} is a Hopf algebra, with antipode $\bar{S}$; thus $\bar{S}$ is an inverse to id_H under the twist convolution defined after 1.4.1. That is, $\sum(\bar{S}h_2)h_1 = \sum h_2(\bar{S}h_1) = \varepsilon(h)1$, all $h \in B$. By 1.5.10 applied to B^{cop}, it follows that $\bar{S}$ is also an anti-algebra and anti-coalgebra morphism. We call $\bar{S}$ a *twisted antipode* for B.

1.5.11 LEMMA. *Let B be a bialgebra. Then B is a Hopf algebra with (composition) invertible antipode $S \Leftrightarrow B^{cop}$ is a Hopf algebra with (composition) invertible antipode $\bar{S}$. In this situation, $S \circ \bar{S} = \bar{S} \circ S = id$* and so $\bar{S} = S^{-1}$.

PROOF. ($\Rightarrow$) Assume that S' is a composition inverse for S; then S' is also an anti-automorphism of B. Now for any $b \in B$,

$$\sum (S'b_2)b_1 = \sum (S'b_2)(S'Sb_1) = S'(\sum (Sb_1)b_2) = S'(\varepsilon(b)1) = \varepsilon(b)1.$$

That is, $S' \times id = u\varepsilon$. Similarly $id \times S' = u\varepsilon$, and so S' is an antipode for B^{cop}.

($\Leftarrow$) This is proved similarly.

Now assume B has both an S and an $\bar{S}$, and choose $b \in B$. Then

$$\begin{aligned} \bar{S}Sb &= \sum \bar{S}S(\varepsilon(b_2)b_1) = \sum \varepsilon(b_2)\bar{S}Sb_1 = \sum b_3(\bar{S}b_2)\bar{S}Sb_1 \\ &= \sum b_3\bar{S}((Sb_1)b_2) = \sum b_2\bar{S}(\varepsilon(b_1)) = \sum \varepsilon(b_1)b_2 = b. \end{aligned}$$

Thus $\bar{S} \circ S = id$, and similarly $S \circ \bar{S} = id$. □

1.5.12 COROLLARY. *If H is commutative or cocommutative, then $S^2 = id$.*

PROOF. In either case, clearly $\bar{S} = S$. □

Contrast 1.5.12 with Examples 1.5.6 and 1.5.8. An old question of Kaplansky asks if H is finite-dimensional and semisimple, must $S^2 = id$? This question is discussed further in Chapter 2.

§1.6. Modules and comodules

As for algebras, we consider modules before dualizing to comodules.

1.6.1 DEFINITION. For a k-algebra A, a (left) *A-module* is a k-space M with a k-linear map $\gamma : A \otimes M \to M$ such that the following diagrams commute:

a)
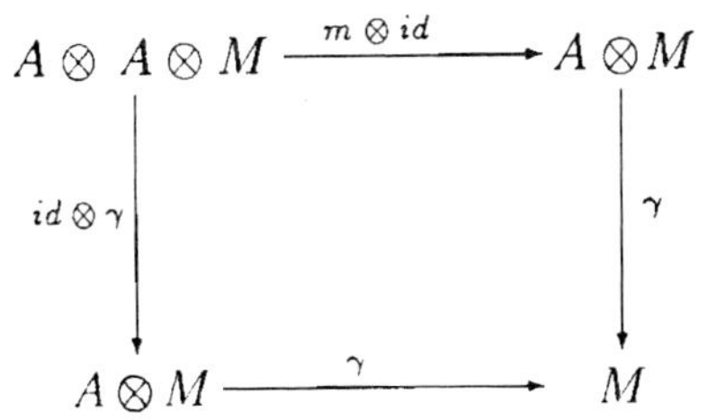

b)
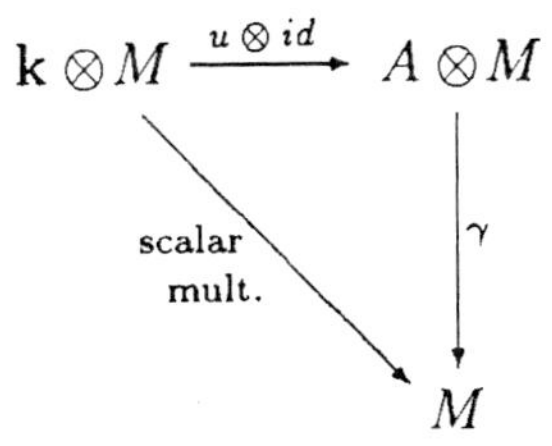

The category of left A-modules is denoted ${}_A\mathcal{M}$.

1.6.2 DEFINITION. For a k-coalgebra C, a (right) *C-comodule* is a k-space M with a k-linear map $\rho : M \to M \otimes C$ such that the following diagrams commute:

a)
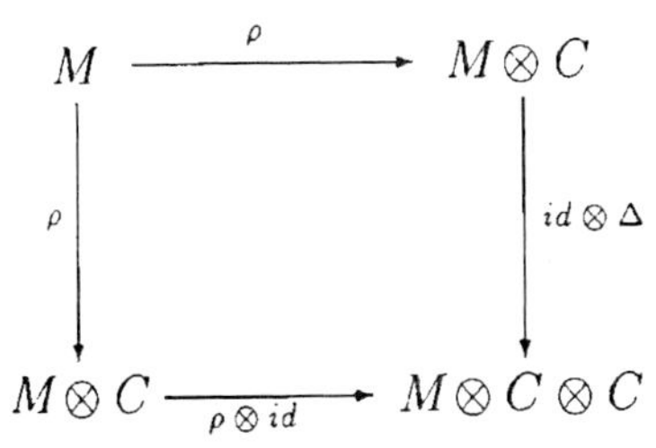

b)
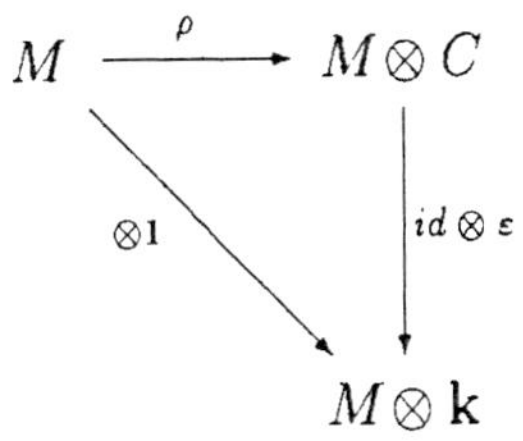

The category of right C-comodules is denoted $\mathcal{M}^C$.

There is also a summation notation for right comodules: we write $\rho(m) =$

$\sum m_{(0)} \otimes m_{(1)} = \sum m_0 \otimes m_1 \in M \otimes C$, preserving the convention that $m_{(i)} \in C$ for $i \neq 0$. Analogously one has left comodules, via some map $\rho' : M \to C \otimes M$, and we use the notation $\rho'(m) = \sum m_{(-1)} \otimes m_{(0)} = \sum m_{-1} \otimes m_0$.

1.6.3 DEFINITION. Let M and N be right C-comodules, with structure maps ρ_M and ρ_N, respectively. A map $f : M \to N$ is a *comodule morphism* if $\rho_N \circ f = (f \otimes id) \circ \rho_M$.

As in §1.2, there is a strong connection between modules and comodules.

1.6.4 LEMMA. *1) If M is a right C-comodule, then M is a left C^*-module. 2) Let M be a left A-module. Then M is a right A°-comodule $\Leftrightarrow$ $\{A \cdot m\}$ is finite-dimensional, for all $m \in M$.*

PROOF(sketch). 1) If $\rho : M \to M \otimes C$ is the comodule map, with $\rho(m) = \sum m_0 \otimes m_1$, and $f \in C^*$, then M becomes a left C^*-module via

$$f \cdot m = \sum \langle f, m_1 \rangle m_0.$$

2) ($\Leftarrow$) Choose $m \in M$ and let $\{m_1 \ldots, m_n\}$ be a basis for $A \cdot m$. Then for all $a \in A$, $a \cdot m = \sum_{i=1}^{n} f_i(a) m_i$, for some $f_i(a) \in \mathbf{k}$. Now $I = \ker(A \to \operatorname{End}_{\mathbf{k}}(A \cdot m))$ is a cofinite-dimensional ideal of A, on which f_i vanishes. Thus $f_i \in A^\circ$, for all $i = 1, \ldots, n$. Then M becomes an A°-comodule via $\rho : M \to M \otimes A^\circ, m \mapsto \sum_i m_i \otimes f_i$.

($\Rightarrow$) This follows from the formula in 1): $A \cdot m$ is spanned by the set $\{m_0\}$ in $\rho(m)$. □

The converse of 1) is false in general; that is, not all left C^*-modules are also C-comodules. A C^*-module M which becomes a C-comodule in the natural way is called *rational.* See [A], [S].

We now consider some basic examples of comodules.

1.6.5 EXAMPLE. For any coalgebra C, $M = C$ is a right comodule using $\rho = \Delta$. As in the Lemma, this determines a left action of C^* on C : for $f \in C^*, c \in C$,

$$f \rightharpoonup c = \sum \langle f, c_2 \rangle c_1.$$

This action can also be expressed in terms of right multiplication in C^*: for $f, g \in C^*$, $c \in C$,

$$\langle g, f \rightharpoonup c \rangle = \sum \langle f, c_2 \rangle \langle g, c_1 \rangle = (g * f)(c) = \langle gf, c \rangle.$$

Thus $\rightharpoonup$ equals the transpose of right multiplication of C^* on itself.

There is also a natural right action of C^* on C, given by $c \leftharpoonup f = \sum \langle f, c_1 \rangle c_2$; as before $\langle g, c \leftharpoonup f \rangle = \langle fg, c \rangle$, and thus $\leftharpoonup$ is the transpose of left multiplication of C^* on itself.

A right subcomodule D of C is a subspace such that $\Delta D \subseteq D \otimes C$; such a D is called a *right coideal* of C. Similarly a left subcomodule E is a subspace such that $\Delta E \subseteq C \otimes E$ and is called a *left coideal.*

In this case, the converse of 1.6.4, 1) is true: every left C^*-submodule D of C is a right C-subcomodule. For, choose $d \in D$ and write $\Delta d = \sum_{i=1}^{n} d_i \otimes d_i'$, where $d_1, \dots, d_m \in D$ and $d_{m+1}, \dots, d_n$ are linearly independent mod D. For any $f \in C^*$, $f \rightharpoonup d = \sum f(d_i')d_i \in D$. By the independence of $d_{m+1}, \dots, d_n$ mod D, $f(d_i') = 0$, all $i > m$. Since f was arbitrary, $d_i' = 0$, all $i > m$, and thus $\Delta d \in D \otimes C$. Thus D is a right C-comodule.

1.6.6 EXAMPLE. Analogously, beginning with an algebra A one can define a left action $\rightharpoonup$ of A on A^* which is the transpose of right multiplication on A. That is, for any $h \in A$, $f \in A^*$, $h \rightharpoonup f$ is that element of A^* such that

$$\langle h \rightharpoonup f, k \rangle = \langle f, kh \rangle$$

for all $k \in A$. If $f \in A^\circ$, then Δf makes sense, and as above we see that $h \rightharpoonup f = \sum \langle f_2, h \rangle f_1$.

We may also define $f \leftharpoonup h$ as the transpose of left multiplication by h on A.

1.6.7 EXAMPLE. Let $C = \mathrm{k}G$. Then M is a (right) $\mathrm{k}G$-comodule $\Leftrightarrow$ M is a G-graded module; that is $M = \oplus_{g \in G} M_g$.

PROOF. ($\Rightarrow$) For any $m \in M$, write $\rho(m) = \sum m_g \otimes g$. Then 1.6.2 a) gives that $(m_g)_h = \delta_{g,h} m_g$ and thus $\rho(m_g) = m_g \otimes g$. Setting $M_g = \{m_g \mid m \in M\}$, this shows the sum is direct. By 1.6.2 b), $\sum m_g = m$, and thus $\oplus M_g = M$.

($\Leftarrow$) Simply set $\rho(m) = m \otimes g$ for each $m \in M_g$. □

Now if we begin with $A = \mathrm{k}G$ and an A-module M, then by Lemma 1.6.4, M is a $(\mathrm{k}G)^\circ$-comodule $\Leftrightarrow$ $G \cdot m$ spans a finite-dimensional space, for all $m \in M$. That is, the G-action is locally finite.

§1.7 Invariants and coinvariants

We specialize in this section to a Hopf algebra H.

1.7.1 DEFINITION. 1) Let M be a left H-module. The *invariants* of H on M are the set

$$M^H = \{m \in M \mid h \cdot m = \varepsilon(h)m, \forall h \in H\}.$$

2) Let M be a right H-comodule. The *coinvariants* of H in M are the set

$$M^{coH} = \{m \in M \mid \rho(m) = m \otimes 1\}.$$

Since we eventually will also consider right modules, perhaps we should distinguish left and right invariants; in that case our notation in 1) should be ${}^HM$. However M^H is currently the standard notation for invariants, whether left or right.

The next lemma follows from the constructions in the proof of 1.6.4.

1.7.2 LEMMA. *1) Let M be a right H-comodule, and consider its left H^*-module structure. Then $M^{H^*} = M^{\mathrm{co}H}$.*

2) Let M be a left H-module such that it is also a right H°-comodule. Then $M^H = M^{\mathrm{co}H^\circ}$.

We review our basic examples:

1.7.3 EXAMPLE. Let $H = \mathrm{k}G$. If M is a left H-module, then $M^H = M^G$, the elements of M fixed by G. If M is a right H-comodule, then $M^{\mathrm{co}H} = M_1$, the identity component of the G-graded module M.

1.7.4 EXAMPLE. Let $H = U(\mathfrak{g})$. If M is a left H-module, then $M^H = \{m \in M \mid x \cdot m = 0, \text{ all } x \in \mathfrak{g}\}$, the "constants" of the action of $\mathfrak{g}$.

1.7.5 EXAMPLE. For any H, consider H as a left H-module via left multiplication. Then

$$H^H = \{t \in H \mid ht = \varepsilon(h)t, \forall\, h \in H\}.$$

The question will be considered in Chapter 2 as to when $H^H \neq 0$.

§1.8 Tensor products of H-modules and H-comodules

The action of a Hopf algebra on the tensor product of modules extends the usual diagonal action for groups.

1.8.1 DEFINITION. Let H be a Hopf algebra, and V and W left H-modules. Then $V \otimes W$ is also a left H-module, via

$$h \cdot (v \otimes w) = \sum (h_1 \cdot v) \otimes (h_2 \cdot w)$$

for all $h \in H, v \in V, w \in W$.

Similarly the tensor product of right H-modules is again a right H-module.

When H is cocommutative, then $V \otimes W \cong W \otimes V$, again extending what we know for groups. However, it is false in general that $V \otimes W \cong W \otimes V$ as H-modules. For an easy counterexample, consider $H = (\mathbf{k}G)^*$ for any finite non-abelian group G. Choose $g, h \in G$ with $gh \neq hg$, and let $V = \mathbf{k}p_g, W = \mathbf{k}p_h$. Then $p_{gh} \cdot (V \otimes W) \neq 0$ but $p_{gh} \cdot (W \otimes V) = 0$, using 1.3.7.

In Chapter 10 we will see that this symmetry property $V \otimes W \cong W \otimes V$ still holds for some of the quantum groups.

We may rewrite 1.8.1 in terms of maps. Thus if $\phi_V : H \otimes V \to V$ and $\phi_W : H \otimes W \to W$ are the two given module actions, then

$$\phi_{V\otimes W} = (\phi_V \otimes \phi_W) \circ (id \otimes \tau \otimes id) \circ (\Delta \otimes id^2) : H \otimes V \otimes W \longrightarrow V \otimes W.$$

Dualizing this definition, we obtain:

1.8.2 DEFINITION. Let H be a Hopf algebra, and V and W right H-comodules with structure maps ρ_V and ρ_W respectively. Then $V \otimes W$ is also a right H-comodule via

$$\rho_{V\otimes W} = (id \otimes m) \circ (id \otimes \tau \otimes id) \circ (\rho_V \otimes \rho_W) : V \otimes W \longrightarrow V \otimes W \otimes H.$$

In terms of elements we have $\rho(v \otimes w) = \sum v_0 \otimes w_0 \otimes v_1 w_1$.

§1.9 Hopf modules

Just as a Hopf algebra is both an algebra and a coalgebra, a Hopf module is both a module and a comodule.

1.9.1 DEFINITION. For a k-Hopf algebra H, a *right H-Hopf module* is a k-space M such that
1) M is a right H-module
2) M is a right H-comodule, via $\rho : M \to M \otimes H$
3) ρ is a right H-module map, where $M \otimes H$ is a right H-module as in 1.8.1, and where H acts on itself by right multiplication.

We may also write 3) as $\sum (m \cdot h)_0 \otimes (m \cdot h)_1 = \sum m_0 \cdot h_1 \otimes m_1 h_2$, all $m \in M, h \in H$.

More generally, in the module part of the definition we may replace H by any subHopfalgebra K of H; M then becomes a *right (H, K) - Hopf module.* The category of all right (H, K) - Hopf modules is denoted $\mathcal{M}_K^H$.

Clearly we may also replace the right action or coaction by a left action or coaction, obtaining three additional categories of (H, K) -Hopf modules: ${}^H\mathcal{M}_K$, ${}_K\mathcal{M}^H$, and ${}_K^H\mathcal{M}$.

1.9.2 EXAMPLE. For any $H, M = H$ is an H-Hopf module using $\rho = \Delta$.

1.9.3 EXAMPLE. Let W be any right H-module. Then $M = W \otimes H$ is a (right) H-Hopf module using $\rho = id \otimes \Delta$.

As a special case of this example, let W be the trivial H-module; that is, $w \cdot h = \varepsilon(h) w$, for all $w \in W, h \in H$. Let $M = W \otimes H$ as before; note that now $(w \otimes k) \cdot h = w \otimes kh$, all $w \in W, h, k \in H$. Such an M is called a *trivial Hopf module*.

Although the following theorem is proved in [A] and [S], it is so important that we give an outline of the proof .

1.9.4 FUNDAMENTAL THEOREM OF HOPF MODULES. *Let $M \in \mathcal{M}_H^H$. Then $M \cong M^{coH} \otimes H$ as right H-Hopf modules, where $M^{coH} \otimes H$ is a trivial Hopf module.*
In particular, M is a free right H-module of rank $= \dim_k M^{coH}$.

PROOF(sketch). We define $\alpha : M^{coH} \otimes H \to M$ by $m' \otimes h \mapsto m' \cdot h$ and $\beta : M \to M \otimes H$ by $m \mapsto \sum m_0 \cdot (S m_1) \otimes m_2$.

First one shows that $\beta(M) \subseteq M^{coH} \otimes H$ by showing that $\rho(\sum m_0 \cdot S m_1) = \sum m_0 \cdot S m_1 \otimes 1$; that is, $\sum m_0 \cdot (S m_1) \in M^{coH}$. This uses 1.9.1, 3). Next, one verifies that $\alpha\beta = id$ and $\beta\alpha = id$ (these are good exercises in summation notation).

Finally, check that α is a right H-comodule map, again using 1.9.1, 3); it is clear that α is a right H-module map. Thus α is an isomorphism of H-Hopf modules. $\square$

A very similar proof works for any left H-Hopf module $M \in {}_H^H\mathcal{M}$. However, for the "mixed" Hopf modules in ${}^H\mathcal{M}_H$ or ${}_H\mathcal{M}^H$, one needs that H has a twisted antipode $\bar{S}$. For example if $M \in {}_H\mathcal{M}^H$, then the map β in 1.9.4 is replaced by $\beta' : M \to M \otimes H$ given by $m \mapsto \sum(\bar{S}m_1) \cdot m_0 \otimes m_2$. Then $M \cong M^{coH} \otimes H$ as before.

1.9.5 EXAMPLE. Let $H = \mathrm{k}G$; we will describe the right $\mathrm{k}G$-Hopf modules. Thus, let M be such a Hopf module. By 1.6.7 we know that $M = \oplus_{g \in G} M_g$, a G-graded k-space, and that $\rho(m_g) = m_g \otimes g$ for $m_g \in M_g$. Also G acts on M, and since ρ is a right $\mathrm{k}G$-map, $\rho(m \cdot h) = \rho(m) \cdot h$, for all $h \in G$. That is, $\rho(m_g \cdot h) = m_g \cdot h \otimes gh$, or $m_g \cdot h = (m_g)_{gh}$, and so the G-action permutes the summands M_g. In particular, $M_1 \cdot g = M_g$. This is precisely what the Fundamental Theorem says: here $M^{coH} = M_1$, and so $M \cong M_1 \otimes \mathrm{k}G$ as $\mathrm{k}G$-Hopf modules implies $M_g \cong M_1 \otimes g$.

The reader may ask at this point why Hopf modules are useful, if in fact they are all trivial. As we shall see in Chapter 2, the difficulty will be to *prove* that a given M is an H-Hopf module, so that the Fundamental Theorem may be applied.

In Chapter 3 we will see that, more generally, any right (H, K)-Hopf module is a free right K-module provided H is finite-dimensional.

Chapter 2

Integrals and Semisimplicity

In this chapter we first show the existence of non-zero integrals in any finite-dimensional Hopf algebra, a basic fact used throughout these notes. We then use integrals to prove a version of Maschke's theorem, and to describe the structure of semisimple restricted enveloping algebras. Finally we briefly discuss cosemisimple Hopf algebras, including a dual Maschke theorem, and Kaplansky's conjecture on involutory antipodes.

We assume throughout this chapter that H is finite-dimensional, unless stated otherwise.

§2.1 Integrals

An integral in H is simply an invariant under left multiplication:

2.1.1 DEFINITION. A *left integral in* H is an element $t \in H$ such that $ht = \varepsilon(h)t$, for all $h \in H$; a *right integral in* H is an element $t' \in H$ such that $t'h = \varepsilon(h)t'$, for all $h \in H$.

$\int_H^l$ denotes the space of left integrals, and $\int_H^r$ the space of right integrals. H is called *unimodular* if $\int_H^l = \int_H^r$.

2.1.2 EXAMPLES.

1) If $H = \mathrm{k}G$, then $t = \sum_{g \in G} g$ generates the space of left and right integrals in H.
2) If $H = (\mathrm{k}G)^*$, then $t = p_1$ generates the space of left and right integrals in H.
3) The Hopf algebra $H = H_4$ of 1.5.6 is not unimodular: it is not difficult to check that $\int_H^l = \mathrm{k}(x + gx)$ but $\int_H^r = \mathrm{k}(x - gx)$.
4) Even if H is cocommutative it may not be unimodular, as the following example shows:

Assume char$\mathrm{k} = 2$ and let $\mathbf{g}$ be the 2-dimensional solvable Lie algebra; that is $\mathbf{g} = \mathrm{k}\langle x, y \mid [xy] = x \rangle$. $\mathbf{g}$ becomes a restricted 2-Lie algebra if we define $x^{[2]} = 0$ and $y^{[2]} = y$ (restricted Lie algebras will be discussed in more detail in 2.3.2). Then $H = u(\mathfrak{g})$ has basis $\{1, x, y, xy\}$. One may check that $\int_H^l = \mathbf{k}xy$ but that $\int_H^r = \mathbf{k}yx$.

Observe that in the above examples, the spaces of integrals are all one-dimensional. A result of Larson and Sweedler says that this is true in general:

2.1.3 THEOREM [LS]. *Let H be any (finite-dimensional) Hopf algebra. Then*
1) $\int_H^l$ and $\int_H^r$ are each one-dimensional
2) the antipode S of H is bijective, and $S(\int_H^\ell) = \int_H^r$
3) H is a cyclic left and right H^*-module
4) H is a Frobenius algebra.

Recall that an algebra A is a *Frobenius algebra* if there exists an associative, non-degenerate bilinear form $(\ ,\ ) : A \otimes A \to \mathbf{k}$.

Although versions of 2.1.3 are proved in [A] and [S], we sketch the proof here for the sake of completeness; it is also a nice application of the Fundamental Theorem of Hopf modules.

The main preliminary step is to show that $M = H^*$ becomes a right H-Hopf module using a particular action and coaction. First, H^* is a left H^*-module via left multiplication, and thus becomes a right H-comodule as in Lemma 1.6.4. That is, if $\{g_1 \cdots, g_n\}$ is a basis of H^* and $f \in H^*$, then there exist $h_1 \cdots, h_n \in H$ such that for any $g \in H^*, gf = \sum_i \langle g, h_i\rangle g_i$. The comodule map $\rho : H^* \to H^* \otimes H$ is then given by $\rho(f) = \sum_i g_i \otimes h_i$. Conversely if $\rho(f) = \sum f_0 \otimes f_1$, then $gf = \sum \langle g, f_1\rangle f_0$.

Second, H^* is also a right H-module via $\leftharpoonup$, as follows: if $f \in H^*$, and $h, \ell \in H$, then $\langle f \leftharpoonup h, \ell\rangle = \langle f, \ell(Sh)\rangle$. Equivalently, using 1.6.6, we have $f \leftharpoonup h = Sh \rightharpoonup f$.

2.1.4 LEMMA. *$M = H^* \in \mathcal{M}_H^H$ using ρ and $\leftharpoonup$ as above.*

PROOF. First we require the technical fact that

$$(*) \qquad g(f \leftharpoonup h) = \sum ((h_2 \rightharpoonup g)f) \leftharpoonup h_1$$

for all $f, g \in H^*$ and $h \in H$. This is a straightforward computation using the fact that H^* is an "H-module algebra" under $\rightharpoonup$ (that is, $h \rightharpoonup fg = \sum (h_1 \rightharpoonup f)(h_2 \rightharpoonup g)$; see 4.1.10).

To prove that $H^* \in \mathcal{M}_H^H$, we must show that ρ is a right H-map; that is, $\rho(f \leftharpoonup h) = \sum (f_0 \leftharpoonup h_1) \otimes f_1 h_2 = \rho(f) \cdot h$. Using the remarks before the statement of the Lemma, this is equivalent to showing that

$$g(f \leftharpoonup h) = \sum \langle g, f_1 h_2 \rangle (f_0 \leftharpoonup h_1),$$

for all $f, g \in H^*, h \in H$. Now using $(*)$,

$$\begin{aligned} g(f \leftharpoonup h) &= \sum ((h_2 \rightharpoonup g)f) \leftharpoonup h_1 \\ &= \sum [\langle h_2 \rightharpoonup g, f_1 \rangle f_0] \leftharpoonup h_1 \\ &= \sum \langle h_2 \rightharpoonup g, f_1 \rangle (f_0 \leftharpoonup h_1) \\ &= \sum \langle g, f_1 h_2 \rangle (f_0 \leftharpoonup h_1). \end{aligned}$$ □

PROOF OF 2.1.3. 1) Since $M = H^* \in \mathcal{M}_H^H$ by the Lemma, $M \cong M^{coH} \otimes H$ by the Fundamental Theorem of Hopf modules, 1.9.4. Since $\dim M^* = \dim H^* = \dim H$, it follows that $\dim M^{coH} = 1$. But by Lemma 1.7.2,

$$(H^*)^{coH} = (H^*)^{H^*} = \{f \in H^* \mid gf = \varepsilon_{H^*}(g)f, \text{ for all } g \in H^*\}.$$

Thus $\dim \int_{H^*}^l = 1$. Replacing H by H^* proves 1).
2) Choose $0 \neq f \in \int_{H^*}^l$. If $h \in \ker S$, then $\alpha(f \otimes h) = Sh \rightharpoonup f = 0$, where α is the map in the proof of the Fundamental Theorem. Since α is injective, $f \otimes h = 0$, and so $h = 0$. Thus S is injective. Since H is finite-dimensional, it is bijective. It is now clear that $S(\int_H^\ell) = \int_H^r$, and thus $\int_H^r$ is also one-dimensional.
3) Again using α and f as in 2), $f \otimes H \cong H^*$. Thus $H^* = f \leftharpoonup H = SH \rightharpoonup f = H \rightharpoonup f$, using 2) for the last equality. Dualizing gives 3).
4) Again choose $0 \neq f \in \int_{H^*}^l$ and define $(h, k) = \langle f, hk \rangle \in \mathbf{k}$. This form is clearly associative and bilinear. To see that it is non-degenerate, it suffices to prove left non-degeneracy since H is finite-dimensional. Assume that for some $a \in H, (a, H) = 0$. Then $0 = \langle f, aH \rangle = \langle H \rightharpoonup f, a \rangle = \langle H^*, a \rangle$ using 3) and the definition of $\rightharpoonup$. Now $a = 0$ by the non-degeneracy of $\langle \ , \ \rangle$. □

2.1.5 EXAMPLE. When $H = \mathbf{k}G$, the bilinear form is determined by $f = p_1 \in \int_{H^*}^l$. Thus for $v = \sum \alpha_g g, w = \sum \beta_h h \in H$,

$$(v, w) = \langle p_1, vw \rangle = \langle p_1, \sum_{g,h} \alpha_g \beta_h gh \rangle = \sum_g \alpha_g \beta_{g^{-1}} \in \mathbf{k}.$$

This form is well-known in work on group rings.

§2.2 Maschke's Theorem

Our first application of integrals is a version of Maschke's theorem for Hopf algebras, again due to Larson and Sweedler. First recall the classical result : for G a finite group, $\mathbf{k}G$ is semisimple $\Leftrightarrow |G|^{-1} \in \mathbf{k}$.

Translating this into the language of integrals, let $t = \sum_{g\in G} g \in \int_{\mathbf{k}G}$. Then $\varepsilon(t) = |G|$, and thus $|G|^{-1} \in \mathbf{k} \Leftrightarrow \varepsilon(t) \neq 0$ in $\mathbf{k}$. Restating Maschke's theorem in this way generalizes to any finite-dimensional Hopf algebra.

2.2.1 THEOREM [LS]. *Let H be any (finite-dimensional) Hopf algebra. Then H is semi-simple $\Leftrightarrow \varepsilon(\int_H^l) \neq 0 \Leftrightarrow \varepsilon(\int_H^r) \neq 0$.*

PROOF: We use the fact that H being semisimple is equivalent to every (left) H-module being completely reducible.

First, assume H is semisimple. Since Ker ε is an ideal of H, we may write $H = I \oplus$ Ker ε for some left ideal $I \neq 0$ of H, by complete reducibility of ${}_H H$. We claim that $I \subseteq \int_H^l$. Choose $z \in I, h \in H$. Since $h - \varepsilon(h)1 \in$ Ker $\varepsilon, hz = (h - \varepsilon(h))z + \varepsilon(h)z = 0 + \varepsilon(h)z$, and thus $z \in \int_H^l$. Since I is one-dimensional, we may choose $0 \neq z \in I$. Then $\varepsilon(z) \neq 0$ as $z \notin$ Ker ε, and so $\varepsilon(\int_H^l) \neq 0$. Similarly we see $\varepsilon(\int_H^r) \neq 0$.

Conversely, first assume $\varepsilon(\int_H^l) \neq 0$. Then we may choose $t \in \int_H^l$ such that $\varepsilon(t) = 1$. Let M be any (left) H-module and let N be an H-submodule. We will show that N has a complement in M. Let $\pi : M \to N$ be any $\mathbf{k}$-linear projection, and define $\tilde{\pi} : M \to N$ by

$$\tilde{\pi}(m) = \sum t_1 \cdot \pi(St_2 \cdot m)$$

for all $m \in M$. We claim that $\tilde{\pi}$ is an H-projection of M onto N.

First, for any $n \in N, \tilde{\pi}(n) = \sum t_1(St_2) \cdot n = \varepsilon(t)n = n$.

Second, $\tilde{\pi}$ is an H-module map. To see this, first note that for any $h \in H$,

$$(*) \qquad \begin{aligned} \Delta t \otimes h &= \sum \Delta(\varepsilon(h_1)t) \otimes h_2 = \sum \Delta(h_1 t) \otimes h_2 \\ &= \sum h_1 t_1 \otimes h_2 t_2 \otimes h_3. \end{aligned}$$

Then for any $m \in M, h \in H$,

$$\begin{aligned}
\tilde{\pi}(h \cdot m) &= \sum t_1 \cdot \pi((St_2) \cdot h \cdot m) \\
&= \sum h_1 t_1 \cdot \pi(S(h_2 t_2) h_3 \cdot m) \quad \text{by } (*) \\
&= \sum h_1 t_1 \cdot \pi(St_2 (Sh_2) h_3 \cdot m) \\
&= \sum h_1 t_1 \cdot \pi(\varepsilon(h_2) St_2 \cdot m) = \quad h \cdot \tilde{\pi}(m).
\end{aligned}$$

Thus $\ker \tilde{\pi}$ is an H-complement for N, and so M is a completely reducible H-module. If $\varepsilon(\int_H^r) \neq 0$, one may use a similar argument with right modules.

□

Observe that the proof of 2.2.1 is actually self-contained and did not require the use of Theorem 2.1.3: the existence of integrals follows from the complete reducibility of H.

We will see a more general version of Maschke's theorem in Chapter 7.

2.2.2 COROLLARY. *Let H be finite-dimensional and semisimple. Then*
1) H is a separable k*-algebra*
2) for any subHopfalgebra K of H such that H is free over K, K is also semisimple.

PROOF. 1) We must show that for any extension field $E \supseteq \mathbf{k}, H \otimes E$ is also semisimple. Note that $H \otimes E$ is also a Hopf algebra, via

$$\begin{aligned}
\Delta(h \otimes \alpha) &\cdot= \Delta h \otimes \alpha \in H \otimes H \otimes E \cong (H \otimes E) \otimes_E (H \otimes E) \\
\varepsilon(h \otimes \alpha) &\cdot= \varepsilon(h)\alpha \in E \\
S(h \otimes \alpha) &\cdot= Sh \otimes \alpha, \text{ all } h \in H, \alpha \in E.
\end{aligned}$$

It follows that $\int_H^l \otimes E = \int_{H \otimes E}^l$. Now apply 2.2.1 to see that $H \otimes E$ is semisimple.

2) Choose $t \in \int_H^l$ with $\varepsilon(t) \neq 0$. Since ${}_K H$ is free, we may write $t = \sum k_i h_i$, with $k_i \in K$ and $\{h_i\}$ a free basis for ${}_K H$. Then for any $k \in K$,

$$\sum (k k_i) h_i = kt = \varepsilon(k)t = \sum (\varepsilon(k) k_i) h_i.$$

By freeness, $k k_i = \varepsilon(k) k_i$, for all i. Thus $k_i \in \int_K^l$, all i.

Now $0 \neq \varepsilon(t) = \sum_i \varepsilon(k_i)\varepsilon(h_i)$ implies that $\varepsilon(k_j) \neq 0$ for some j. Thus K is semisimple by 2.2.1. □

In fact we will see in Chapter 3 that finite-dimensional Hopf algebras are always free over subHopfalgebras, and so 2.2.2, 2) always holds.

Next we make some comments on the relationship between left and right integrals. For any $0 \neq t \in \int_H^l$, also $th \in \int_H^l$ for any $h \in H$. Since $\int_H^l$ is one-dimensional, it follows that $th = \alpha(h)t$, for some $\alpha(h) \in \mathbf{k}$. Moreover, clearly $\alpha \in \mathrm{Alg}(H, \mathbf{k})$ and so is a group-like element of H^*. Finally, if we had begun with some $0 \neq t' \in \int_H^r$, then $ht' = \alpha^{-1}(h)t'$. For, since $St' \in \int_H^l$, $(St')Sh = \alpha(Sh)St'$, and so $ht' = \alpha(Sh)t'$. But $\langle \alpha, Sh \rangle = \langle S^*\alpha, h \rangle = \langle \alpha^{-1}, h \rangle$.

2.2.3 DEFINITION. The element $\alpha \in G(H^*)$ constructed above is called the *distinguished group-like element* of H^*.

Clearly H is unimodular $\Leftrightarrow \alpha = \varepsilon$.

2.2.4 COROLLARY. *If H is semisimple, then H is unimodular.*

PROOF. By 2.2.1, we may choose $t \in \int_H^l$ with $\varepsilon(t) \neq 0$. Then for any $h \in H$,

$$\alpha(h)\varepsilon(t)t = \alpha(h)t^2 = (th)t = t(ht) = \varepsilon(h)t^2 = \varepsilon(h)\varepsilon(t)t.$$

Since $\varepsilon(t) \neq 0$, we must have $\alpha(h) = \varepsilon(h)$ for all h, and so H is unimodular. □

§2.3 Commutative semisimple Hopf algebras and restricted enveloping algebras

In this section we apply the results of §2.2 to commutative semisimple Hopf algebras. An easy example of such a Hopf algebra is $H = (kG)^*$ for G a finite group. Our first result was proved independently by D. Harrison (unpublished) about 1960 (see [L 67]) and also by Cartier [Ca]. It says that this is the only example, up to extension of the base field.

2.3.1 THEOREM. *Let H be a finite-dimensional commutative semisimple Hopf algebra. Then there exists a group G and a separable extension field E of $\mathbf{k}$ such that $H \otimes E \cong (EG)^*$ as Hopf algebras.*

PROOF. By Wedderburn's theorem $H \cong \oplus_i E_i$, where the E_i are fields containing $\mathbf{k}$. Moreover each E_i is separable over $\mathbf{k}$ by 2.2.2, 1). Thus we may choose E to be a common extension field of all the E_i, which is separable over k. It follows that $H \otimes E \cong E^{(n)}$, where $n = \dim H$. Thus without loss of generality we may assume that $H \cong \mathrm{k}^{(n)}$.

Let $\{p_i \mid i = 1 \cdots, n\}$ be a basis of orthogonal idempotents in H and let $\{g_i\}$ be the corresponding dual basis in H^*. Since each $g_i \in \mathrm{Alg}\ (H, \mathrm{k})$, it follows that each $g_i \in G(H^*)$. Thus H^* has a basis of group-like elements. It follows that the set G of all g_i is a group (since distinct group-like elements are linearly independent) and thus $H^* \cong \mathrm{k}G$. □

We now specialize to the case of restricted Lie algebras and their enveloping algebras; a good reference is [J]. First we review the definition.

2.3.2 DEFINITION. Let k have characteristic $p \neq 0$. A Lie algebra $\mathfrak{g}$ over k is *restricted* if there exists a map $\mathfrak{g} \to \mathfrak{g}, x \mapsto x^{[p]}$ satisfying, for all $x, y \in \mathfrak{g}, \alpha \in \mathrm{k}$:

1) $(\alpha x)^{[p]} = \alpha^p x^{[p]}$,
2) $(x+y)^{[p]} = x^{[p]} + y^{[p]} + \sum_{i=1}^{p-1} s_i(x, y)$, where $is_i(x, y)$ is the coefficient of λ^{i-1} in $x(ad(\lambda x + y))^{p-1}$,
3) $[xy^{[p]}] = x(ad\, y)^p$.

For example, if A is any algebra over a field of char $p \neq 0$, then $\mathfrak{g} = A^-$ is restricted , using $[ab] = ab - ba$ and $a^{[p]} = a^p$, the usual p^{th} power in A.

If $\mathfrak{g}$ is restricted and $U(\mathfrak{g})$ the usual enveloping algebra, let B be the ideal in $U(\mathfrak{g})$ generated by all $x^p - x^{[p]}, x \in \mathfrak{g}$, and define $u(\mathfrak{g}) = U(\mathfrak{g})/B$. $u(\mathfrak{g})$ is called the *restricted enveloping algebra*, or *u-algebra*, of $\mathfrak{g}$. Just as for $U(\mathfrak{g}), \mathfrak{g}$ imbedds into $u(\mathfrak{g})^-$ via $x \mapsto \bar{x} = x + B$; under this imbedding, $\bar{x}^{[p]}$ is the usual p^{th} power. A version of the PBW theorem holds for $u(\mathfrak{g})$: given a basis for $\mathfrak{g}$, the ordered monomials in this basis, where the exponent of each basis element is bounded by $p - 1$, form a basis for $u(\mathfrak{g})$. Consequently if $\dim \mathfrak{g} = n$, then $\dim u(\mathfrak{g}) = p^n$.

$u(\mathfrak{g})$ becomes a Hopf algebra via the Δ, ε, and S induced from $U(\mathfrak{g})$, as one can check that B is actually a Hopf ideal .

We can now give an elementary proof, due to Chin [Ch 92], of an old theorem of Hochschild.

2.3.3 THEOREM [H 54]. *Let* $\mathfrak{g}$ *be a finite-dimensional restricted Lie algebra* *of characteristic $p \neq 0$. Then $u(\mathbf{g})$ is semisimple $\Leftrightarrow$ $\mathbf{g}$ is abelian and $\mathbf{g} = \mathbf{k}\mathbf{g}^p$.*

PROOF. Let E be the algebraic closure of $\mathbf{k}$. Since $u(\mathbf{g} \otimes E) = u(\mathbf{g}) \otimes E$ is semisimple $\Leftrightarrow$ $u(\mathbf{g})$ is semisimple, and $(\mathbf{g} \otimes E)^p = \mathbf{g} \otimes E \Leftrightarrow \mathbf{g} = \mathbf{k}\mathbf{g}^p$, it will suffice to prove that if $\mathbf{k}$ is algebraically closed, then $u(\mathbf{g})$ is semisimple $\Leftrightarrow$ $\mathbf{g}$ is abelian and $\mathbf{g} = \mathbf{g}^p$.

($\Leftarrow$) When $\mathbf{g}$ is abelian, $H = u(\mathbf{g})$ is commutative and the p-map is a semilinear transformation. Also $\mathbf{g} = \mathbf{g}^p$ implies that $H = H^p$ and so ${}^p : H \to H$ is injective. Thus H has no non-zero nilpotent elements so is semisimple.

($\Rightarrow$) Assume $u(\mathbf{g})$ is semisimple. We first show that for any $x \in \mathbf{g}, x \in \langle x \rangle^p$, where $\langle x \rangle$ denotes the restricted subalgebra of $\mathbf{g}$ generated by x. By the restricted PBW theorem, $H = u(\mathbf{g})$ is free over $K = u(\langle x \rangle)$, and thus K is semisimple by 2.2.2, 2). Thus there exists $t \in \int_K$ with $\varepsilon(t) \neq 0$.

If n is the dimension of $\langle x \rangle$, then x satisfies a polynomial of the form $0 = f(x) = \sum_{i=0}^{n} a_i x^{p^i}$; in fact f is the minimal polynomial of x in $u(\langle x \rangle)$by the restricted PBW theorem. Thus $t \in u(\langle x \rangle)$ can be written uniquely as $t = g(x) = \sum_{j=0}^{p^n-1} b_j x^j$, for $a_i, b_j \in \mathbf{k}, a_n \neq 0$. Note $0 \neq b_0 = \varepsilon(t)$. But $xt = \varepsilon(x)t = 0$ and thus $f(x)$ divides $xg(x)$. Comparing degrees implies $xg(x) = \alpha f(x)$ for some $\alpha \in \mathbf{k}$; also $a_0 \neq 0$ since $b_0 \neq 0$. Thus $x = \sum_{i=1}^{n} c_i x^{p^i}$ with $c_i = -a_i/a_0 \in \mathbf{k} = \mathbf{k}^p$, and hence $x \in \langle x \rangle^p$.

This proves that $\mathbf{g} = \mathbf{g}^p$. Moreover $a_0 \neq 0$ says that x satisfies a separable polynomial; thus $ad\, x$ satisfies a separable polynomial and hence its action on $\mathbf{g}$ is completely reducible. Let $y \in \mathbf{g}$ be an eigenvector for $ad\, x$. Then $ad\, x$ acts on the commutative ring $u(\langle y \rangle)$ and $ad\, x$ annihilates $\langle y \rangle^p$. Since $y \in \langle y \rangle^p$ by the above, $(ad\, x)y = 0$. Since $\mathbf{g}$ is spanned by eigenvectors for $ad\, x, (ad\, x)\mathbf{g} = 0$, and thus x is central. Since x was arbitrary, $\mathbf{g}$ is abelian. □

2.3.4 LEMMA. *Let $\mathbf{g}$ be a Lie algebra (respectively, restricted Lie algebra) of characteristic $p \neq 0$. If $f \in U(\mathbf{g})^\circ$ (resp., $u(\mathbf{g})^\circ$) is an algebra morphism, then $f^p = \varepsilon$.*

PROOF. Since $\{1\} \cup \mathbf{g}$ generates $H = U(\mathbf{g})$, it suffices to show that $f^p(1) = 1$ and that $f^p(x) = 0 = \varepsilon(x)$, for all $x \in \mathbf{g}$. Recall that the product in $U(\mathbf{g})^\circ$ is Δ^*, for Δ the comultiplication in H. Thus $f(1) = 1$ implies $f^n(1) = 1$, for all

$n \geq 0$. If $x \in \mathfrak{g}$, we claim that $f^n(x) = nf(x)$, for all $n \geq 1$. We proceed by induction on n. Then $f^n(x) = (f * f^{n-1})(x) = f(x)f^{n-1}(1) + f(1)f^{n-1}(x)$, since $\Delta x = x \otimes 1 + 1 \otimes x$. By induction, $f^n(x) = f(x) \cdot 1 + 1(n-1)f(x) = nf(x)$, proving the claim. But then $f^p(x) = pf(x) = 0 = \varepsilon(x)$. □

2.3.5 COROLLARY. *Let char* $\mathbf{k} = p \neq 0$, *and let* $\mathfrak{g}$ *be a restricted Lie algebra of dim* $n < \infty$ *such that* $u(\mathfrak{g})$ *is semisimple. Then for some finite separable field extension* $E \supset \mathbf{k}$, $u(\mathfrak{g}) \otimes E \cong (EG)^*$, *where* $G \cong (\mathbf{Z}_p)^n$.

PROOF. By Hochschild's theorem, $H = u(\mathfrak{g})$ is commutative. By 2.3.1, $u(\mathfrak{g}) \otimes E \cong (EG)^*$ for $E \supset \mathbf{k}$ as desired. Moreover $|G| = \dim u(\mathfrak{g}) = p^n$. By Lemma 2.3.4, since $G(H^*) = \mathrm{Alg}(H, \mathbf{k})$, every non-identity element of G has order p. Thus $G \cong \mathbf{Z}_p \times \cdots \times \mathbf{Z}_p$. □

§2.4 Cosemisimplicity and integrals on H

We discuss (without proofs) a dual version of Maschke's theorem, which holds even if H is infinite-dimensional. We require some definitions.

2.4.1 DEFINITION. Let C be any coalgebra.
1) C is *simple* if it has no proper subcoalgebras.
2) C is *cosemisimple* if it is a direct sum of simple subcoalgebras.

We will see in §5.1 that when C is finite-dimensional, it is cosemisimple $\Leftrightarrow$ C^* is a semisimple algebra. Moreover C is simple $\Leftrightarrow$ C^* is a finite-dimensional simple algebra (thus we should really call C "cosimple" to be consistent). For example, $C = \mathbf{k}G$ is cosemisimple, since each $g \in G = G(C)$ generates a one-dimensional subcoalgebra $\mathbf{k}g$, which is simple. When G is finite, $(\mathbf{k}G)^* \cong \mathbf{k}^{(n)}$, a direct sum of copies of $\mathbf{k}$, which is certainly semisimple.

2.4.2 DEFINITION. Let C be a coalgebra, and M a right C-comodule.
1) M is *simple* if it has no proper subcomodules.
2) M is *completely reducible* if it is a direct sum of simple subcomodules.

2.4.3 LEMMA. *Let C be any coalgebra. Then C is cosemisimple* $\Leftrightarrow$ *every right (left) C-comodule is completely reducible.*

The lemma is exactly [S, 14.0.1]; the proof uses 1.6.4 and 5.1.4 to reduce the question to classical results about modules. See also 5.1.6.

Now let H be any Hopf algebra. We generalize the definition of integral given in 2.1.1.

2.4.4 DEFINITION. Let H be a Hopf algebra. An element $T \in H^*$ is a *left integral on* H if for all $f \in H^*$,

$$fT = \langle f, 1_H \rangle T.$$

The space of left integrals on H will be denoted by $\mathcal{I}_H^\ell$. Right integrals are defined similarly.

Comparing with 2.1.1 when H is finite-dimensional, we see that an integral *on* H is the same as an integral *in* H^*, since $\varepsilon_{H^*}(f) = \langle f, 1_H \rangle$.

2.4.5 EXAMPLE. This example is the motivating influence for the terminology "integral". Let G be a compact topological group, and let $H = \mathcal{R}_{\mathbf{C}}(\mathcal{G})$ be the Hopf algebra of continuous complex-valued representative functions; that is, $H = \{f \in R_{\mathbf{C}}(G) | f : G \to \mathbf{C} \text{ is continuous}\}$. Let μ denote Haar measure on G, and consider the Haar integral $\int_G f(x) d\mu(x)$. It is translation-invariant in the sense that for all $y \in G$, $\int f(xy) d\mu(x) = \int f(yx) d\mu(x) = \int f(x) d\mu(x)$. It follows that the map from H to $\mathbf{G}$ given by $f \mapsto \int f(x) d\mu(x)$ is an integral in the sense of 2.4.4.

2.4.6 DUAL MASCHKE THEOREM. *If H is any Hopf algebra, then the following are equivalent:*
1) H is cosemisimple (as a coalgebra)
2) There exists a left integral T on H satisfying $\langle T, 1_H \rangle = 1$.

This theorem is [S,14.0.3] and [A, 3.3.2]. Of course when H is finite-dimensional, it is just 2.2.1. More is known: if $\mathbf{k}$ has characteristic 0 and H is commutative, then H is cosemisimple $\Leftrightarrow \mathcal{I}_H \neq 0$ [A, 3.3.11].

For an affine algebraic group G, the Hopf algebra $\mathcal{O}(G)$ of regular functions on G is cosemisimple $\Leftrightarrow$ G itself is linearly reductive; that is, its (finite-dimensional) polynomial representations are completely reducible. When $\mathbf{k}$ has characteristic 0, all the "classical" groups are linearly reductive (see [H81, Chap. V]). In particular $\mathcal{O}(GL_n(\mathbf{k}))$ as in 1.5.7 is cosemisimple in characteristic 0, and so non-zero integrals exist. It is also known that $\mathcal{O}_q(GL_n(\mathbf{k}))$ is cosemisimple; see for example [Tk 92b].

Integrals can also be described for some of the quantum groups. This is done for $\mathcal{O}_q(sl(2))$ in [MMN]; see a discussion in [Tk 92b].

§2.5 Kaplansky's conjecture and the order of the antipode

We now return to finite-dimensional Hopf algebras. By 2.1.3, we know that the antipode of such a Hopf algebra H must be injective. In fact more is known:

2.5.1 THEOREM [R 76]. *Let H be any finite-dimensional Hopf algebra. Then the antipode S has finite order.*

Moreover, S^4 has a very special form. Let $\alpha \in H^*$ be the distinguished group-like element of H^*, as in 2.2.3; analogously, there exists a distinguished group-like element $a \in H$ coming from the left integrals in H^*. Now $ad\,\alpha \in$ Aut H^*, and so $(ad\,\alpha)^* \in$ Aut H. Then it is proved in [R 76] that $S^{-4} = (ad\,a) \circ (ad\,\alpha)^*$.

As noted in 1.5.6, examples exist in which the order of S is $2n$, for any $n > 1$, although if H is commutative or cocommutative, then $S^2 = id$ by 1.5.12. In 1975, Kaplansky conjectured that the antipode of any finite-dimensional cosemisimple Hopf algebra has order 2 [K]. Equivalently one could ask the question for semisimple Hopf algebras, since $S^2 = id_H \Leftrightarrow (S^*)^2 = id_{H^*}$. In two recent papers, Larson and Radford solve this conjecture in characteristic 0. They prove:

2.5.2 THEOREM [LR 88]. *If char* $\mathrm{k} = 0$, *then finite-dimensional cosemisimple Hopf algebras are semisimple.*

2.5.3 THEOREM [LR 87]. *Let H be a finite-dimensional semisimple and cosemisimple Hopf algebra over a field* k *of characteristic* 0 *or of characteristic $p > (\dim H)^2$. Then $S^2 = id$.*

The methods of proof include a detailed study of the trace function Tr of the left regular represention L of H^*. For example, Tr defines a left integral $t \in H$ by $\langle f, t\rangle = Tr(L(f) \circ (S^*)^2)$ for all $f \in H^*$, and $t \neq 0 \Leftrightarrow H$ is cosemisimple.

Kaplansky's conjecture remains open in characteristic p.

Chapter 3

Freeness Over Subalgebras

The main object of this chapter is to prove the important theorem of Nichols and Zoeller, that a finite-dimensional Hopf algebra H is free over any subHopfalgebra K; this is a kind of "Lagrange's Theorem" for Hopf algebras. In addition recent work of Schneider actually strengthens the Nichols - Zoeller theorem by showing that a very nice basis exists for H over K.

Next we consider adjoint actions and normal subHopf algebras; when H is faithfully flat over K we can give another characterization of normal. We also briefly discuss ideals and quotients.

In the infinite-dimensional case, it has been known for a long time that H need not be free over K [OSch 74], although it is true in various special cases. We review some of these results at the end of the chapter, and consider the more general question of the faithful flatness of H over K.

§3.1. The Nichols - Zoeller theorem

We follow fairly closely the proof in [NZ 89]. We first consider an arbitrary (finite-dimensional) Frobenius algebra A. The (right) *principal indecomposable modules* of A are those which appear in A itself considered as a right A-module; call them $P_1, \ldots, P_t$.

Our first lemma is well-known; 1) is [CuR, 59.3] and 2) is an easy consequence of 1).

3.1.1 LEMMA. *Let A be a Frobenius algebra, and let W be a finitely-generated right A-module. Then*

1) W is faithful $\Leftrightarrow$ each P_i appears as a summand of W

2) there exists a positive integer r such that $W^{(r)} \cong F \oplus E$ as A-modules, where F is free and E is not faithful.

Before proceeding, we recall that a k-algebra A is called *augmented* if it has a non-zero algebra map $\varepsilon : A \to \mathrm{k}$. In particular, any Hopf algebra is an augmented algebra.

The next proposition was pointed out to us by H.-J. Schneider; it simplfies the [NZ]-argument.

3.1.2 PROPOSITION. *Let K be a finite-dimensional augmented algebra and W a finitely-generated right K-module. If there exists $r > 0$ such that $W^{(r)}$ is free over K, then W is free.*

PROOF. By assumption $W^{(r)} \cong K^{(n)}$ for some $n \geq 1$. Since K is augmented, via ε, $\mathbf{k}$ becomes a left K-module via $k \cdot \alpha = \varepsilon(k)\alpha$, for all $k \in K, \alpha \in \mathbf{k}$. It follows that

$$(W \otimes_K \mathbf{k})^{(r)} \cong (K \otimes_K \mathbf{k})^{(n)} \cong \mathbf{k}^{(n)}.$$

Thus $n = rt$, where $t = \dim_{\mathbf{k}}(W \otimes_K \mathbf{k})$, and so $W^{(r)} \cong K^{(n)} \cong (K^{(t)})^{(r)}$. Now by the Krull-Schmidt theorem, $W \cong K^{(t)}$, a free K-module. □

3.1.3. THEOREM. *Let K be a finite-dimensional Hopf algebra and W a finitely-generated right K-module. Suppose there exists a finitely-generated faithful right K-module V such that $W \otimes V \cong W^{(\dim V)}$ as right K-modules. Then W is free over K.*

PROOF. By Theorem 2.1.3, K is a Frobenius algebra. Thus by 3.1.1, there exists $r > 0$ such that $W^{(r)} \cong F \oplus E, F$ free, E not faithful. Thus $W^{(r)} \otimes V \cong (W \otimes V)^{(r)} \cong W^{(r \dim V)}$. By 3.1.2, it suffices to show $W^{(r)}$ is free. Thus we may replace W by $W^{(r)}$ and assume that $W \cong F \oplus E$. Setting $t = \dim V$, our hypothesis gives

$$(*) \qquad F^{(t)} \oplus E^{(t)} \cong W^{(t)} \cong W \otimes V \cong (F \otimes V) \oplus (E \otimes V).$$

Also by 3.1.1, $V^{(s)} \cong F' \oplus E'$, for some s, and we may replace V by $V^{(s)}$. Thus we may assume $V \cong F' \oplus E', F'$ free, E' not faithful. Note that $F' \neq (0)$ since V is faithful.

Now consider $K \otimes V$; it is a right K-module via the diagonal action and a left K-comodule via $\Delta \otimes id$, and moreover is in ${}^K\mathcal{M}_K$ as in 1.9.3. Thus it is free over K by the Fundamental Theorem of Hopf modules, 1.9.4, and so $K \otimes V \cong K^{(t)}$. Consequently since F is a free K-module, $F \otimes V \cong F^{(t)}$. Using this fact and the Krull-Schmidt theorem in $(*)$ gives

$$(**) \qquad E^{(t)} \cong E \otimes V \cong E \otimes (F' \oplus E') \cong (E \otimes F') \oplus (E \otimes E').$$

Now if $E \neq 0$, then $E \otimes K \in \mathcal{M}_K^K$ as before (using $id \otimes \Delta$) and so it is free over K. Since F' is a free K-module, it follows that $E \otimes F'$ is also free over K. But then using $(**)$, $E^{(t)}$, and so E, is a faithful K-module, a contradiction. Thus $E = 0$ and $W \cong F$, free. □

3.1.4. LEMMA. *Let H be a Hopf algebra with bijective antipode, let K be a subHopfalgebra, and let $M \in \mathcal{M}_K^H$. Then $M \otimes H \cong M^{(\dim H)}$ as right K-modules.*

PROOF. Let H_0 denote H considered as a trivial right K-module. Since $M \otimes H_0 \cong M^{(\dim H)}$, it will suffice to prove $M \otimes H \cong M \otimes H_0$ as right K-modules. Let $\rho : M \to M \otimes H$ denote the comodule structure map.

Define $\phi : M \otimes H_0 \to M \otimes H$ by $m \otimes h \mapsto (1 \otimes h)\rho(m) = \sum m_0 \otimes hm_1$. Then ϕ is a right K-module map, since ρ is a right K-map. Also, define $\psi : M \otimes H \to M \otimes H_0$ by $m \otimes h \mapsto \sum m_0 \otimes h\bar{S}m_1$, where $\bar{S}$ is the inverse of S. Then it is easy to see that ϕ and ψ are inverses, and so ϕ is an isomorphism. □

3.1.5. THEOREM. *Let H be a finite-dimensional Hopf algebra and let K be a subHopfalgebra. Then every right (H, K)-Hopf module is free as a right K-module. In particular H is free as a right K-module.*

PROOF. First assume that $M \in \mathcal{M}_K^H$ is finitely generated. Since the antipode of H is bijective, 3.1.4 implies that $M \otimes H \cong M^{(\dim H)}$ as right K-modules. We may then apply 3.1.3 with $H = V$ to see that M is free. The general case, of any $M \in \mathcal{M}_K^H$, follows from the next lemma:

3.1.6. LEMMA. *Let H be a bialgebra and K a finite-dimensional subbialgebra. If every finite-dimensional $M \in \mathcal{M}_K^H$ is free as a right K-module, then every $M \in \mathcal{M}_K^H$ is free as a right K-module.*

PROOF. We follow the proof of [R 77, Prop. 1]. Choose $0 \neq M \in \mathcal{M}_K^H$. Call a subset $L \subseteq M$ a *partial basis* if L is a basis for LK as a K-module and LK is an H-subcomodule of M. Let $\mathcal{L}$ denote the set of partial bases of M. By convention we assume that (0) is free on the empty set; thus $\mathcal{L} \neq \phi$. If $\mathcal{L}$ is ordered by inclusion, Zorn's lemma applies, and thus there exists a maximal partial basis L. We claim that $LK = M$.

If $LK \neq M$, consider $M' = M/LK \neq 0$ and let $\pi : M \to M'$ be the natural projection. Since M' is an H-comodule, it contains a non-zero finite-dimensional subcomodule V' by 5.1.1. Then $V = \pi^{-1}(V')$ is a subcomodule of M, LK is a proper subset of VK, and $VK/LK \cong V'K$. Since $V'K \in \mathcal{M}_K^H$ and is finite-dimensional as a K-module, it is free over K by hypotheses. Thus L can be extended to a larger partial basis, a contradiction. Therefore $LK = M$, and M is free. □

§3.2 Applications: Hopf algebras of prime dimension and semisimple subHopfalgebras

We give a few immediate consequences of 3.1.5. First is the more familiar statement of Lagrange's theorem:

3.2.1 COROLLARY. *If H is a finite-dimensional Hopf algebra and K a subHopfalgebra, then* $\dim K$ *divides* $\dim H$. *In particular if* $\dim H = p$, *a prime, then H has no proper subHopfalgebras.*

The corollary provides some positive evidence for the following conjecture, which has been open since 1974 [K].

3.2.2 CONJECTURE. Let k be algebraically closed of char 0, and let H be a Hopf algebra of dimension p, a prime. Then $H \cong \mathrm{k}\mathbf{Z}_p$, the group algebra over $\mathbf{Z}_p$.

It suffices to show that H (or H^*) contains a non-trivial group-like element g; for then $K = \mathrm{k}\langle g\rangle = H$ by 3.2.1. If no such elements exist, then $\alpha = \varepsilon$, where α is the distinguished group-like element as in 2.2.3. Thus one may assume that both H and H^* are unimodular.

ADDED IN PROOF: Zhu [Zh] has just announced a proof of the conjecture.

We can improve 2.2.2.

3.2.3. COROLLARY. *If H is a finite-dimensional semisimple Hopf algebra, then every subHopfalgebra is also semisimple.*

PROOF. 2.2.2 and 3.1.5. □

It is false, however, that subHopfalgebras of unimodular Hopf algebras are unimodular; in fact any finite-dimensional Hopf algebra may be imbedded in a unimodular one by 10.3.7 and 10.3.12.

3.3. A normal basis for H over K

Observe that 3.1.5 does not recover the full strength of Lagrange's theorem, for in that case, when L is a subgroup of the group G, $\mathbf{k}G$ has as a basis over $\mathbf{k}L$ a set of left (or right) coset representatives of L in G. Let $\{x_1 \dots, x_n\}$ be such a set of left coset representatives. Then the images $\bar{x}_1, \dots, \bar{x}_n$ of the x_i form a basis of the quotient space $\mathbf{k}G/(\mathbf{k}L)^+G$, where for any Hopf algebra H, we define $H^+ = H \cap (\text{Ker } \varepsilon)$. Since $(\mathbf{k}L)^+G$ is a subcoalgebra of $\mathbf{k}G$, in fact $\mathbf{k}G/(\mathbf{k}L)^+G$ is a coalgebra; it is also a trivial left L-module. Thus for the case of a group algebra $\mathbf{k}G$, we obtain

$$\mathbf{k}G \cong \mathbf{k}L \otimes \mathbf{k}G/(\mathbf{k}L)^+G$$

as left $\mathbf{k}L$-modules and as right $\mathbf{k}G/(\mathbf{k}L)^+G$-comodules. In fact this formulation extends to arbitrary finite-dimensional Hopf algebras, as shown by Schneider:

3.3.1 THEOREM [Sch 92]. *Let $K \subset H$ be finite-dimensional Hopf algebras. Then*
1) $H \cong K \otimes H/K^+H$ as left K-modules and right H/K^+H-comodules
2) $H \cong H/HK^+ \otimes K$ as right K-modules and left H/HK^+-comodules.

We describe the conclusions of the theorem by saying that H has a right (or left) *normal basis* over K; we will consider normal bases in more detail in Chapter 8. In fact Schneider's result is a corollary to a more general result about Galois extensions and crossed products; see Theorem 8.4.6 and Corollary 8.4.7. His proof also uses 3.1.5 itself.

Masuoka has extended 3.1.5 to the more general situation in which K is only assumed to be a right coideal subalgebra of H; that is, K is a subalgebra of H and $\Delta K \subseteq K \otimes H$. He proves in that setting that H is left and right free over $K \Leftrightarrow K$ is a Frobenius algebra $\Leftrightarrow$ every object in $\mathcal{M}_K^H$ or in ${}_K\mathcal{M}^H$ is a free K-module [Ma 92] (the first of these equivalences was also shown by K. Hoffman, unpublished). Later Masuoka and Doi [MaD] and Koppinen [Ko 93] showed that the above conditions are equivalent to the existence of a normal basis as above; that is, $H \cong H/HK^+ \otimes K$.

3.4. The adjoint action, normal subHopfalgebras, and quotients

In this section we do not assume that H is finite-dimensional.

3.4.1 DEFINITION. Let H be any Hopf algebra.

1) The *left adjoint action* of H on itself is given by

$$(ad_l h)(k) = \sum h_1 k (S h_2),$$

for all $h, k \in H$.

2) The *right adjoint action* of H on itself is given by

$$(ad_r h)(k) = \sum (S h_1) k h_2,$$

for all $h, k \in H$.

3) A subHopfalgebra K of H is called *normal* if both

$$(ad_l H)(K) \subseteq K \text{ and } (ad_r H)(K) \subseteq K.$$

In the case of $H = \mathbf{k}G$ and $g \in G$, then $(ad_l g)(k) = gkg^{-1}$, all $k \in \mathbf{k}G$, and if $H = U(\mathfrak{g})$ and $x \in \mathfrak{g}$, then $(ad_l x)(k) = xk - kx$, all $k \in U(\mathfrak{g})$. Thus in these cases we get the usual classical adjoint actions.

For normal subgroups of groups, there are several other characterizations of normal, such as the fact that every left coset is a right coset, or that kernels of homomorphisms are normal subgroups. We consider the analogs of these questions for Hopf algebras.

First we need some notation. Let $\pi : H \to \bar{H}$ be a morphism of Hopf algebras; then H becomes both a right and a left $\bar{H}$-comodule, via $\rho = (id \otimes \pi) \circ \Delta$ and $\rho' = (\pi \otimes id) \circ \Delta$ respectively. Let $H^{co\bar{H}}$ denote the coinvariants for ρ and ${}^{co\bar{H}}H$ the coinvariants for ρ'.

3.4.2 LEMMA. *Let K be a subHopfalgebra of H.*

1) If K is normal, then $HK^+ = K^+H$, $I = HK^+$ is a Hopf ideal of H, and $\pi : H \to H/HK^+$ is a morphism of Hopf algebras.

2) Let $\pi : H \to \bar{H}$ be a morphism of Hopf algebras, and consider H as a right and left $\bar{H}$-comodule as above. Then $H^{co\bar{H}}$ is ad_ℓ-stable and ${}^{co\bar{H}}H$ is ad_r-stable.

PROOF. Consider the identity

$$(*) \qquad ha = \sum h_1 a\varepsilon(h_2) = \sum h_1 a(Sh_2)h_3 = \sum (ad_\ell h_1)(a)h_2.$$

If K is normal and $a \in K$, then $(ad_l h_1)(a) \in K$ and so $ha \in KH$. Moreover if $\varepsilon(a) = 0$, then $\varepsilon((ad_l h_1)(a)) = 0$ and thus $HK^+ \subseteq K^+H$. The other containment follows using $ad_r H$. It follows that $I = HK^+$ is an ideal (it is always a coideal) and $SI = I$. Thus I is a Hopf ideal, and π a Hopf morphism, proving 1).

2) Choose $a \in H^{co\bar{H}}$, so that $\rho(a) = \sum a_1 \otimes \bar{a}_2 = a \otimes \bar{1}$. For any $h \in H$,

$$\begin{aligned}
\rho((ad_l h)a) &= \rho(\sum h_1 a(Sh_2)) = \sum (h_1)_1 a_1 (Sh_2)_1 \otimes (\bar{h}_1)_2 \bar{a}_2 (S\bar{h}_2)_2 \\
&= \sum h_1 a(Sh_4) \otimes \bar{h}_2(S\bar{h}_3) = \sum h_1 \varepsilon(h_2) a(Sh_3) \otimes \bar{1} \\
&= (ad_l h)(a) \otimes \bar{1}.
\end{aligned}$$

Thus $(ad_l h)(a) \in H^{co\bar{H}}$. The argument is similar on the left. □

The converse of 3.4.2, 1) is open in general. However it is true for "nice" extensions. Before proving this, we review some definitions. Recall that a ring extension $A \subset B$ is (left) *faithfully flat* if for any right A-module map $f : M \to N, f$ is injective $\Leftrightarrow f \otimes id_B : M \otimes_A B \to N \otimes_A B$ is injective. See [W, Chap. 13] for a discussion of faithful flatness.

Given two maps $f, g : M \to N$, the *equalizer* of f and g is $\ker(f,g) = \{m \in M | f(m) = g(m)\}$. The equalizer diagram $L \xrightarrow{h} M \overset{f}{\underset{g}{\rightrightarrows}} N$ is *exact* if $\operatorname{Im} h = \ker(f,g)$ and h is injective.

We can now prove our converse. We follow the argument in [Sch 92, 1.2 and 1.3].

3.4.3 PROPOSITION. *Let K be a subHopfalgebra of H such that H is left or right faithfully flat over K, and such that $HK^+ = K^+H$. Let $\bar{H} = H/HK^+$ and consider H as a right and left $\bar{H}$ comodule as before. Then*
1) $K = H^{co\bar{H}} = {}^{co\bar{H}}H$
2) K is a normal subHopfalgebra of H.

PROOF. The proof proceeds by comparing two equalizer diagrams. First, the diagram $K \subset H \overset{f}{\underset{g}{\rightrightarrows}} H \otimes_K H$, where $f(h) = h \otimes 1$ and $g(h) = 1 \otimes h$, is exact by faithful flatness [W, 13.1 Theorem]. Now $HK^+ = K^+H$ implies that $\bar{H} = H/HK^+$ is a Hopf algebra and thus the natural map $H \to \bar{H}$ is a morphism of Hopf algebras. Then the diagram $H^{co\bar{H}} \subset H \rightrightarrows H \otimes \bar{H}$, where the two maps on the right are given by $h \mapsto h \otimes \bar{1}$ and $h \mapsto \sum h_1 \otimes \bar{h}_2$, is exact by the definition of $\bar{H}$-coinvariants.

Finally we tie these two diagrams together. Define a map

$$\beta : H \otimes_K H \to H \otimes \bar{H}, \text{ via } x \otimes y \mapsto \sum xy_1 \otimes \bar{y}_2$$

(this is the Galois map studied in Chapter 8). β has a well-defined inverse, namely $x \otimes \bar{y} \mapsto \sum x(Sy_1) \otimes y_2$, and thus β is bijective. It is also easy to check that $K \subseteq H^{co\bar{H}}$. Thus we have a commutative diagram

$$\begin{array}{ccccc} K & \subset & H & \rightrightarrows & H \otimes_K H \\ i\downarrow & & \| & & \downarrow\beta \\ H^{co\overline{H}} & \subset & H & \rightrightarrows & H \otimes \overline{H} \end{array}$$

By exactness and the bijectivity of β, we must have $K = H^{co\bar{H}}$.

Using $\bar{H} \otimes H$ and repeating the argument, we obtain that $K = {}^{co\bar{H}}H$. Now by Lemma 3.4.2, part 2), it follows that K is normal. □

3.4.4 COROLLARY. *Let H be finite-dimensional and K a subHopfalgebra. Then K is normal $\Leftrightarrow$ $HK^+ = K^+H$.*

PROOF. By the Nichols - Zoeller theorem, H is free over K, so certainly faithfully flat. □

The question as to when H is faithfully flat over K will be discussed in the next section.

A related question, as to when Hopf ideals are of the form $HK^+ = K^+H$ for some normal subHopf algebra K of H, is more difficult. We first need the dual of 3.4.1.

3.4.5 DEFINITON. Let H be any Hopf algebra.

1) The *left adjoint coaction* of H on itself is given by

$$\rho_\ell : H \to H \otimes H \text{ via } h \mapsto \sum h_1 Sh_3 \otimes h_2$$

2) The *right adjoint coaction* of H on itself is given by

$$\rho_r : H \longrightarrow H \otimes H \text{ via } h \mapsto \sum h_2 \otimes (Sh_1)h_3$$

3) A Hopf ideal I of H is called *normal* if both

$$\rho_\ell(I) \subseteq H \otimes I \text{ and } \rho_r(I) \subseteq I \otimes H$$

(that is, I is a subcomodule of H under ρ_ℓ and ρ_r). If I is normal, the canonical surjection $\pi : H \to H/I$ is called *conormal.*

To see that this definition is indeed the dual of 3.4.1, note that $\rho_\ell : H \to H \otimes H$ can be written as

$$\rho_\ell = (m \otimes id) \circ (id \otimes \tau) \circ (id^2 \otimes S) \circ \Delta_2.$$

Considering $ad_\ell : H \otimes H \to H$ as a left action via $h \otimes k \mapsto \sum h_1 k S h_2$, we see that

$$ad_\ell = m^2 \circ (id^2 \otimes S) \circ (id \otimes \tau) \circ (\Delta \otimes id).$$

Thus ad_ℓ and ρ_ℓ are formal duals; the same is true for ad_r and ρ_r.

Moreover, a conormal surjection $\pi : H \to H/I$ is the formal dual of a normal subHopfalgebra $i : K \hookrightarrow H$. For, $\rho_\ell(I) \subseteq H \otimes I$ means that $(id \otimes \pi) \circ \rho_\ell : H \to H \otimes H/I$ factors through $\pi : H \to H/I$. Dualizing gives that $ad_\ell \circ (id \otimes i) : H \otimes K \to H$ factors through $i : K \hookrightarrow H$. That is, K is ad_ℓ-stable. Thus if H is finite-dimensional, a Hopf ideal I of H is normal $\Leftrightarrow (H/I)^*$ is a normal subHopfalgebra of H^*.

Of course if H is cocommutative, I is always normal.

Normal Hopf ideals also arise in the context of affine k-groups. For, if G is an affine group scheme (that is, a functor from commutative k-algebras to the category of groups which is represented as a set-valued functor), and $N \subset G$ a closed subgroup scheme of G represented by the surjective map $\pi : H \to H/I$ of commutative Hopf algebras, then N is a normal subgroup scheme (that is, $N(A)$ is a normal subgroup of $G(A)$ for all commutative k-algebras A) $\Leftrightarrow I \subset H$ is a normal Hopf ideal [DG, Tk 72, W].

The question on Hopf ideals can now be formulated more precisely : let H be any Hopf algebra. When are

$$\left\{ K \,\middle|\, \begin{matrix} K \text{ a normal subHopf-} \\ \text{algebra of } H \end{matrix} \right\} \underset{\psi}{\overset{\phi}{\rightleftarrows}} \left\{ I \,\middle|\, \begin{matrix} I \text{ a normal Hopf} \\ \text{ideal of } H \end{matrix} \right\},$$

where $\phi(K) = HK^+$ and $\psi(I) = {}^{coH/I}H$, inverse bijections?

Both maps ϕ and ψ are well-defined, as is noted in [Sch 93b]. In general, ϕ and ψ induce inverse bijections when restricted to those K such that H is right faithfully flat over K (respectively, to those I such that H is an injective right H/I-comodule) [Sch 93b, depending on Tk 79]. It is not known whether these additional conditions on K and I are always satisfied; they are satisfied in the situation of the following basic theorem (see also §3.5).

3.4.6 THEOREM. *Let H be any Hopf algebra. Then ϕ and ψ are inverse bijections if either H is commutative or if the coradical H_0 of H is cocommutative.*

PROOF. If H is commutative this is the fundamental theorem on affine **k**-groups [DG, III, §3, no. 7] or [W, 16.3]; a purely algebraic proof was given in [Tk 72]. For H cocommutative the result is shown in [Ga, 4.2] in the language of schemes; if the coradical H_0 of H is cocommutative it is shown in [Ma 91].

□

Here the *coradical* H_0 of H is just the sum of the simple subcoalgebras of H; it will be discussed in more detail in §5.1.

More generally, one may ask when ϕ and ψ give inverse bijections between non-normal objects. In [Ma 91] it is shown that when H_0 is cocommutative, ϕ and ψ give a correspondence between the right coideal subalgebras K of H such that $G(K \otimes \bar{\mathbf{k}})$ is a group (here $\bar{\mathbf{k}}$ is the algebraic closure of $\mathbf{k}$) and the coideals I of H which are also left ideals. This generalizes an old result of [Ne] for cocommutative H, in which case the K's as above are simply the subHopfalgebras of H. An easier proof of [Ne], which in fact extends to the situation when only H_0 is cocommutative, is given in [Sch 90a].

§3.5 Freeness and faithful flatness in the infinite-dimensional case

In this last section we briefly discuss some of the main results when H is infinite-dimensional. First, H is not always free over K, even if H is both commutative and cocommutative, by an example of Oberst and Schneider.

Before giving their example, we need a lemma.

3.5.1 LEMMA. *Let $\mathbf{k} \subset E$ be a finite Galois field extension with Galois group*

G, and let H be a Hopf algebra over E. Assume that G acts on H as semi-linear automorphisms (that is, $g\cdot(\alpha h) = (g\cdot\alpha)(g\cdot h)$ for $g \in G, \alpha \in E, h \in H$). Then H^G is a Hopf algebra over $\mathbf{k}$.

PROOF (sketch). First, $H \cong H^G \otimes_{\mathbf{k}} E$ as algebras by a standard result on descent theory for algebras; see, for example, [J, Chap. 10, Sec. 2]. Next, using the fact that the action of G commutes with Δ_H and that $H \otimes_E H \cong (H^G \otimes_{\mathbf{k}} H^G) \otimes_{\mathbf{k}} E$ as vector spaces, one sees that Δ_H induces a comultiplication on H^G [RTW, Prop. 1]. □

3.5.2 EXAMPLE [OSch 74]. Let $F \subset E$ be a Galois field extension of degree 2, with Galois group $G = \{1, \sigma\}$. Let σ act on $\mathbf{Z}$ by $z \mapsto -z$. Then G acts on the group algebra $E\mathbf{Z}$ by acting on both E and $\mathbf{Z}$. Let $H = (E\mathbf{Z})^G$ and $K = (E(n\mathbf{Z}))^G \subset H$. If n is even, then H is not free over K.

PROOF. H and K are F-Hopf algebras by the Lemma. Writing $\mathbf{Z}$ multiplicatively as $\{x^m \mid m \in \mathbf{Z}\}$, we have $H = (E[x, x^{-1}])^G$ and $K = (E[x^n, x^{-n}])^G$. Now $\{1, x, \ldots, x^{n-1}\}$ is an $E[x^n, x^{-n}]$-basis of $E[x, x^{-1}]$. Assume that there exists a K-basis $\{a_0, \ldots, a_{n-1}\}$ of H. Then there exists $A \in GL_n(E[x^n, x^{-n}])$ such that $(a_0, \ldots, a_{n-1}) = (1, x^{-1}, \ldots, x^{n-1})A$. Applying σ, we see

$$\begin{aligned}(a_0, \ldots, a_{n-1}) &= (1, x^{-1}, \ldots, x^{-(n-1)})\sigma(A) \\ &= (1, x, \ldots, x^{n-1}) \begin{bmatrix} 1 & 0 & \ldots & 0 \\ 0 & 0 & \ldots & x^{-n} \\ \vdots & & & \vdots \\ 0 & x^{-n} & \ldots & 0 \end{bmatrix} \sigma(A).\end{aligned}$$

Thus $\det A = \pm x^{-n(n-1)}\sigma(\det A)$. Thus $ax^{nk} = \pm x^{-n(n-1)}\sigma(a)x^{-nk}$ for some $0 \neq a \in E, k \in \mathbf{Z}$. But then $k = -(n-1) - k$, a contradiction. □

Positive results can be obtained by assuming some special property of the subHopfalgebra K. We mention two recent ones:

3.5.3 THEOREM. *H is free over the finite-dimensional subHopfalgebra K if either*

1) K is semisimple [NZ 92], or

2) *K is normal in H* [Sch 93b].

The situation for faithful flatness may be more interesting at this point, since it is open in general, and is almost as good a property as freeness in many applications. We formally ask :

3.5.4 QUESTION. Is H always left and right faithfully flat over any subHopf-algebra K?

It is known to be true if H is commutative [DG, see W, 14.1, or Tk 72] or if the coradical H_0 of H is cocommutative [Tk 72].

For other recent work concerning faithful flatness, see [Sch 92] and [MaW].

Chapter 4

Actions of Finite-Dimensional Hopf Algebras and Smash Products

In this chapter we study actions of a Hopf algebra H on a $\mathbf{k}$-algebra A, with the object of trying to extend to Hopf algebras facts which are known for actions of finite groups. In particular we are interested in classical questions such as the integrality of A over the subalgebra of invariants A^H, the finite generation of A^H, and the A^H-module structure of A. The question of when $A^H \subset A$ is a Galois extension will be considered in Chapter 8.

Most of the results we present here depend in one way or another on the smash (or semi-direct) product $A\#H$; this is the generalization for Hopf algebras of the skew group ring $A * G$, which has proved very useful in studying group actions. The close relationship between A^H and $A\#H$ can be expressed in terms of a Morita context.

§4.1 Module algebras, comodule algebras, and smash products

4.1.1 DEFINITION. An algebra A is a *(left) H-module algebra* if
1) A is a (left) H-module, via $h \otimes a \mapsto h \cdot a$
2) $h \cdot (ab) = \sum (h_1 \cdot a)(h_2 \cdot b)$
3) $h \cdot 1_A = \varepsilon(h) 1_A$

for all $h \in H, a, b \in A$.

We say that H *measures* A if only 2) and 3) are satisfied.

If we write the multiplication and unit of A as $m_A : A \otimes A \to A$ and $u_A : \mathbf{k} \to A$ as in 1.1.1, then 2) and 3) say that m_A and u_A are H-module maps, using 1.8.1. This enables us to dualize 4.1.1.

4.1.2 DEFINITION. An algebra A is a *(right) H-comodule algebra* if
1) A is a (right) H-comodule, via $\rho : A \to A \otimes H$
2) m_A and u_A are right H-comodule maps.

More concretely, this says that $\rho(ab) = \sum a_0 b_0 \otimes a_1 b_1$, all $a, b \in A$, and that $\rho(1) = 1 \otimes 1$, using 1.8.2.

When H is finite-dimensional, it follows from the constructions in 1.6.4 that A is a left H-module algebra $\Leftrightarrow$ A is a right H^*-comodule algebra.

4.1.3 DEFINITION. Let A be a left H-module algebra. Then the *smash product algebra* $A\#H$ is defined as follows, for all $a, b \in A, h, k \in H$:
1) as k-spaces, $A\#H = A \otimes H$. We write $a\#h$ for the element $a \otimes h$
2) multiplication is given by

$$(a\#h)(b\#k) = \sum a(h_1 \cdot b)\#h_2 k. \tag{4.1.4}$$

It is easy to see that $A \cong A\#1$ and $H \cong 1\#H$; for this reason we frequently abbreviate the element $a\#h$ by ah. In this notation, we may write $ha = \sum (h_1 \cdot a)h_2$ using (4.1.4).

4.1.5 EXAMPLE. For any H and any A, we have the trivial action $h \cdot a = \varepsilon(h)a$, all $h \in H, a \in A$. In this case $A\#H \cong A \otimes H$ as algebras.

4.1.6 EXAMPLE. Let $H = \mathrm{k}G$, and let A be an H-module algebra. Since $\Delta g = g \otimes g$ for $g \in G$, 4.1.1, 2) implies that $g \cdot (ab) = (g \cdot a)(g \cdot b)$ for all $a, b \in A$, and thus g acts as an endomorphism of A. In addition, 4.1.1, 1) implies each g acts as an automorphism of A, since $g^{-1}g = 1$ (more generally, for any H and any $g \in G(H), g$ must act as an automorphism of A). Thus we have a (group) homomorphism $G \to \mathrm{Aut}_{\mathrm{k}} A$. Conversely any such map makes A into a kG-module algebra.

In this case $A\#\mathrm{k}G = A * G$, the skew group ring; here multiplication is just $(ag)(bh) = a(g \cdot b)gh$, for all $a, b \in A, g, h \in G$.

4.1.7 EXAMPLE. Again let $H = \mathrm{k}G$, but now let A be a kG-comodule algebra. We already know from 1.6.7 that $A = \oplus_{g \in G} A_g$, a G-graded vector space, and that for $a_g \in A_g, \rho(a_g) = a_g \otimes g$. Thus $\rho(a_g b_h) = a_g b_h \otimes gh$; that is, $a_g b_h \in A_{gh}$. Thus $A_g A_h \subseteq A_{gh}$, all $g, h \in G$, and also $1_A \in A_1$. Thus A is a G-graded algebra.

When G is finite, A is G-graded (a kG-comodule algebra) $\Leftrightarrow$ A is a $(\mathrm{k}G)^*$-module algebra. It is easy to see here that $p_g \cdot A = A_g$; that is, the $\{p_g\}$ act as projections. Using 1.3.7, we see that multiplication in $A\#(\mathrm{k}G)^*$ is given by

$$(a\#p_g)(b\#p_h) = \sum_{uv=g} a(p_u \cdot b)\#p_v p_h = ab_{gh^{-1}}\#p_h.$$

4.1.8 EXAMPLE. For any H, consider $x \in P(H)$, the primitive elements, and let A be an H-module algebra. Since $\Delta x = x \otimes 1 + 1 \otimes x$, $x \cdot (ab) = (x \cdot a)b + a(x \cdot b)$; that is, x acts as a k-derivation of A and we have a Lie homomorphism $P(H) \to \mathrm{Der}_{\mathrm{k}} A$.

More generally, consider $x \in P_{g,h}(H)$, where $g, h \in G(H)$. Since $\Delta x = x \otimes g + h \otimes x$, $x \cdot (ab) = (x \cdot a)(g \cdot b) + (h \cdot a)(x \cdot b)$. We say that x acts as a *g, h-derivation* of A; it is also called a *skew derivation.*

When $H = U(\mathbf{g})$, the action of $\mathbf{g}$ determines that of $U(\mathbf{g})$. In this case $A\#H$ is sometimes called the differential polynomial ring. In particular when $\mathbf{g} = \mathrm{k}x$, the one-dimensional Lie algebra, and x acts as a derivation δ of A, then $A\#U(\mathbf{g}) = A[x;\delta]$, the usual Ore extension in which $xa = ax + \delta(a)$. A special case of this construction is the first Weyl algebra $\mathbf{A}_1$. In this case $A = \mathrm{k}[y]$, the polynomial ring, and $\delta = \frac{d}{dy}$. Then $\mathbf{A}_1 = \mathrm{k}[y][x;\delta] = \mathrm{k}\langle\, x, y \mid xy - yx = 1 \rangle$.

4.1.9 EXAMPLE. For any H, H itself is an H-module algebra for both the left and right adjoint actions (see 3.4.1). ad_l and ad_r are examples of inner actions; see Chapter 6.

For either adjoint action $H\#H \cong H \otimes H$, even though the action may be nontrivial. We return to this fact in 7.3.3.

4.1.10 EXAMPLE. The action of H on H^* given in 1.6.6 gives H^* the structure of an H-module algebra (this fact has already been used in 2.1.4). For, recall that for $h \in H$ and $f \in H^*, h \rightharpoonup f$ was given by $\langle\, h \rightharpoonup f, l \rangle = \langle\, f, lh \rangle$, for all $l \in H$. Thus if also $g \in H^*$,

$$\begin{aligned}\langle\, h \rightharpoonup (fg), l \rangle &= \langle\, fg, lh \rangle = \sum \langle\, f, l_1 h_1 \rangle \langle\, g, l_2 h_2 \rangle \\ &= \sum \langle\, h_1 \rightharpoonup f, l_1 \rangle \langle\, h_2 \rightharpoonup g, l_2 \rangle = \sum \langle\, (h_1 \rightharpoonup f)(h_2 \rightharpoonup g), l \rangle.\end{aligned}$$

It follows that $h \rightharpoonup (fg) = \sum (h_1 \rightharpoonup f)(h_2 \rightharpoonup g)$, proving 4.1.1, 2).

Similarly, H is an H^0-module algebra under $\rightharpoonup$ as in 1.6.5. When H is finite-dimensional, we will see in 9.4.3 that $H\#H^* \cong \mathrm{End}_{\mathrm{k}}(H)$; moreover from 8.5.2, ${}_{H\#H^*}\mathcal{M}$ may be identified with the Hopf module category ${}_H\mathcal{M}^H$. Recently $H\#H^*$ has been called the *Heisenberg double*, denoted $\mathcal{H}(H)$, in the quantum groups literature.

4.1.11 EXAMPLE. We give a newer example using quantum groups. When

$H = U_q(sl(2))$ and $A = \mathbf{C}[X]$, it is shown in [MSm] that A is an H-module algebra in an essentially unique way. From the relations for $U_q(sl(2))$ given in the Appendix, we note that $K \in G(H)$ and that $E, F \in P_{K^{-1},K}(H)$. By studying the possible skew derivations of $\mathbf{C}[X]$, the following is proved: up to isomorphism of module algebras and automorphisms of $U_q(sl(2))$, there are two possibilities for the action of H on $\mathbf{C}[X]$:

(1) K acts as $\sigma : X \mapsto q^{-2}X$, $E = \delta\sigma^{-1}$, and $F = -q^{-4}X^2\delta\sigma^{-1}$, or

(2) K acts as $\sigma : X \mapsto q^2X$, $E = -q^4X^2\delta\sigma^{-1}$, and $F = \delta\sigma^{-1}$,

where δ is the $1,\sigma$-derivation determined by $X \mapsto 1$.

§4.2 Integrality and affine invariants: the commutative case

In this section we prove a generalization of the well-known fact that if a finite group G acts on a commutative ring A, then A is integral over A^G.

4.2.1 THEOREM. *Let H be finite-dimensional cocommutative and let A be a commutative H-module algebra. Then A is integral over A^H.*

The theorem is a consequence of an old result of Grothendieck on quotients [DGa, p.309]; a direct proof for the special case of 4.2.1 has been pointed out to us by H.-J. Schneider. We present here instead a recent proof of W. Ferrer-Santos which uses the exterior algebra, although we have changed the coaction arguments in [F-S] to action arguments. As a preliminary step, we prove:

4.2.2 PROPOSITION. *Let $M \in {}_{A\#H}\mathcal{M}$ be free of finite rank as a left A-module, and let $f : M \to M$ be an $A\#H$-module map. Then χ_f, the characteristic polynomial of f in $End_A(M)$, has coefficients in A^H.*

PROOF. Let n be the A-rank of M. We first consider the case when $n = 1$. Thus, we may write $M = Am_0$, some $m_0 \in M$. Then for f as above, $f(m_0) = am_0$, some $a \in A$, and thus $f(m) = am$ for any $m \in M$ since A is commutative. In fact $a = \det f$. We claim that $a \in A^H$.

Now for any $h \in H$ and $m \in M$, $f(h \cdot m) = h \cdot f(m)$, and so $a(h \cdot m) =$

$h \cdot (am) = \sum(h_1 \cdot a)(h_2 \cdot m)$. But then

$$\begin{aligned}(h \cdot a)m &= \sum(h_1 \cdot a)\varepsilon(h_2)m = \sum(h_1 \cdot a)(h_2 Sh_3 \cdot m) \\ &= \sum(h_{11} \cdot a)(h_{12} \cdot (Sh_2 \cdot m)) = \sum h_1 \cdot (a(Sh_2 \cdot m)) \\ &= \sum a\, h_1 \cdot (Sh_2 \cdot m) \qquad \text{by the above} \\ &= \varepsilon(h)am.\end{aligned}$$

Using $m = m_0$ and the freeness of M, we see $h \cdot a = \varepsilon(h)a$, for all $h \in H$. Thus $a \in A^H$, proving the case $n = 1$.

We proceed to the general case. Consider the nth exterior power $N = \wedge_A^n M$ of M over A. Recall that $N = T_A^n M/I$, the quotient of the nth tensor power of M by the submodule I generated by the symmetric tensors $\{m \otimes m \mid m \in M\}$. f induces the A-module map $\wedge^n f : N \to N$, with the same determinant as f, and N is free of rank 1 as an A-module. Moreover, since H is cocommutative it stabilizes I; thus there is an induced action of H (and so of $A\#H$) on N, and $\wedge^n f$ is an $A\#H$-map. We may thus apply the $n = 1$ case to see that $\det \wedge^n f = \det f \in A^H$.

Let t be an indeterminate. Then $A[t]$ and $M[t]$ become $A\#H$-modules by letting H act trivially on t, and f extends to an $A\#H$-map $\tilde{f} : M[t] \to M[t]$. But now also $t \cdot id - \tilde{f}$ is an $A\#H$-map, and thus $(\det t \cdot id - \tilde{f}) = \chi_f(t) \in A[t]^H = A^H[t]$ by the above arguments. □

Note that the proof of the Proposition did not require H to be finite-dimensional.

PROOF OF 4.2.1. Let $M = A\#H$; M is a free left A-module of rank $n = \dim H$. Let $f = r_a$, right multiplication by $a \in A$; f is a left $A\#H$-map. Thus by 4.2.2, χ_{r_a} has coefficients in A^{H^a}. Since $\chi_{r_a}(a) = 0$, a is integral over A^H. □

4.2.1 can be applied to the question of finite generation.

4.2.3 DEFINITION. A k-algebra is called k-*affine* if it is finitely generated as a k-algebra.

4.2.4 ARTIN-TATE LEMMA. *Let $B \subset A$ be commutative* k-*algebras. If A*

is k*-affine and integral over* B*, then* B *is* k*-affine.*

The lemma is classical. It is the remaining fact we need to prove a generalization of E. Noether's classical theorem on affine invariants for group actions.

4.2.5 THEOREM [F-S]. *Let* A *be a commutative algebra which is* k*-affine, and which is an* H*-module algebra for* H *a finite-dimensional cocommutative Hopf algebra. Then* A^H *is* k*-affine.*

PROOF. 4.2.1 and 4.2.4. □

Note that cocommutativity was used only in the proof of 4.2.2. Thus we ask:

4.2.6 QUESTION. If A is a commutative k-algebra and H any finite dimensional Hopf algebra such that A is an H-module algebra, is A integral over A^H?

§4.3 Trace functions and affine invariants: the non-commutative case

If the algebra A is non-commutative, there is little hope of imitating the arguments in §4.2, since integrality fails much of the time, even for actions of finite groups. If we use a non-commutative definition of integrality due to Schelter, then some positive results are known: a theorem of Quinn says that if a finite group G acts on a k-algebra A such that $|G|^{-1} \in$ k, then A is Schelter-integral over A^G [Q 89]. In general an extension $R \subset S$ is *Schelter-integral* if each $s \in S$ satisfies some monic polynomial $\rho(x)$ in the free product $R \coprod \mathrm{k}[x]$. That is, $\rho(x) = x^n + q(x)$, where the total x-degree of each monomial in q is less than n. Thus we ask:

4.3.1 QUESTION. If A is an H-module algebra and H is finite-dimensional and semisimple, is A Schelter-integral over A^H?

Some partial results are known on this problem; see [C 86], [Q 91].

Fortunately it is possible to obtain some results on affine invariants without using integrality. However, we do have to generalize the notion of trace for a group action. Recall that if a finite group G acts on an algebra A,

$$tr : A \to A^G \qquad \text{is given by} \qquad a \mapsto \sum_{g \in G} g \cdot a.$$

Note that we can rephrase this by writing $tr(a) = t \cdot a$, where $t = \sum g$ is an integral in kG. This formulation generalizes:

4.3.2 LEMMA. *Let H be finite-dimensional acting on A and choose $0 \neq t \in \int_H^l$. Then the map $\hat{t} : A \to A$ given by $\hat{t}(a) = t \cdot a$ is an A^H-bimodule map with values in A^H.*

The proof of the lemma is straightforward.

4.3.3 DEFINITION. A map $\hat{t} : A \to A^H$ as in 4.3.2 is called a (left) *trace function* for H on A.

Our next lemma comes from [CFM]; it generalizes a well-known fact for group actions.

4.3.4 LEMMA. *Let H be finite-dimensional acting on A and assume that $\hat{t} : A \to A^H$ is surjective. Then there exists a non-zero idempotent $e \in A\#H$ such that $e(A\#H)e = A^H e \cong A^H$ as algebras.*

PROOF. First, we check that for $h \in H, a \in A$, and $t \in \int_H^l$, $hat = (h \cdot a)t$ in $A\#H$. For, $(1\#h)(a\#t) = \sum (h_1 \cdot a)\#h_2 t = \sum (h_1 \cdot a)\#\varepsilon(h_2)t = (h \cdot a)\#t$. Now since $\hat{t}$ is surjective, there exists $c \in A$ with $t \cdot c = 1$. Define $e = tc$; then $e^2 = tctc = (t \cdot c)tc = tc = e$.

For any $a \in A, h \in H$, we then have

$$e(ah)e = tcahtc = \varepsilon(h)tcatc = \varepsilon(h)\, t \cdot (ca)tc \in A^H e.$$

Conversely if $a \in A^H$, then $t \cdot (ca) = (t \cdot c)a = a$, and thus $ae = atc = tcatc$. That is, $e(A\#H)e = A^H e$. Finally, this is algebra-isomorphic to A^H, since $(ae)(be) = atcbtc = abe$ as above. □

4.3.5 COROLLARY. *Let H be finite-dimensional acting on A and assume that $\hat{t}$ is surjective. If A is left or right Noetherian, then so is A^H.*

PROOF. If A is left Noetherian, then clearly so is $A\#H$; on the right, this follows from 4.4.3, 1) or 7.2.11. Now it is true for any Noetherian ring S

that eSe is Noetherian. Alternatively, we could show directly that A^H is Noetherian by blowing up a chain of (left) ideals in A^H to A, using the fact that A is Noetherian to get a finite chain, and then applying $\hat{t}$ to recover the original chain, which is thus finite. □

We note that the Corollary is false if $\hat{t}$ is not surjective, even for group actions. The next proposition from [MS 81] was used for group actions.

4.3.6 PROPOSITION. *Let S be a* k*-algebra and e a non-zero idempotent in S. If S is* k*-affine and (left) Noetherian, then eSe is* k*-affine.*

PROOF. Since S is left Noetherian, SeS is a finitely-generated left ideal of S. We claim that eS is (left) finite over eSe. For, write $SeS = \sum_{i=1}^{n} Sx_i$, and write $x_i = \sum_{i,j} v_{ij}ew_{ij}$. Choose $r \in S$. Then $er \in e(SeS)$, and so $er = e(\sum_i s_i x_i) = \sum es_i v_{ij} ew_{ij}$. Thus the set $\{ew_{ij}\}$ generates eS as an eSe-module, proving the claim. For simplicity, rewrite the generators as $\{ew_i\}$.

Now since S is k-affine, it is generated by some finite set $\{t_j\}$. Write $et_j = \sum_i ey_{ij}ew_i$ and $ew_k t_j = \sum_i ez_{ijk}ew_i$. Then the set $\{ew_ie, ey_{ij}e, ez_{ijk}e\}$ generates eSe as a k-algebra. For, every element of eSe is a linear combination of terms of the form $et_{j_1}t_{j_2}\ldots t_{j_m}e$, and it is not difficult to see that every such term may be expressed in terms of the given generators. For example,

$$et_1t_2e = (\sum_i ey_{i1}ew_i)t_2e = \sum_i ey_{i1}e(\sum_m ez_{m2i}ew_m)e. \qquad \square$$

We can now generalize [MS 81, Theorem 1].

4.3.7 THEOREM. *Let A be a left Noetherian ring which is an affine* k*-algebra, let H be finite-dimensional, and assume that A is an H-module algebra such that the trace $\hat{t} : A \to A^H$ is surjective. Then A^H is* k*-affine.*

PROOF. 4.3.4 and 4.3.6, using that $S = A\#H$ is left Noetherian since it is a finite A-module. □

Contrasting this result with 4.2.5, we see that we do not require that H be cocommutative, although we require that $\hat{t}$ be surjective and that A be Noetherian (this last hypothesis is automatic in the commutative case, of course, by the Hilbert Basis Theorem). Examples are given in [MS 81] which show that 4.3.7 can fail for group actions if A is not Noetherian or if $\hat{t}$ is not surjective. Note that it is always true that $\hat{t}$ is surjective if H is semisimple. For then, we may choose $t \in \int_H^l$ with $\varepsilon(t) = 1$. It follows that $\hat{t}(1) = t \cdot 1 = 1$, and so $\hat{t}(A) = A^H$.

The dual notion of "surjective trace" is also important. We first need a definition due to Doi [D 85].

4.3.8 DEFINITION. Let A be a right H-comodule algebra. Then a (right) *total integral* for A is a right H-comodule map $\phi : H \to A$ such that $\phi(1) = 1$.

Recall from 2.1.3, 3) that if $0 \neq T \in \int_{H^*}^l$, then the map $\theta : H \to H^*$ given by $h \mapsto (h \rightharpoonup T)$ is a left H-module isomorphism. Setting $t = \theta^{-1}(\varepsilon)$, so that $t \rightharpoonup T = \varepsilon$, we claim that $t \in \int_H^l$ (an observation of Radford). For, if $h \in H$, then $ht = h\theta^{-1}(\varepsilon) = \theta^{-1}(h \rightharpoonup \varepsilon) = \theta^{-1}(\varepsilon(h)\varepsilon) = \varepsilon(h)\theta^{-1}(\varepsilon) = \varepsilon(h)t$.

The next lemma appears in [CF 92], although a special case is in [D 85].

4.3.9 LEMMA. *Let A be a left H-module algebra, where H is finite-dimensional, and consider A as a right H^*-comodule algebra. Then $\hat{t} : A \to A^H$ is surjective $\Leftrightarrow$ there exists a total integral $\phi : H^* \to A$.*

PROOF. First assume $\hat{t}$ is surjective and choose $c \in A$ with $t \cdot c = 1$. With θ as above, define $\phi : H^* \to A$ by $f \mapsto \theta^{-1}(f) \cdot c$; ϕ is a left H-module map since θ is, and thus ϕ is a right H^*-comodule map. Moreover $\phi(1_{H^*}) = \phi(\varepsilon) = t \cdot c = 1$, and so ϕ is a total integral for A.

Conversely, assume that $\phi : H^* \to A$ is a total integral and set $c = \phi(T)$. Then $t \cdot c = t \cdot \phi(T) = \phi(t \rightharpoonup T) = \phi(\varepsilon) = 1$, and so $\hat{t}$ is surjective. □

Doi also defines the notion of "trace" for a right H-comodule algebra A such that there exists a (right) comodule map $\phi : H \to A$. Namely, if $a \in A$, set $tr(a) = \sum a_0 \phi(Sa_1)$; it is easy to check that $tr(a) \in A^{coH}$, and that tr is the identity on A^{coH} if ϕ is a total integral [D 90].

We return to trace functions in §4.5.

§4.4 Ideals in $A \# H$ and A as an A^H-module

In this section we look a bit more closely at how the structure of $A \# H$ influences the relationship between A and A^H when H is finite-dimensional. In particular we are interested in when A is a finitely-generated A^H-module.

We first consider the situation when $\hat{t}$ is surjective as in the last section.

In this case things work fairly well; we are able to extend arguments given by Lorenz and Passman for the action of a finite group G with $|G|^{-1} \in k$; see [M 80, 7.5 and 7.6].

Let $e = tc$ be the idempotent in 4.3.4, and fix a basis $\{h_1, \ldots, h_n\}$ of H. For any right A-module V, let $W = V \otimes_A (A\#H)$ be the induced $A\#H$-module. We will compare the lattice of A^H-submodules of V, denoted $\mathcal{L}(V_{A^H})$, with the lattice $\mathcal{L}(W_{A\#H})$ of $A\#H$-submodules of W. Define

$$\sigma : \mathcal{L}(V_{A^H}) \to \mathcal{L}(W_{A\#H}) \text{ via } U \mapsto (U \otimes e)(A\#H)$$

and

$$\mu : \mathcal{L}(W_{A\#H}) \to \mathcal{L}(V_{A^H}) \text{ via } \sum_i v_i \otimes h_i \mapsto \sum_i \varepsilon(h_i)v_i,$$

for any $w = \sum_i v_i \otimes h_i \in W$; μ is well-defined since any $w \in W$ has a unique representation in this form. μ is an A^H-module map, since $ha = ah$ for all $a \in A^H$, and thus if $X \in \mathcal{L}(W_{A\#H}), X^\mu \in \mathcal{L}(V_{A^H})$. Clearly both σ and μ preserve inclusions.

4.4.1 LEMMA. *Let e, V, W, σ and μ be as above. Then for any $U \in \mathcal{L}(V_{A^H})$, $U^{\sigma\mu} = U$. Thus σ is an injection of $\mathcal{L}(V_{A^H})$ into $\mathcal{L}(W_{A\#H})$.*

PROOF. First, note that if $v \otimes e = 0$, then $v \otimes et = v \otimes t = 0$ and so $v = 0$. Thus $v_1 \otimes e = v_2 \otimes e$ implies $v_1 = v_2$. Now for any $w = \sum_i v_i \otimes h_i \in W$, note $we = \sum_i v_i \otimes h_i tc = \sum_i \varepsilon(h_i)v_i \otimes tc = \mu(w) \otimes e$. For $U \in \mathcal{L}(V_{A^H}), U^\sigma e = U \otimes e(A\#H)e = U \otimes A^H e = U \otimes e$ by 4.3.4. Thus if $w \in U^\sigma, we = \mu(w) \otimes e \in U \otimes e$, and so $\mu(w) \in U$; that is, $U^{\sigma\mu} \subseteq U$. If $u \in U$, then $w = u \otimes e \in U^\sigma$, and so $u \otimes e = we = \mu(w) \otimes e$. It follows that $u = \mu(w)$ and so $U = U^{\sigma\mu}$. □

4.4.2 THEOREM. *Let H be finite-dimensional and let A be an H-module algebra such that $\hat{t}$ is surjective. If A is right Noetherian, then A is a right Noetherian A^H-module.*

PROOF. We apply the Lemma with $V = A$. Then $W = A \otimes_A A\#H \cong A\#H$; since $A\#H \cong H \otimes A$ as right A-modules (see 4.4.3, 1) or 7.2.11) and A is right Noetherian, W is a Noetherian $A\#H$-module. Now $\mathcal{L}(V_{A^H}) \hookrightarrow \mathcal{L}(W_{A\#H})$ by 4.4.1; thus A is a Noetherian A^H-module. □

When $\hat{t}$ is not surjective, the situation is not so nice; nevertheless, some things can be said. We first need a small technical lemma.

4.4.3 LEMMA. *Let A be an H-module algebra and choose $0 \neq t \in \int_H^l$. Then for all $a \in A, h \in H$,*
1) $ah = \sum h_2(\bar{S}h_1 \cdot a)$
2) $hat = (h \cdot a)t$ and $tah = t(\bar{S}h^\alpha \cdot a)$, where α is the distinguished group-like element in H^* and $h^\alpha = \alpha \rightharpoonup h$
3) $(t) = AtA$ is an ideal of $A\#H$.

PROOF. 1) This is straightforward using properties of $\bar{S}$ (see §1.5) and the definition of multiplication in $A\#H$.
2) The first identity is shown in the proof of 4.3.4. For the second, use 1) and the fact that $th = \alpha(h)t$; recall α is defined in 2.2.3. Thus

$$tah = \sum th_2(\bar{S}h_1 \cdot a) = t(\bar{S}(\sum \alpha(h_2)h_1) \cdot a) = t(\bar{S}(\alpha \rightharpoonup h) \cdot a).$$

3) This follows from 2). □

How large (t) is in $A\#H$ influences whether or not A is a finitely-generated A^H-module. The next result is from [BeM 86].

4.4.4 PROPOSITION. *Consider $(t) = AtA$ in $A\#H$ as above. Then*
1) For any $a \in (t) \cap A$, there exist $\{b_i\}, \{c_i\} \in A$ such that for all $d \in A$,

$$ad = \sum_{i=1}^{n} b_i\hat{t}(c_i d).$$

That is, $aA \subseteq \sum_{i=1}^n b_i A^H$. This says that $I = (t) \cap A$ is "Shirshov locally finite" over A^H.
2) If $(t) = A\#H$, then A is a finitely-generated (right) A^H-module.
3) If $I = (t) \cap A$ contains a regular element of A, then A is a (right) A^H-submodule of a finite free A^H-module.

PROOF. 1) Write $a = \sum_{i=1}^n b_i t c_i \in AtA$, for some $b_i, c_i \in A$. Then $ad = \sum b_i t c_i d = \sum b_i\, t_1 \cdot (c_i d)\#t_2$. Applying $id \otimes \varepsilon$ to $A\#H$, we see that $ad = \sum b_i t \cdot (c_i d)$.
2) We may use $a = 1$ in 1).
3) Let a be a regular element in I, and let $\{b_i\}, \{c_i\}$ be as in 1). Define $\phi : A \to \oplus_{i=1}^n A^H$ by $d \mapsto (\hat{t}(c_i d))_i$. If $\phi(d) = 0$, then $\hat{t}(c_i d) = 0$, for all i. Thus $ad = 0$ by 1), and $d = 0$ since a is regular. Thus ϕ is injective. □

Note that an immediate consequence of 4.4.4 is that if $A\#H$ is a simple ring, then A is a finitely-generated A^H-module, and thus A is Noetherian provided A^H is Noetherian. This last fact is true more generally, as the following result of [BeM 86] shows; we state it without proof. Recall that a ring R is called *semiprime* if it has no non-zero nilpotent ideals.

4.4.5 THEOREM. *Assume that $A\#H$ is semiprime and that every non-zero ideal of $A\#H$ intersects A non-trivially. Then*
1) A^H is a (right) Goldie ring $\Leftrightarrow$ A is (right) Goldie
2) if A^H is (right) Noetherian or Artinian, so is A.

The proof of 4.4.5 uses 4.4.4 as well as the next lemma; we see that the semiprimeness of $A\#H$ has consequences for the trace function:

4.4.6 LEMMA. *Assume that $A\#H$ is semiprime, and choose $0 \neq t \in \int_H^l$. If I is any non-zero left or right H-stable ideal of A, then $\hat{t}(I) \neq 0$.*

PROOF. If $\hat{t}(I) = 0$, then $tIt = 0$ by 4.4.3, 2). Thus if (for example) I is a left ideal, then $J = It$ is a left ideal of $A\#H$ such that $J^2 = 0$. Since $A\#H$ is semiprime, $J = 0$ and thus $I = 0$ a contradiction. If I is a right ideal, the same argument works using $J = tI$. □

From the above results, it is clear that it would be very useful to know when $A\#H$ is semiprime. If A is semiprime, it is known that $A\#H$ is semiprime in the following cases: if $H = \mathrm{k}G$ and $|G|^{-1} \in \mathrm{k}$ [FM], or if $H = (\mathrm{k}G)^*$ [CM]. In both of these cases, H is semisimple. This suggests:

4.4.7 QUESTION. *If H is finite-dimensional and semisimple, and A is a semiprime H-module algebra, is $A\#H$ semiprime?*

This question is mentioned explicitly in [CF 86]. In that paper, a first step is taken for general H. Namely, with H as above, it is proved that if A is semisimple Artinian, then also $A\#H$ is semisimple Artinian. The method of proof is to use a variant of the Larson-Sweedler averaging function $\tilde{\pi}$ as in 2.2.1 to show that an $A\#H$-submodule N of an $A\#H$-module has an $A\#H$-complement. To show $\tilde{\pi}$ is an H-map follows as before; what is new is to show that it is also an A-map.

A number of other results are known on this question; however one may ask the same question for crossed products, and at this writing anything known for $A\#H$ is also known for $A\#_\sigma H$. Thus we postpone a discussion of this problem until §7.4.

We close this section with an example in which the ideal (t) is not very well-behaved.

4.4.8 EXAMPLE. Let $H = H_4$ as in 1.5.6 with $\mathbf{k} = \mathbf{R}$, and let $A = \mathbf{C}$. Then $\mathbf{C}$ is an H-module algebra via g acting as complex conjuation and x as the $1, g$ - derivation δ sending i to 1. It was shown in [M 92] that this action is well-defined.

We claim that $B = \mathbf{C}\#H = (t) \oplus (t')$, where $t = x + gx \in \int_H^l$ and $t' = x - gx \in \int_H^r$ as in 2.1.2. First, B has $\mathbf{R}$-basis $\{1, i, g, ig, x, ix, gx, igx\}$, and relations $gi = -ig, xi + ix = 1$, and $xg = -gx$. It is easy to verify that $e_1 = it/2, e_2 = ti/2, e_3 = it'/2$, and $e_4 = t'i/2$ are orthogonal idempotents in B whose sum is 1. Now one may check that $Bt = Be_1$, and so $BtB = Be_1 \oplus Be_2$. Moreover $e_1 + e_2$ is a central idempotent in B. Similarly $Bt'B = Be_3 \oplus Be_4$ and $e_3 + e_4$ is a central idempotent in B. Since $\sum e_i = 1, B = BtB \oplus Bt'B = (t) \oplus (t')$.

Thus $(t) \cap A = (0)$, and so 4.4.4 and 4.4.5 do not apply. With a little more work, one sees that each Be_i is a simple B-module. Thus (t) and (t') are both four-dimensional algebras and B is semisimple Artinian. These last computations were done jointly with M. Cohen.

§4.5 A Morita context relating $A\#H$ and A^H

It is clear from the previous two sections that there is a close relationship between $A\#H$ and A^H; here we formalize that relationship via a Morita context. A reference for Morita contexts is [Am] or [McR, §3.6]. The basic idea is to find a relationship between two rings via their modules, which is weaker than Morita equivalence (in which the two module categories are equivalent), but is still strong enough to enable the rings to share some properties. The work presented here comes from [CFM]; it extends earlier work of [CF 86], who considered the case when H was unimodular. That paper in turn was extending earlier work on group actions, going back to [CHR].

We first review what it means for two rings R and S to be connected by a Morita context. One needs a left R, right S-bimodule ${}_RM_S$, a left S, right R-bimodule ${}_SN_R$, and two bilinear maps

$$[\ ,\ ] : N \otimes_R M \to S \quad \text{and} \quad (\ ,\ ) : M \otimes_S N \to R,$$

such that

1) $[\ ,\ ]$ is an S-bimodule map which is middle R-linear
2) $(\ ,\ )$ is an R-bimodule map which is middle S-linear
3) for all $m, m' \in M$ and $n, n' \in N$, "associativity" holds; that is

$$m' \cdot [n,m] = (m',n) \cdot m \quad \text{and} \quad [n,m] \cdot n' = n \cdot (m,n').$$

We show that such a set-up exists with $R = A^H$ and $S = A\#H$, for any finite-dimensional H, using $M = N = A$. Now A is a left (or right) A^H-module simply by left (or right) multiplication. A is a left $A\#H$ module in the standard way, that is

$$(a\#h) \cdot b = a(h \cdot b) \tag{4.5.1}$$

for all $a, b \in A$ and $h \in H$. To define a right $A\#H$ action which will work is a trickier matter. Again let $\alpha \in H^*$ be the distinguished group-like element as in 2.2.3. Since $\alpha \in G(H^*)$, its left action on H is an automorphism of H. We write

$$h^\alpha = \alpha \rightharpoonup h = \sum \alpha(h_2)h_1.$$

By [R 76], $t^\alpha = St$ for $0 \neq t \in \int_H^l$; in particular H is unimodular if and only if $St = t$. We define our new right action of $A\#H$ on A by

$$a \leftarrow (b\#h) = \bar{S}h^\alpha \cdot (ab) \tag{4.5.2}$$

for all $a, b \in A$ and $h \in H$. This is the usual right action but with h "twisted" by α. In this notation, the second part of 4.4.3, 2) becomes $tah = t(a \leftarrow h)$. We now have

4.5.3 THEOREM. *Consider A as a left (respectively right) A^H-module via left (right) multiplication, as a left $A\#H$-module via 4.5.1, and as a right $A\#H$-module via 4.5.2. Then $M = {}_{A^H}A_{A\#H}$ and $N = {}_{A\#H}A_{A^H}$, together with the maps*

$$[\ ,\ ] : A \otimes_{A^H} A \to A\#H \text{ given by } [a,b] = atb$$

$$(\ ,\ ) : A \otimes_{A\#H} A \to A^H \text{ given by } (a,b) = \hat{t}(ab)$$

give a Morita context for A^H and $A\#H$.

PROOF (sketch). Most of the desired properties can be checked in a straightforward manner. The one fact needing some work is that (,) is middle $A\#H$-linear: for $a, b, c \in A, h \in H$,

$$\begin{aligned}
(c \leftarrow ah, b) &= (\bar{S}h^\alpha \cdot (ca), b) = \hat{t}((\bar{S}h^\alpha \cdot (ca))b) \\
&= t \cdot ((\sum \alpha(h_2)\bar{S}h_1 \cdot (ca))b) \\
&= \sum th_2 \cdot ((\bar{S}h_1 \cdot (ca))b) \\
&= t \cdot \sum (h_2\bar{S}h_1 \cdot (ca))(h_3 \cdot b) \\
&= t \cdot \sum (\varepsilon(h_1)ca)(h_2 \cdot b) \\
&= t \cdot ((cah) \cdot b) \\
&= (c, (ah) \cdot b).
\end{aligned}$$

□

Note that the trace function $\hat{t}$ and the ideal $(t) = AtA$ which we have studied in the last two sections play an important role here: for, $\hat{t}(A) = (A, A)$, and $(t) = [A, A]$. This observation gives us a criterion for Morita equivalence:

4.5.4 COROLLARY. *Let A be an H-module algebra, where H is finite dimensional, and choose $0 \neq t \in \int_H^l$. If both $\hat{t} : A \to A^H$ is surjective and $(t) = A\#H$, then $A\#H$ is Morita-equivalent to A^H.*

PROOF. For Morita equivalence via the context in 4.5.3, we need that the two maps [,] and (,) are surjective. However this is precisely our hypothesis. □

We give a few ring-theoretic consequences:

4.5.5 COROLLARY [BeM 86]. *Let A be an H-module algebra, where H is finite dimensional, such that $A\#H$ is a simple ring. Then $A\#H$ is Morita-equivalent to $A^H \Leftrightarrow \hat{t}$ is surjective $\Leftrightarrow A^H$ is simple.*

PROOF. Since $A\#H$ is simple, $(t) = A\#H$ and so [,] is surjective. Thus if $\hat{t}$ is surjective, Morita equivalence follows by 4.5.4. Conversely, Morita equivalence implies that A^H is simple since being simple is a Morita invariant. Finally $\hat{t}(A)$ is an ideal of A^H, and is non-zero by 4.4.6. Thus $\hat{t}(A) = A^H$ if A^H is simple, and so $\hat{t}$ is surjective. □

In particular, $\hat{t}$ is surjective whenever H is semisimple, and thus $A\#H$ simple implies A^H is simple in that case.

We note that it is false in general that $A\#H$ simple implies A^H is simple, even for the case of group actions. However, primeness behaves somewhat better; recall that a ring is *prime* if the product of non-zero ideals is non-zero.

4.5.6. COROLLARY [CFM]. *Let A be an H-module algebra, where H is finite-dimensional. Then $A\#H$ is a prime ring $\Leftrightarrow$ A is a left and right faithful $A\#H$-module and A^H is a prime ring.*

PROOF. This result is [CF 86, Theorem 14] in the case H is unimodular, and their proof extends to the general case by using 4.5.3. They use work of Nicholson and Watters on prime Morita contexts; a Morita context $[R,\ {}_RM_S,\ {}_SN_R,\ S]$ is prime when the ring of matrices $C = \begin{bmatrix} R & M \\ N & S \end{bmatrix}$ is prime; the primeness of C depends on that of R and S and on the non-degeneracy of $[\ ,\ ]$ and $(\ ,\ )$. □

Under the hypothesis that A is left and right $A\#H$-faithful, a number of other results now follow from [Am]; for example, A^H is primitive $\Leftrightarrow$ $A\#H$ is primitive.

We note that Example 4.4.8 shows that A^H and $A\#H$ are certainly not always Morita equivalent, and that A is not always a faithful $A\#H$-module.

Chapter 5

Coradicals and Filtrations

In this chapter we return to the basics and study the structure of coalgebras. A major topic is the coradical filtration $\{C_n\}$ of a coalgebra C; this filtration is very useful as a vehicle for inductive arguments. Here we will prove a theorem of Heyneman-Sweedler-Radford, which says that a coalgebra morphism is injective provided it is injective on C_1, and the Taft-Wilson theorem, which for pointed coalgebras describes C_n in terms of "higher degree" skew primitive elements. We then prove the classical results describing the structure of pointed cocommutative Hopf algebras in characteristic 0. We close by giving an application to the classification of semisimple cocommutative Hopf algebras in characteristic p.

§5.1 Simple subcoalgebras and the coradical

Throughout C denotes an arbitrary coalgebra. Our first result says that coalgebras are always "locally finite"; in [S] this is called the Fundamental Theorem on Coalgebras. We follow the (easier) proof in [W].

5.1.1 FINITENESS THEOREM. *Let C be a coalgebra.*
1) Given any (right) C-comodule M and any finite subset $\{m_i\} \subset M$, there exists a finite-dimensional subcomodule N of M such that $m_i \in N$, $\forall i$.
2) Given any finite subset $\{c_i\} \subset C$, there exists a finite-dimensional subcoalgebra D of C such that $c_i \in D, \forall i$.

PROOF. 1) Since a sum of subcomodules is again a subcomodule, it suffices to show that each $m \in M$ lies in a finite-dimensional subcomodule. Let $\{c_i\}$ be a basis for C. If $\rho : M \to M \otimes C$ is the comodule structure map, write $\rho(m) = \sum_i w_i \otimes c_i$, where all but finitely many of the w_i are zero. Also write $\Delta c_i = \sum \alpha_{ijk} c_j \otimes c_k$. Then

$$\sum \rho(w_i) \otimes c_i = (\rho \otimes id)\rho(m) = (id \otimes \Delta)\rho(m) = \sum w_i \otimes \alpha_{ijk} c_j \otimes c_k.$$

Comparing the coefficients of c_k we see that $\rho(w_k) = \sum w_i \otimes \alpha_{ijk} c_j$. Thus the subspace N spanned by m and the w_i is a subcomodule.

2) By 1) applied to $M = C, \rho = \Delta$, the given $\{c_i\}$ are contained in a finite-dimensional subspace V with $\Delta V \subseteq V \otimes C$. Let $\{v_j\}$ be a basis of V, with $\Delta v_j = \sum_i v_i \otimes c_{ij}$. By coassociativity it follows that $\Delta c_{ij} = \sum_k c_{ik} \otimes c_{kj}$. Thus the span D of $\{v_j\}$ and $\{c_{ij}\}$ satisfies $\Delta D \subseteq D \otimes D$. By construction $V \subseteq D$, proving 2). □

From 2.4.1 and 2.4.2, we recall that a coalgebra is *simple* if it has no proper subcoalgebras, and that a comodule is *simple* if it has no proper subcomodules.

5.1.2 COROLLARY. *Let C be any coalgebra. Then*
1) every simple subcoalgebra of C is finite-dimensional, and
2) every simple C-comodule is finite-dimensional.

Next we wish to characterize simple coalgebras in terms of their duals. To do this we first recall a little linear algebra. For any vector space V and subspace $W \subseteq V$, $W^\perp = \{f \in V^* \mid \langle f, W\rangle = 0\}$; for any subspace $U \subseteq V^*$, $U^\perp = \{v \in V \mid \langle U, v\rangle = 0\}$. Note that $W^{\perp\perp} = W$, but that it can happen that $U^{\perp\perp}$ properly contains U if V is infinite-dimensional. For any $U_1, U_2 \subseteq V^*$, one has

$$(U_1 \otimes U_2)^\perp = V \otimes U_2^\perp + U_1^\perp \otimes V \quad \text{in } V \otimes V. \tag{5.1.3}$$

Recall from 1.6.5 that a subspace $D \subseteq C$ is a right (left) coideal if $\Delta D \subseteq D \otimes C$ (resp. $\Delta D \subseteq C \otimes D$).

5.1.4 LEMMA. *Let C be any coalgebra. Then*
1) D is a right (left) coideal of $C \Leftrightarrow D^\perp$ is a right (resp. left) ideal of C^*
2) if I is a right (left) ideal of C^*, then $I^\perp$ is a right (resp. left) coideal of C. The converse holds if C is finite-dimensional.

Consequently, if D is a subcoalgebra of C, then
3) D is a simple subcoalgebra $\Leftrightarrow D^*$ is a (finite-dimensional) simple algebra $\Leftrightarrow D^\perp$ is a maximal ideal of C^* of finite codimension.

PROOF. 1) Suppose D is a right coideal, and choose $a \in D^\perp, b \in C^*$. Then $\langle ab, d\rangle = \langle a \otimes b, \Delta d\rangle \subseteq \langle a, D\rangle\langle b, C\rangle = 0$, for all $d \in D$. Hence $ab \in D^\perp$, and $D^\perp$ is a right ideal of C^*. Since $D^{\perp\perp} = D$, the converse will follow from 2).

2) Let I be a right ideal of C^*. To show that $I^\perp$ is a right coideal of C, it suffices to show that it is a left C^*-module under $\rightharpoonup$, using 1.6.5. But

$\langle I, C^* \rightharpoonup I^\perp \rangle = \langle IC^*, I^\perp \rangle \subseteq \langle I, I^\perp \rangle = 0$. Thus $C^* \rightharpoonup I^\perp \subseteq I^\perp$, and so $I^\perp$ is a right coideal. For the converse, when C is finite-dimensional we have $I^{\perp\perp} = I$ and we may apply the first part of 1).

The left-handed versions can be proved similarly, using $\leftharpoonup$.

For 3), D is a subcoalgebra $\Leftrightarrow$ D is both a left and right coideal. Thus the first equivalence follows from 1), 2), and 5.1.2. The second follows from the fact that $D^\perp$ is the kernel of the restriction map $\phi : C^* \to D^*$, and so $D^* \cong C^*/D^\perp$. □

As noted in §2.4, simple coalgebras should really be called "cosimple" in view of 5.1.4.

5.1.5 DEFINITION. Let C be a coalgebra.
1) The *coradical* C_0 of C is the sum of all simple subcoalgebras of C.
2) C is *pointed* if every simple subcoalgebra is one-dimensional.
3) C is *connected* if C_0 is one-dimensional.

Necessarily a one-dimensional subcoalgebra is of the form $\mathbf{k}g$, for $g \in G(C)$. Thus C is pointed $\Leftrightarrow$ $C_0 = \mathbf{k}G(C)$. It is also easy to see that a sum of simple subcoalgebras is in fact a direct sum. Thus the definition given in 2.4.1 of C being *cosemisimple*, that C is a direct sum of simple coalgebras, is equivalent to saying that $C = C_0$, its coradical.

5.1.6 EXAMPLES. Consider our basic Hopf algebras $\mathbf{k}G$ and $U(\mathfrak{g})$. Clearly $C = \mathbf{k}G$ is pointed and $C = C_0$. Now for $C = U(\mathfrak{g})$, in fact C is connected and $C_0 = \mathbf{k}1$. We will return to this in §5.5.

5.1.7 REMARK. If C is finite-dimensional, then it follows immediately from 5.1.4, 3) and the definition of C_0 that $C_0^\perp = \mathrm{Jac}(C^*)$, the Jacobson radical of C^*. We shall see in 5.2.9 that this fact is true for any coalgebra C.

We can give another characterization of C_0 in terms of submodules. Recall from 1.6.5 the left action of C^* on C, and the fact that the C^*-submodules of C are precisely the right coideals of C. For any ring R and left R-module M, the *socle* of M, $\mathrm{Soc}(M)$, is the sum of the simple submodules of M. The next result is due to [Gr].

5.1.8 COROLLARY. *Under the left action of C^* on C,* $\mathrm{Soc}(C) = C_0$, *the* *coradical. Thus any simple subcoalgebra D is a direct sum of simple C^*-submodules, and conversely any simple C^*-submodule of C is a summand of such a D.*

PROOF. We use 5.1.4. For any simple subcoalgebra D, $D^\perp = \cap_{i=1}^n I_i$, a finite intersection of maximal right ideals of C^*. Thus $D = \sum I_i^\perp$, a (direct) sum of minimal right coideals of C, by 5.1.4. As noted above, each $I_i^\perp$ is a simple left C^*-submodule by 1.6.5. Thus $C_0 \subseteq \text{Soc}(C)$. Conversely, let V be a simple left C^*-submodule of C; thus V is a minimal right coideal. Thus $V^\perp$ is a maximal right ideal of C^*, and $V^\perp$ has finite codimension since $C^*/V^\perp \cong V^*$, which is finite-dimensional by 5.1.1. Now $V^\perp \supset I$, a maximal ideal of C^* of finite codimension. Then $V = V^{\perp\perp} \subset I^\perp$ and so $C^* \neq V^\perp \supset I^{\perp\perp} \supseteq I$. Since I is maximal, $I = I^{\perp\perp}$; thus $I^\perp$ is a simple subcoalgebra by 5.1.4, and V is a summand of $I^\perp$ as above. Thus $V \subseteq C_0$, and so Soc $(C) \subseteq C_0$. □

We close this section with two lemmas we will need later on.

5.1.9 LEMMA. *If D is a subcoalgebra of C, then $D_0 = D \cap C_0$.*

PROOF. Clearly $D_0 \subseteq D \cap C_0$. For the converse, write $C_0 = \sum_\alpha \oplus T_\alpha$, where each T_α is a simple subcoalgebra; we claim $D \cap C_0 = \sum_\alpha \oplus (D \cap T_\alpha)$ (it will then follow that $D \cap C_0 \subseteq D_0$, as it's a sum of simple subcoalgebras of D). Choose $d \in D \cap C_0$ and write $d = \sum_{i=1}^n t_i$, where $t_i \in T_{\alpha_i}$. For each i, choose $f_i \in C_0^*$ such that $f|_{T_{\alpha_i}} = \varepsilon$ and $f|_{T_\beta} = 0$ for $\beta \neq \alpha_i$. Then $f_i \rightharpoonup d = \sum \langle f_i, d_2 \rangle d_1 \in D \cap C_0$ since $D \cap C_0$ is a subcoalgebra. But $f_i \rightharpoonup d = t_i$ by construction; thus each $t_i \in D \cap C_0$, and $d \in \sum \oplus (D \cap T_\alpha)$. □

5.1.10 LEMMA. *For any coalgebras C and D, $(C \otimes D)_0 \subseteq C_0 \otimes D_0$. If also C and D are pointed (respectively connected), then equality holds and $C \otimes D$ is pointed (resp. connected).*

PROOF. It suffices to show that if $0 \neq X$ is a subcoalgbra of $C \otimes D$, then $X \cap (C_0 \otimes D_0) \neq 0$. To show this, we may assume C and D are finite-dimensional: for, choose $x = \sum c_i \otimes d_i \neq 0$ in X, and replace C and D, respectively, by the subcoalgebras C' generated by the $\{c_i\}$ and D' generated by the $\{d_i\}$.

Now consider $(C \otimes D)^* \cong C^* \otimes D^*$. By 5.1.7, $C_0^\perp = \text{Jac}(C^*)$, the Jacobson radical of C^*, and $D_0^\perp = \text{Jac}(D^*)$. Since C^*, D^* are finite-dimensional, there exists $n > 0$ such that $(C_0^\perp)^n = (D_0^\perp)^n = 0$. Then $I = C_0^\perp \otimes D^* + C^* \otimes D_0^\perp$ is an ideal of $C^* \otimes D^*$ with $I^{2n} = 0$; thus $I \subseteq \mathrm{Jac}(C^* \otimes D^*) \subseteq P$, for any maximal ideal P. Thus $I^\perp = C_0 \otimes D_0$ contains all simple subcoalgebras of $C \otimes D$, and so $X \cap (C_0 \otimes D_0) \neq 0$.

If also C and D are pointed, then it is clear that $C_0 \otimes D_0 \subseteq (C \otimes D)_0$ since $G(C) \otimes G(D) \subseteq G(C \otimes D)$. Thus $C \otimes D$ is pointed. □

§ 5.2. The coradical filtration

In fact the coradical C_0 is the bottom piece of a filtration of C. We define C_n inductively as follows: for each $n \geq 1$, define

(5.2.1) $$C_n = \Delta^{-1}(C \otimes C_{n-1} + C_0 \otimes C)$$

The basic properties of $\{C_n\}$ are given in the next theorem; it is [S, 9.0.4 and 9.1.7] and [A, 2.4.1].

5.2.2 THEOREM. *For all $n \geq 0$, the $\{C_n\}$ are a family of subcoalgebras of C satisfying*

1) $C_n \subseteq C_{n+1}$ and $C = \bigcup_{n \geq 0} C_n$

2) $\Delta C_n \subseteq \sum_{i=0}^{n} C_i \otimes C_{n-i}$.

PROOF (sketch). The proof uses the "wedge" of two subspaces X, Y of C. This is defined by

$$X \wedge Y = \text{Ker } (C \xrightarrow{\Delta} C \otimes C \xrightarrow{\pi_X \otimes \pi_Y} C/X \otimes C/Y)$$

where π_X and π_Y are the canonical quotient maps. It has the following properties [S, 9.0.0]:

(5.2.3) $$X \wedge Y = \Delta^{-1}(C \otimes Y + X \otimes C)$$

(5.2.4) $$X \wedge Y = (X^\perp Y^\perp)^\perp, \text{ where the product } X^\perp Y^\perp \text{ is in } C^*$$

(5.2.5) $$(X \wedge Y) \wedge Z = X \wedge (Y \wedge Z).$$

We also define $\bigwedge^0 X = \{0\}, \bigwedge^1 X = X$, and $\bigwedge^n X = (\bigwedge^{n-1} X) \wedge X$, all $n \geq 1$. It follows from these facts that if X and Y are subcoalgebras, then

$X \wedge Y$ is also a subcoalgebra. For, $X^\perp$ and $Y^\perp$ are ideals of C^* by 5.1.4. Thus if I is the ideal of C^* generated by $X^\perp Y^\perp$, then $I^\perp = (X^\perp Y^\perp)^\perp = X \wedge Y$ by 5.2.4. Thus $X \wedge Y$ is a subcoalgebra by 5.1.4. Note also that $X + Y \subseteq X \wedge Y$.

In this notation, $C_n = \bigwedge^{n+1} C_0$; thus the $\{C_n\}$ are an ascending chain of subcoalgebras. By 5.2.5, in fact $C_n = (\bigwedge^i C_0) \wedge (\bigwedge^{n+1-i} C_0)$, for all $0 \leq i \leq n+1$. Thus for $1 \leq i \leq n$,

$$(*) \qquad \Delta C_n \subseteq C \otimes \wedge^{n+1-i} C_0 + \wedge^i C_0 \otimes C = C \otimes C_{n-i} + C_{i-1} \otimes C.$$

Also $(*)$ holds for $i = 0$ and $i = n+1$, since C_n is a subcoalgebra by the above. Now 2) follows from the fact [S, 9.1.5] that for any vector space $V$and any ascending chain of subspaces $\{0\} = V_0 \subset V_1 \subset \cdots \subset V_n \subset \cdots$,

$$\bigcap_{i=0}^{n+1} (V \otimes V_{n+1-i} + V_i \otimes V) = \sum_{i=1}^{n+1} V_i \otimes V_{n+2-i}.$$

For, apply this identity with $V_i = C_{i-1}$.

It remains to show that $\bigcup_{n \geq 0} C_n = C$. By 5.1.1, C is the union of its finite-dimensional subcoalgebras, and thus it suffices to show that for any finite-dimensional subcoalgebra D, $D \subseteq C_n$ for some n. To see this, first note that $D_0 = C_0 \cap D$ by 5.1.9. Now D^* has Jacobson radical $D_0^\perp$ by 5.1.7 (where here we mean the $\perp$ in D^*) and thus $(D_0^\perp)^n = 0$ for some n, since D is finite-dimensional. Using 5.2.4 it follows that $\bigwedge_D^n D_0 = ((D_0^\perp)^n)^\perp = 0^\perp = D$, where $\bigwedge_D$ is the wedge for the coalgebra D. Since $\bigwedge_D^n D_0 \subseteq \bigwedge_C^n D_0$, it follows that $D \subseteq \bigwedge_C^n C_0 = C_{n-1}$. The theorem is proved. □

Any set of subspaces of C satisfying conditions 1) and 2) in the theorem is called a *coalgebra filtration* of C. The coradical filtrations of $U(\mathfrak{g})$ and $U_q(\mathfrak{g})$ will be considered in §5.5; for another example see 5.4.8. Here we give an example from combinatorics.

5.2.7 EXAMPLE. Let P be a locally finite poset, and let $\mathcal{I}$ be the set of intervals in P. The *incidence coalgebra* C of P is given as follows: C is the $\mathbf{k}$-space with basis the elements of $\mathcal{I}$. For any interval $x_0 < x_1 < \cdots < x_n$, where $x_i \in P$, one defines $\Delta[x_0, x_n] = \sum_{i=0}^{n} [x_0, x_i] \otimes [x_i, x_n]$. The intervals of length 0 correspond to the points of P, and are group-like elements in C. In this example, C_n is the span of all intervals of length n [JR].

If H is a Hopf algebra, then a set $\{A_n\}$ of subspaces of H is a *Hopf algebra filtration* if it is a coalgebra filtration, an algebra filtration (that is, $A_nA_m \subseteq A_{n+m}$, for all $n, m \geq 0$) and $SA_n \subseteq A_n$ for all n. These conditions are precisely what is needed for the associated graded space $gr(H) = \oplus_{n\geq 0} A_n/A_{n-1}$ to be a Hopf algebra, where we set $A_{-1} = (0)$. The question as to when the coradical filtration of a Hopf algebra is a Hopf algebra filtration is given in the following lemma.

5.2.8 LEMMA. *Let $\{H_n\}$ be the coradical filtration of the Hopf algebra H. Then $\{H_n\}$ is a Hopf algebra filtration $\Leftrightarrow$ H_0 is a subHopfalgebra of H.*

PROOF. ($\Rightarrow$) This is immediate. ($\Leftarrow$) First, $SH_0 \subseteq H_0$ since H_0 is a Hopf subalgebra. It then follows by induction on n that $SH_n \subseteq H_n$, using the definition of H_n and the fact that $\Delta Sh = \sum Sh_2 \otimes Sh_1$. Thus it remains to show that $H_nH_m \subseteq H_{n+m}$, all $n, m \geq 0$. If $m = 0$, this follows by induction on n:

$$\begin{aligned}\Delta H_nH_0 &\subseteq (H_0 \otimes H_n + H_n \otimes H_{n-1})(H_0 \otimes H_0)\\ &\subseteq H_0^2 \otimes H + H \otimes H_{n-1}H_0 \subseteq H_0 \otimes H + H \otimes H_{n-1}\end{aligned}$$

(here we have used that $H_0^2 \subseteq H_0$ since H_0 is a subalgebra). Thus $H_nH_0 \subseteq \Delta^{-1}(H_0 \otimes H + H \otimes H_{n-1}) = H_n$. Similarly $H_0H_m \subseteq H_m$, for all $m \geq 0$. We may now assume, by induction on both n and m, that $H_{n-1}H_m \subseteq H_{n+m-1}$ for all m and that $H_nH_{m-1} \subseteq H_{n+m-1}$ for all n. But then

$$\begin{aligned}\Delta H_nH_m &\subseteq (H_0 \otimes H_n + H_n \otimes H_{n-1})(H_0 \otimes H_m + H_m \otimes H_{m-1})\\ &\subseteq H_0^2 \otimes H_nH_m + H_nH_0 \otimes H_{n-1}H_m + H_0H_m \otimes H_nH_{m-1}\\ &\quad + H_nH_m \otimes H_{n-1}H_{m-1}\\ &\subseteq H_0 \otimes H + H \otimes H_{n+m-1}.\end{aligned}$$

Thus $H_nH_m \subset H_{n+m}$, and $\{H_n\}$ is a Hopf algebra filtration. □

In particular the lemma applies when H is pointed, for then $H_0 = \mathbf{k}G$, a subalgebra. However it is false in general that H_0 is a subalgebra. For, if H_0 is a subalgebra and H is finite-dimensional, then $\mathrm{Jac}(H^*) = H_0^{\perp}$ is a coideal of H^*, and thus the tensor product of two semisimple H^*-modules is always semisimple since it is annihilated by $\mathrm{Jac}(H^*)$. But if G is a finite group and char $\mathbf{k} = p \neq 0$, then $\mathbf{k}G$-modules have this property only if G has a unique p-Sylow subgroup. Thus when $H = (\mathbf{k}G)^*$, H_0 is not always a subHopfalgebra.

By similar considerations about the tensor product of completely reducible G-modules being completely reducible, for G an affine algebraic group, it can be shown that if H is a commutative affine Hopf algebra over a field of characteristic 0, then H_0 is a subHopfalgebra [A, 4.6.1].

We can now prove the result mentioned in 5.1.7; namely for any coalgebra C, it is true that $C_0^{\perp}$ is the Jacobson radical of C^*.

5.2.9 PROPOSITION. *Let* $J = C_0^{\perp}$ *in* C^*. *Then*
1) $J = Jac(C^*) = \bigcap_\alpha M_\alpha$, *where the* M_α *are maximal ideals of* C^* *of finite codimension*
2) $C_n = (J^{n+1})^{\perp}$, *for all* $n \geq 0$
3) $\bigcap_{n\geq 0} J^n = (0)$.

PROOF. Write $C_0 = \sum D_\alpha$ where the D_α are simple subcoalgebras of C. Then by 5.1.4, each $M_\alpha = D_\alpha^{\perp}$ is a maximal ideal of C^* of finite codimension. Then $J = (\sum_\alpha D_\alpha)^{\perp} = \bigcap_\alpha D_\alpha^{\perp} = \bigcap_\alpha M_\alpha$, and thus $J \supseteq \mathrm{Jac}(C^*)$. For the other containment, we first show 2), by induction on n.

Now $C_0 = C_0^{\perp\perp} = J^{\perp}$, so the result is true when $n = 0$. Assume true for $n-1$ and choose $c \in C$. Then

$$\begin{aligned}
\langle J^{n+1}, c\rangle = 0 \quad &\Leftrightarrow \quad \langle J \otimes J^n, \Delta c\rangle = 0 \Leftrightarrow \Delta c \in (J \otimes J^n)^{\perp} \\
&\Leftrightarrow \quad \Delta c \in C \otimes (J^n)^{\perp} + J^{\perp} \otimes C, \text{ by (5.1.3)} \\
&\Leftrightarrow \quad \Delta c \in C \otimes C_{n-1} + C_0 \otimes C, \text{ by induction} \\
&\Leftrightarrow \quad c \in C_n
\end{aligned}$$

proving 2).

We now return to 1). Assume $f \in J$. Since $\langle f^{n+1}, C_n\rangle = 0 \ \forall n \geq 0, g = \sum_{n=0}^{\infty} f^n$ is defined on all of C, where $f^0 = \varepsilon$. But $g = (\varepsilon - f)^{-1}$ in C^*; that is, $\varepsilon - f$ is invertible for all $f \in J$. It follows that $J \subseteq Jac(C^*)$.

Finally 3) follows from 2), since $J^n \subseteq (J^n)^{\perp\perp}$ and $C = \bigcup_{n\geq 0} C_n$. □

A slightly more complicated version of the argument in 5.2.9 proves a very useful lemma of Takeuchi.

5.2.10 LEMMA [Tk 71]. *Let C be a coalgebra and A an algebra. Then $f \in Hom(C, A)$ is (convolution) invertible $\Leftrightarrow$ $f\mid_{C_0}$ is invertible in $Hom(C_0, A)$.*

PROOF. ($\Rightarrow$) This direction is trivial. ($\Leftarrow$) Let $g \in \mathrm{Hom}(C_0, A)$ be the $*$-inverse of f on C_0. Extend g to a map $g' \in \mathrm{Hom}(C, A)$ by setting $g' \equiv 0$ on some vector space complement of C_0 in C. Now set $\gamma = u \circ \varepsilon - f * g'$; then $\gamma \mid C_0 = 0$. By induction on n, it follows that $\gamma^{n+1} = 0$ on C_n. Thus $\sum_{n=0}^{\infty} \gamma^n$ is well-defined on C, and is an inverse of $f * g'$. Thus $g' * \sum_{n=0}^{\infty} \gamma^n$ is a right inverse for f. Similarly, using $u \circ \varepsilon - g' * f$, we see that f has a left inverse. Thus f is invertible in $\mathrm{Hom}(C, A)$. □

The Corollary seems to be "folklore".

5.2.11 COROLLARY. *Let H be a Hopf algebra with cocommutative coradical H_0. Then the antipode of H is bijective.*

PROOF. Consider the bialgebra H^{cop} as in 1.3.11; it has the opposite coalgebra structure from H. Clearly $(H^{cop})_0 = H_0$, since the coopposite of a simple subcoalgebra is simple. Then $id \in \mathrm{Hom}(H^{cop}, H^{cop})$ has a convolution inverse on H_0, namely S itself, since H_0 is cocommutative. By 5.2.10, $S|_{H_0}$ extends to an antipode $\bar{S}$ on H^{cop}. Thus $\bar{S}$ is an inverse for S on H by 1.5.11.
□

We close this section by noting that Lemma 5.1.9 can be extended to the entire filtration $\{C_n\}$ [HR, 2.3.7].

5.2.12 LEMMA. *If D is a subcoalgebra of C, then $D_n = D \cap C_n$, for all $n \geq 0$.*

PROOF (sketch). The proof goes by induction on n, the case $n = 0$ being 5.1.9. The result will then follow from the identity, true for any subspaces V, W of C, that

$$(C \otimes V + W \otimes C) \cap (D \otimes D) = D \otimes (V \cap D) + (W \cap D) \otimes D,$$

by setting $V = C_{n-1}$ and $W = C_0$. □

5.1.8 can also be extended to C_n; one can show that $C_n = \mathrm{Soc}^{n+1}(C)$, the $(n+1)^{st}$ term in the socle series for C as a left C^*-module (recall that for any ring R and left R-module M, the socle series is defined inductively by $\mathrm{Soc}^{n+1}(M) = \mathrm{Soc}(M/\mathrm{Soc}^n(M))$. The dual version, for H acting on $C = H^0$, is shown in [ChM].

§5.3 Injective coalgebra maps

The object of this section is to prove the following theorem, due to Heyneman and Radford [HR]. It improves an earlier result of [HS].

5.3.1 THEOREM. *Let C and D be coalgebras and $f : C \to D$ a coalgebra morphism such that $f|_{C_1}$ is injective. Then f is injective.*

Before proving the theorem we need some lemmas.

5.3.2 LEMMA. *Let C be a connected coalgebra, with $G(C) = \{1\}$. Then*
1) $C_1 = \mathbf{k}1 \oplus P(C)$, where $P(C)$ is the set of primitive elements of C
2) for any $n \geq 1$ and $c \in C_n, \Delta c = c \otimes 1 + 1 \otimes c + y$, where $y \in C_{n-1} \otimes C_{n-1}$.

PROOF. 2) By 5.2.2, $\Delta c \in C_n \otimes C_0 + C_0 \otimes C_n + \sum_{i=1}^{n-1} C_i \otimes C_{n-i}$. Since $C_0 = \mathbf{k}1$, we may write $\Delta c = a \otimes 1 + 1 \otimes b + w$, for $w \in C_{n-1} \otimes C_{n-1}$. Now $c = (id \otimes \varepsilon)\Delta c = a + \varepsilon(b)1 + (id \otimes \varepsilon)w \in a + C_0 + C_{n-1}$; thus $a - c = c' \in C_{n-1}$. Similarly $b - c = c'' \in C_{n-1}$. It follows that $\Delta c = c \otimes 1 + 1 \otimes c + y$, for $y = w + c' \otimes 1 + 1 \otimes c'' \in C_{n-1} \otimes C_{n-1}$.

For 1), choose $c \in C_1$. Then $\Delta c = c \otimes 1 + 1 \otimes c + \alpha(1 \otimes 1)$, some $\alpha \in \mathbf{k}$, by 2). Using $c = (id \otimes \varepsilon)\Delta c$ gives $\alpha = -\varepsilon(c)$; it follows that $c - \varepsilon(c)1 \in P(C)$. Thus $C_1 = \mathbf{k}1 + P(C)$. The sum is direct since $\varepsilon(P(C)) = 0$. □

5.3.3 LEMMA. *If C is connected and $f : C \to D$ is a coalgebra map such that $f|_{P(C)}$ is injective, then f is injective.*

PROOF. We will show that $f|_{C_n}$ is injective for all $n \geq 0$. Since f is a coalgebra map, $\varepsilon(f(1)) = \varepsilon(1) = 1$ and so $f(1) \neq 0$. Also $\varepsilon(f(P(C))) = 0$ and thus $f(\mathbf{k}1) + f(P(C))$ is a direct sum. It follows that $f|_{C_1}$ is injective, using 5.3.2, 1).

Now assume $f|_{C_n}$ is injective and choose $x \in C_{n+1}$. By 5.3.2, 2), $\Delta x = x \otimes 1 + 1 \otimes x + y$, for $y \in C_n \otimes C_n$. Thus $\Delta f(x) = (f \otimes f)\Delta x = f(x) \otimes f(1) + f(1) \otimes f(x) + (f \otimes f)(y)$. If $x \in \operatorname{Ker} f$, then $(f \otimes f)y = 0$. But $f \otimes f$ is injective on $C_n \otimes C_n$, and so $y = 0$. But then $x \in P(C) \subseteq C_1$ and we know f is injective on C_1; thus $x = 0$, and f is injective. □

5.3.4 LEMMA. *Let C be any coalgebra and $\{A_n\}$ a coalgebra filtration of C.*

Then $A_0 \supseteq C_0$.

PROOF. It suffices to show that if D is any non-zero subcoalgebra of C, then $D \cap A_0 \neq 0$. Since $\bigcup_{n \geq 0} A_n = C$, we may choose n minimal such that $D \cap A_n \neq 0$; we claim $n = 0$. Choose $0 \neq d \in D \cap A_n$. Then 1) $\Delta d \in \sum_{i=0}^n A_i \otimes A_{n-i}$ and 2) $\Delta d \in D \otimes D$. If $\Delta d \in C \otimes A_0$, then $d = (\varepsilon \otimes id)\Delta d \in A_0$, done. If not, choose $f \in C^*$ with $f \in A_0^\perp$ and $0 \neq \langle id \otimes f, \Delta d \rangle = \bar{d}$. By 1), $\bar{d} \in A_{n-1}$, and by 2) $\bar{d} \in D$. Thus $\bar{d} \in D \cap A_{n-1}$, a contradiction. Thus $D \cap A_0 \neq 0$. $\square$

5.3.5 COROLLARY. *If $f: C \to D$ is a surjective coalgebra map, then $f(C_0) \supset D_0$. Thus if C is pointed, $f(C_0) = D_0$ and D is pointed.*

PROOF. Let $A_n = f(C_n)$, all $n \geq 0$; it is easy to check that $\{A_n\}$ is a coalgebra filtration of D. Thus $D_0 \subseteq A_0 = f(C_0)$ by 5.3.4. $\square$

We recall the "wedge" $\wedge$ which was defined in 5.2.2.

5.3.6 LEMMA. *Let $f : C \to D$ be a surjective coalgebra map and let W_1, W_2 be subspaces of C such that $Ker\, f \subseteq W_1 \cap W_2$. Then*

$$f(W_1 \wedge W_2) = f(W_1) \wedge f(W_2).$$

PROOF. Define $f_i : C/W_i \to D/f(W_i)$, $i = 1, 2$ as the maps induced from f on the quotients. Then $Ker\, f \subseteq W_i$ implies f_1, f_2, and so $f_1 \otimes f_2$ are bijective. Now define α and β as the compositions

$$\alpha: \quad C \to C \otimes C \to C/W_1 \otimes C/W_2 = \tilde{C}$$

$$\beta: \quad D \to D \otimes D \to D/f(W_1) \otimes D/f(W_2) = \tilde{D}.$$

From the definition of $\wedge$ in 5.2.2, $Ker\, \alpha = W_1 \wedge W_2$ and $Ker\, \beta = f(W_1) \wedge f(W_2)$. Thus we must show $Ker\, \beta = f(Ker\, \alpha)$. Since $f_1 \otimes f_2$ is bijective and f is surjective, this follows from the commutative diagram

$$
\begin{array}{ccc}
C & \xrightarrow{\;f\;} & D \\
\downarrow{\scriptstyle \alpha} & & \downarrow{\scriptstyle \beta} \\
\tilde{C} & \xrightarrow{\;f_1 \otimes f_2\;} & \tilde{D}
\end{array}
$$

$\square$

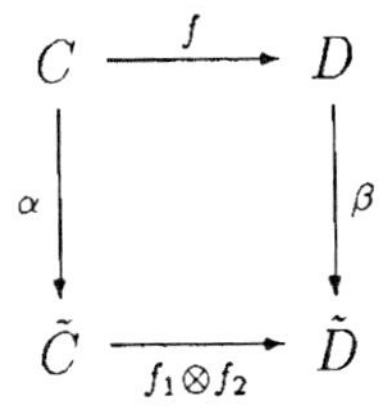

5.3.7 DEFINITION. Let C be any coalgebra, and let $C^+ = \operatorname{Ker} \varepsilon$. Then the *associated connected coalgebra* of C is $R = R(C) = C/C_0^+$.

We must show that $R(C)$ is in fact connected. Let $\pi : C \to R(C)$ be the canonical quotient map.

5.3.8 LEMMA. *For all $n \geq 0$, $R(C)_n = \pi(C_n)$; in particular $R(C)$ is connected.*

PROOF. Write $R = R(C)$ for simplicity. Now $\pi(C_0) \supseteq R_0$ by 5.3.5; however $\pi(C_0) = C_0/C_0^+$ is one-dimensional. Thus $\pi(C_0) = R_0$ and R is connected. Now apply 5.3.6 with $f = \pi$; since $\operatorname{Ker} \pi = C_0^+ \subseteq C_0 \cap C_{n-1}$ for all $n \geq 1$, it follows that $\pi(C_0 \wedge C_{n-1}) = \pi(C_0) \wedge \pi(C_{n-1})$ for all $n \geq 1$. By induction $\pi(C_{n-1}) = R_{n-1}$; thus $\pi(C_n) = R_0 \wedge R_{n-1} = R_n$. □

PROOF of 5.3.1. It suffices to show that if N is a coideal in C such that $N \cap C_1^+ = 0$, where $C_1^+ = C_1 \cap \operatorname{Ker} \varepsilon$, then $N = 0$. For, if we apply this with $N = \operatorname{Ker} f$, then $N \cap C_1^+ = 0$ implies $N = 0$.

Let R be the associated connected coalgebra and $\pi : C \to R$ the quotient map, as above, and assume that $N \cap C_1^+ = 0$. We claim that $\pi(N) \cap R_1^+ = 0$. To see this, it suffices to show $\pi(N) \cap \pi(C_1^+) = 0$ since $\pi(C_1^+) = R_1^+$ by 5.3.8. Choose $r \in \pi(N) \cap \pi(C_1^+)$; write $r = \pi(n) = \pi(c)$, for $n \in N, c \in C_1^+$. Then $n - c \in \operatorname{Ker} \pi = C_0^+ \subseteq C_1^+$, and so $n \in N \cap C_1^+ = 0$. Thus $r = \pi(n) = 0$, proving the claim.

Now consider $g : R \to R/\pi(N) = \pi(C)/\pi(N) = \pi(C/N)$. R is connected, and g is injective on R_1^+ since $\pi(N) \cap R_1^+ = 0$. Thus by 5.3.3, g is injective on R. Thus $\operatorname{Ker} g = \pi(N) = 0$. Then $N \subseteq \operatorname{Ker} \pi = C_0^+ \subseteq C_1^+$, and thus $N = N \cap C_1^+ = 0$. □

§5.4 The coradical filtration of pointed coalgebras

As noted after 5.1.5, if C is pointed then $C_0 = \mathbf{k}G$, where $G = G(C)$, the group-like elements. Recall that for $g, h \in G$,

$$P_{g,h}(C) = \{c \in C \mid \Delta c = c \otimes g + h \otimes c\},$$

the set of g, h-primitives of C. Note that $\mathbf{k}(g - h) = P_{g,h}(C) \cap P_{h,g}(C) \cap C_0$. For each pair $g, h \in G$, let $P'_{g,h}(C)$ be a subspace of $P_{g,h}(C)$ such that

$$P_{g,h}(C) = \mathbf{k}(g - h) \oplus P'_{g,h}(C)$$

(this subspace is not unique).

The following theorem of Taft and Wilson generalizes 5.3.2.

5.4.1 THEOREM [TW]. *Let C be a pointed coalgebra, with $G = G(C)$. Then*
1) $C_1 = \mathbf{k}G \oplus (\oplus_{g,h\in G} \ P'_{g,h}(C))$
2) for any $n \geq 1$ and $c \in C_n$,

$$c = \sum_{g,h\in G} c_{g,h}, \text{ where } \Delta c_{g,h} = c_{g,h} \otimes g + h \otimes c_{g,h} + w,$$

for some $w \in C_{n-1} \otimes C_{n-1}$.

The proof we give of 5.4.1 uses the methods of [R78] and [R82]; see also [Ml]. In fact the full strength of 5.4.1, 1) is not completely proved in [TW]. We first need a preliminary result; it is [A, 2.3.11]. A coalgebra C is said to have a *separable coradical* if for every simple subcoalgebra D of C, D^* is a separable k-algebra. The coradical is always separable if C is pointed or if k is algebraically closed.

5.4.2 THEOREM. *Let C be a coalgebra with separable coradical. Then there exists a coideal I of C such that $C = I \oplus C_0$ as* k*-spaces. That is, there exists a coalgebra projection of C onto C_0.*

PROOF. We first show that if C is finite-dimensional and D is a subcoalgebra, then a projection π from D to $D_0 = D \cap C_0$ can be extended to a projection from C to C_0. Let $\alpha = \pi' \circ i$ be the composite of the imbedding $i : C_0 \hookrightarrow C$ and the quotient map $\pi' : C \to C/\mathrm{Ker}\ \pi = E$. Then $\mathrm{Ker}\ \alpha = \{0\}$ and $\mathrm{Im}\ \alpha \subseteq E_0$. By 5.3.5, $E_0 \subseteq \mathrm{Im}\ \alpha$; thus $E_0 = \mathrm{Im}\ \alpha$.

Then $\alpha^* : E^* \to (C_0)^*$ is a surjective algebra morphism, and by hypothesis $(C_0)^*$ is a separable k-algebra. Thus by the Wedderburn Principal Theorem, $E^* = B \oplus \mathrm{Jac}(E^*)$ where $B \cong (C_0)^*$ as algebras. Then there exists an algebra map $\phi : (C_0)^* \to E^*$ such that $\alpha^* \circ \phi = id_{C_0^*}$. Now set $\tilde{\pi} = \phi^* \circ \pi'$; it is a projection from C to C_0 and extends π.

Now let $\mathcal{F}$ be the set of all pairs (F, π), where F is a subcoalgebra of C and $\pi : F \to F_0$ a projection. Since (C_0, id) is such a pair, $\mathcal{F} \neq \emptyset$. Moreover, $\mathcal{F}$ is an ordered set via

$$(F', \pi') \leq (F, \pi) \Leftrightarrow F' \subset F \text{ and } \pi\ |_{F'} = \pi'.$$

We apply Zorn's Lemma to $\mathcal{F}$ and obtain a maximal element (F, π); we claim that $F = C$.

Suppose $F \neq C$. Then there exists $c \in C, c \notin F$; let D be the subcoalgebra generated by c. Then $\mathrm{Im}(\pi \mid_{F \cap D}) = (F \cap D)_0$. Since D is finite-dimensional, the argument at the beginning of the proof applied to $F \cap D \subset D$ shows that there exists a projection $\pi_1 : D \to D_0$ which extends $\pi \mid_{F \cap D}$. Since (D, π_1) and (F, π) have projections which agree on $F \cap D$, they extend to a projection $\pi_2 : F + D \to (F + D)_0$. This contradicts the maximality of F. Thus $F = C$. □

Now, assume C is pointed and fix a coideal I satisfying $C = I \oplus C_0$ as in 5.4.2. For each $x \in G = G(C)$, define $e_x \in C^*$ as follows:

(5.4.3) $$e_x(I) = 0 \text{ and } e_x(y) = \delta_{x,y} \text{ for all } y \in G.$$

It is easy to verify that $\sum_{x \in G} e_x = \varepsilon$ and that the $\{e_x\}$ are orthogonal idempotents in C^*. Now for each $c \in C$ and $x, y \in G$, define:

(5.4.4) $${}^x c = c \leftharpoonup e_x, \quad c^y = e_y \rightharpoonup c, \quad \text{and } {}^x c^y = ({}^x c)^y = {}^x(c^y),$$

where we recall the definition of $\rightharpoonup$ and $\leftharpoonup$ from 1.6.5. Let ${}^x C^y = \{{}^x c^y \mid c \in C\}$. Then it is clear that $c = \sum_{x,y \in G} {}^x c^y$ and that

(5.4.5) $$I = \cap_{x \in G} \operatorname{Ker} e_x = \oplus_{x,y} ({}^x C^y)^+,$$

where recall that $V^+ = V \cap \operatorname{Ker} \varepsilon$ for any subspace V of C. Note that I and the components ${}^x C^y$ are not unique, since the complement obtained in 5.4.2 using the Wedderburn Principal Theorem was not unique. If $x = y$, then ${}^x C^x = ({}^x C^x)^+ \oplus \mathrm{k}x$, and if $x \neq y$, then ${}^x C^y = ({}^x C^y)^+$. One may also check the following formula, from [R 78, p. 285]:

(5.4.6) $$\Delta\, {}^x c^y = \sum_{z \in G} {}^x c_1^z \otimes {}^z c_2^y.$$

PROOF of 5.4.1. We first prove 2). Let $I_n = I \cap C_n$; then in fact $C_n = I_n \oplus C_0$. It will suffice to prove the result for $c \in I_n$. Now by 5.2.2, 2), C_n is stable under $\rightharpoonup$ and $\leftharpoonup$. It then follows as for 5.4.5 that $I_n = \oplus_{x,y} I_n \cap ({}^x C^y)^+$. Thus we may assume $c \in I_n \cap ({}^x C^y)^+$. In that case, using 5.2.2 and the facts that

$C_n = I_n \oplus C_0$ and $C_0 \subseteq C_{n-1}$, we have

$$\begin{aligned}\Delta c &\in C_n \otimes C_0 + C_0 \otimes C_n + \sum_{i=1}^{n-1} C_i \otimes C_{n-i} \\ &\subseteq I_n \otimes C_0 + C_0 \otimes I_n + C_{n-1} \otimes C_{n-1}.\end{aligned}$$

Thus we may write

$$(*) \qquad \Delta c = \sum_{g \in G} c_g \otimes g + \sum_{h \in G} h \otimes c_h + \sum_i v_i \otimes w_i$$

for $c_g, c_h \in I_n$ and $v_i, w_i \in C_{n-1}$. Apply 5.4.6 to $(*)$, using that $c = {}^x c^y$:

$$\Delta c = \sum_{g,z} {}^x c_g^z \otimes {}^z g^y + \sum_{h,z} {}^x h^z \otimes {}^z c_h^y + \sum_{i,z} {}^x v_i^z \otimes {}^z w_i^y.$$

Now ${}^z g^y = \delta_{z,g,y} y$ and ${}^x h^z = \delta_{x,h,z} x$; moreover $C^* \rightharpoonup C_n = C_n$ and $C_n \leftharpoonup C^* = C_n$ gives ${}^x v_i^z = v'_{i,z} \in C_{n-1}$ and ${}^z w_i^y = w'_{i,z} \in C_{n-1}$. Thus $(*)$ becomes

$$(**) \qquad \Delta c = {}^x(c_y)^y \otimes y + x \otimes {}^x(c_x)^y + \sum_{i,z} v'_{i,z} \otimes w'_{i,z}.$$

Applying $(\varepsilon \otimes id) \circ \Delta = id$, we obtain

$$c = \varepsilon({}^x(c_y)^y)y + \varepsilon(x){}^x(c_x)^y + \sum_{i,z} \varepsilon(v'_{i,z})w'_{i,z} = {}^x(c_x)^y + v,$$

for some $v \in C_{n-1}$, since $\varepsilon(I_n) = 0$. Similarly $c = {}^x(c_y)^y + u$, some $u \in C_{n-1}$. Substituting in $(**)$ gives

$$\Delta c = c \otimes y + x \otimes c + t, \text{ where } t \in C_{n-1} \otimes C_{n-1}.$$

This proves 2).

We now finish 1). From the remarks before the proof, we know that $C_1 = C_0 \oplus I_1$ and that $I_1 = \oplus_{x,y} {}^x(C_1)^{y+}$. We claim that $P_{y,x}(C) = \mathbf{k}(y - x) \oplus {}^x(C_1)^{y+}$. To see this, choose $w \in {}^x(C_1)^{y+}$; by 2),

$$\Delta w = w \otimes y + x \otimes w + \sum_{g,h \in G} \alpha_{g,h}(g \otimes h),$$

since $C_0 = \mathbf{k}G$. Using 5.4.6 again, since $w = {}^x w^y$, we see that

$$\Delta w = w \otimes y + x \otimes w + \alpha_{x,y}(x \otimes y).$$

Since $\varepsilon(w) = 0$, using $(\varepsilon \otimes id) \circ \Delta = id$ we see that $\alpha_{x,y} = 0$; that is $w \in P_{y,x}(C)$. Thus $\mathrm{k}(y - x) \oplus {}^x(C_1)^{y+} \subseteq P_{y,x}(C)$.

Conversely, choose any skew-primitive element $w \in P_{y,x}(C)^+$, so that $\Delta w = w \otimes y + x \otimes w$ and $\varepsilon(w) = 0$. For any $g, h \in G$, (5.4.6) gives

$$\Delta\, {}^g w^h = \delta_{h,y}\, {}^g w^h \otimes y + \delta_{g,x} x \otimes {}^g w^h.$$

Thus if $h \neq y$ and $g \neq x$, ${}^g w^h = 0$, and if $h = y$ and $g = x$, then ${}^x w^y \in P_{y,x}$, and so $\varepsilon({}^x w^y) = 0$. If $h = y$ but $g \neq x$, then $\Delta\, {}^g w^y = {}^g w^y \otimes y$; applying $id = (\varepsilon \otimes id) \circ \Delta$ gives ${}^g w^y = \varepsilon({}^g w^y) y$. Since ${}^g y^y = \delta_{g,y} y$, ${}^g w^y = 0$ unless $g = y$, and then ${}^y w^y = \alpha y$, some $\alpha \in \mathrm{k}$. Similarly ${}^x w^h = 0$ unless $h = x$, in which case ${}^x w^x = \beta x$, for $\beta \in \mathrm{k}$. Thus $w = \sum_{g,h} {}^g w^h = {}^x w^y + \alpha y + \beta x$. Since w and ${}^x w^y \in P_{y,x}(C)$, also $\alpha y + \beta x \in P_{y,x}(C)$, and so $\beta = -\alpha$. Thus $w \in {}^x(C_1)^{y+} + \mathrm{k}(y - x)$, proving the claim.

Now consider any complement $P'_{g,h}$ of $\mathrm{k}(g - h)$ in $P_{g,h}$. It is clear that $C_1 = \mathrm{k}G + \sum_{g,h} P'_{g,h}$, since $C_1 = C_0 \oplus (\oplus_{x,y}\, {}^x(C_1)^{y+})$ and ${}^x(C_1)^{y+} \subseteq P_{y,x}$. Thus it suffices to show the sum is direct. If not, choose $w \in \mathrm{k}G$ and $w_{g,h} \in P'_{g,h}$ such that $\alpha w + \sum_{g,h} \alpha_{g,h} w_{g,h} = 0$, for $\alpha, \alpha_{g,h} \in \mathrm{k}$. Now $w_{g,h} \in P_{g,h} = \mathrm{k}(g - h) + {}^h(C_1)^{g+}$ by the above, and so we may write $w_{g,h} = \beta_{g,h}(g - h) + u_{g,h}$, for $u_{g,h} \in {}^h(C_1)^{g+}$. The dependence relation now implies that $\alpha w + \sum_{g,h} \alpha_{g,h}\beta_{g,h}(g - h) = \sum_{g,h} -\alpha_{g,h} u_{g,h} \in C_0 \cap (\oplus_{g,h}\, {}^h(C_1)^{g+}) = 0$. Thus all $\alpha_{g,h} = 0$ since $\oplus\, {}^h(C_1)^{g+}$ is also direct. But then $\alpha = 0$. This proves 1). □

As a corollary, we show that a stronger version of 5.3.1 holds for pointed coalgebras. This fact has recently been shown independently in [Tk 92a] and in [R 93b].

5.4.7 COROLLARY. *Let $f : C \to D$ be a coalgebra map. Assume that C is pointed and that f is injective on each subspace $P_{g,h}(C)$ of g,h-primitives. Then f is injective on C.*

PROOF. By 5.3.1, it suffices to show that f is injective on C_1. First choose $g \neq h$ in $G = G(C)$. Then $0 \neq g - h \in P_{g,h}(C)$ and thus $0 \neq f(g - h) = f(g) - f(h)$. Thus f is injective on the set G. However, each $f(g) \in G(D)$ and distinct group-like elements are independent. Thus in fact f is injective on C_0. Let $P'_{g,h}(C)$ be a vector space complement of $\mathrm{k}(g - h)$ in $P_{g,h}(C)$ as in

5.4.1; since $P_{g,h}(C)$ imbeds into $P_{f(g),f(h)}(D)$ by assumption we may choose a vector space complement $P'_{f(g),f(h)}(D)$ of $k(f(g)-f(h))$ in $P_{f(g),f(h)}(D)$ so that $P'_{g,h}(C) \hookrightarrow P'_{f(g),f(h)}(D)$. Since C is pointed, 5.4.1 gives

$$C_1 = kG(C) \oplus (\oplus_{g,h} P'_{g,h}(C)) \hookrightarrow kG(D) \oplus (\oplus_{f(g),f(h)} P'_{f(g),f(h)}(D)) \subset D_1.$$

Thus f is injective on C_1. □

5.4.8 EXAMPLE. Let C be the $n \times n$ matrix coalgebra; that is, C has a basis of n^2 elements X_{ij}, for $1 \leq i, j \leq n$, and coalgebra structure given by $\Delta X_{ij} = \sum_k X_{ik} \otimes X_{kj}$ and $\varepsilon(X_{ij}) = \delta_{ij}$. Note that C is just the degree 1 part of $\mathcal{O}(M_n(k))$ as in 1.3.8, and also that $C^* \cong M_n(k)$, the algebra of $n \times n$ matrices over k. Since C^* is a simple algebra, C is a (co)simple coalgebra and thus $C_0 = C$.

Now let I be the k-span of $\{X_{ij} \mid i > j\}$; it is straightforward to verify that I is a coideal of C, and thus $D = C/I$ is a coalgebra. Let $\{\bar{X}_{ij} \mid i \leq j\}$ denote the basis for D, and note that D^* is the subalgebra of C^* of upper triangular matrices. Now for each $i \leq j$,

$$\Delta \bar{X}_{ij} = \bar{X}_{ii} \otimes \bar{X}_{ij} + \bar{X}_{i,i+1} \otimes \bar{X}_{i+1,j} + \ldots + \bar{X}_{ij} \otimes \bar{X}_{jj}.$$

Thus $G(D) = \{\bar{X}_{ii}, \text{ all } i\}$ and D is pointed with $D_0 = kG(D)$. Each $\bar{X}_{i,i+1}$ is $\bar{X}_{ii}, \bar{X}_{i+1,i+1}$-primitive, and $D_1 = D_0 + \sum_i k\bar{X}_{i,i+1}$; continuing, for each $m \leq n-1$ we have that D_m is the span of all $\bar{X}_{ij}$ with $j \leq i+m$. An easy way to see this is to use the fact that $D_m = (J^{m+1})^{\perp}$, where $J = \mathrm{Jac}(D^*)$, and the fact that J consists of the strictly upper triangular matrices.

Thus in this example we can see the coradical filtration explicitly. One can also see that elements in D_m have the form described in the Taft-Wilson theorem 5.4.1: for

$$\Delta \bar{X}_{i,i+m} = \bar{X}_{i,i} \otimes \bar{X}_{i,i+m} + \bar{X}_{i,i+m} \otimes \bar{X}_{i+m,i+m} + w,$$

where $w \in D_{m-1} \otimes D_{m-1}$. The example also shows that 5.3.5 can not be improved, since if $\pi : C \to D$ is the quotient map, $\pi(C_0) = D$, which properly contains D_0. Thus the coradical filtration is not preserved in homomorphic images.

§5.5 Examples: $U(\mathfrak{g})$ and $U_q(\mathfrak{g})$

We first consider enveloping algebras $U(\mathfrak{g})$ and show that they are always connected. Moreover, when $\mathbf{k}$ has characteristic 0, the coradical filtration of $U(\mathfrak{g})$ is just the obvious filtration by degree.

We need an elementary lemma.

5.5.1 LEMMA. *Let H be a Hopf algebra which contains subspaces $A_0 \subset A_1$ with A_0 a subalgebra with 1 such that*
1) A_1 generates H as an algebra, and as an algebra and $A_0A_1, A_1A_0 \subset A_1$
2) $\Delta A_0 \subseteq A_0 \otimes A_0$, and $\Delta A_1 \subseteq A_1 \otimes A_0 + A_0 \otimes A_1$.
Then, if we set $A_n = (A_1)^n$ for all $n \geq 1, \{A_n\}$ is a coalgebra filtration of H and $A_0 \supseteq H_0$. If also $A_0 = H_0$, then $A_n \subseteq H_n$, for all n.

PROOF. Since $1 \in A_0$, it follows that $\Delta A_n \subseteq (A_1 \otimes A_0 + A_0 \otimes A_1)^n \subseteq \sum_{k=0}^{n}(A_1 \otimes A_0)^k(A_0 \otimes A_1)^{n-k} \subseteq \sum_{k=0}^{n} A_k \otimes A_{n-k}$. Moreover $\cup_{n\geq 0}A_n = A$ by 1). Thus $\{A_n\}$ is a coalgebra filtration. By 5.3.4, $A_0 \supseteq H_0$.

Now assume $A_0 = H_0$. By induction, we may assume $A_{n-1} \subseteq H_{n-1}$. Then the above computation gives

$$\Delta A_n \subseteq H_0 \otimes A_n + A_n \otimes A_{n-1} \subseteq H_0 \otimes H + H \otimes H_{n-1},$$

and so $A_n \subseteq H_n$ by 5.2.1. □

Now let $\mathfrak{g}$ be a Lie algebra and fix an ordered basis $\{x_\lambda \mid \lambda \in \Lambda\}$ for $\mathfrak{g}$. We will use "multi-index" notation for the PBW basis monomials of $U(\mathfrak{g})$. That is, we consider functions $\mathfrak{n} : \Lambda \to \mathbf{Z}_{\geq 0}$ with finite support, say $\mathfrak{n}(\lambda) \neq 0 \Leftrightarrow \lambda \in \{\lambda_1, \cdots, \lambda_m\}$, where $\lambda_1 < \lambda_2 < \cdots < \lambda_m$. Then $x^{\mathfrak{n}}$ denotes the basis monomial $x_{\lambda_1}^{\mathfrak{n}(\lambda_1)} x_{\lambda_2}^{\mathfrak{n}(\lambda_2)} \cdots x_{\lambda_m}^{\mathfrak{n}(\lambda_m)}$, and any $a \in U(\mathfrak{g})$ may be written as $a = \sum_{\mathfrak{n}} \alpha_{\mathfrak{n}} x^{\mathfrak{n}}$, where $\alpha_{\mathfrak{n}} \in \mathbf{k}$.

In addition, $\mathfrak{m} \leq \mathfrak{n}$ means $\mathfrak{m}(\lambda) \leq \mathfrak{n}(\lambda)$ for all $\lambda \in \Lambda, \mathfrak{n}! = \prod_{\lambda\in\Lambda} \mathfrak{n}(\lambda)!$, and $\binom{\mathfrak{n}}{\mathfrak{m}} = \prod_{\lambda\in\Lambda} \binom{\mathfrak{n}(\lambda)}{\mathfrak{m}(\lambda)}$. Set $|\mathfrak{n}| = \sum_{\lambda\in\Lambda} \mathfrak{n}(\lambda)$.

In this notation, one has for all basis monomials $x^{\mathfrak{n}}$,

$$\Delta x^{\mathfrak{n}} = \sum_{0\leq \mathfrak{m} \leq \mathfrak{n}} \binom{\mathfrak{n}}{\mathfrak{m}} x^{\mathfrak{m}} \otimes x^{\mathfrak{n}-\mathfrak{m}}. \tag{5.5.2}$$

5.5.3 PROPOSITION. 1) *For any $\mathfrak{g}$, $H = U(\mathfrak{g})$ is connected with $H_0 = \mathbf{k}1$.*

2) *If* k *has characteristic* 0, *then* $P(H) = \mathrm{g}$ *and so* $H_1 = \mathrm{g} \oplus \mathrm{k}1$.
3) *If* k *has characteristic* $p \neq 0$, *then* $P(H) = \hat{\mathrm{g}}$, *the restricted Lie algebra spanned by all* $\{x^{p^k} \mid x \in \mathrm{g}, k \geq 0\}$, *and so* $H_1 = \hat{\mathrm{g}} \oplus \mathrm{k}1$.
Moreover $U(\mathrm{g}) = u(\hat{\mathrm{g}})$, *the restricted enveloping algebra of* $\hat{\mathrm{g}}$.
4) *In any characteristic,* $H_n = (H_1)^n$ *for all* $n \geq 1$.

PROOF. 1) Let $A_0 = \mathrm{k}1$ and $A_1 = \mathrm{g} \oplus \mathrm{k}1$. Then $A_0 \subset A_1$ satisfy the hypotheses of the Lemma and so $A_0 \supseteq H_0$. But clearly $A_0 \subseteq H_0$; thus $A_0 = H_0$ and H is connected.

2) Assume char $\mathrm{k} = 0$; we also show 4) in this case. Let $A_1 = \mathrm{g} \oplus \mathrm{k}1$ as in 1); then the lemma also gives $A_n \subseteq H_n$ for all n, where $A_n = (A_1)^n$. Thus we must show $H_n \subseteq A_n$ for $n \geq 1$. We proceed by induction on n; the case $n = 0$ is part 1). Thus assume true for $n-1$, and choose $a \in H_n$. Thus $\Delta a \in H \otimes H_{n-1} + H_0 \otimes H = H \otimes A_{n-1} + 1 \otimes H$. Write $a = \sum_{\mathrm{m}} \alpha_{\mathrm{m}} x^{\mathrm{m}}$ and let x^{t} be a monomial in a of maximal degree $|\mathrm{t}|$. By 5.5.2,

$$\Delta x^{\mathrm{t}} = \sum_{0<\mathrm{s}\leq\mathrm{t}} \binom{\mathrm{t}}{\mathrm{s}} x^{\mathrm{s}} \otimes x^{\mathrm{t}-\mathrm{s}} + 1 \otimes x^{\mathrm{t}}.$$

Since char $\mathrm{k} = 0$, $\binom{\mathrm{t}}{\mathrm{s}} \neq 0$, and so $\alpha_{\mathrm{t}} \binom{\mathrm{t}}{\mathrm{s}} x^{\mathrm{s}} \otimes x^{\mathrm{t}-\mathrm{s}}$ is a non-trivial term in Δa. When $|\mathrm{s}| > 0$, it is not in $1 \otimes H$, and thus must be in $H \otimes A_{n-1}$. Thus $|\mathrm{t}-\mathrm{s}| \leq n-1$, for all $0 < \mathrm{s} < \mathrm{t}$. In particular we may choose s with $|\mathrm{s}| = 1$; it follows that $|\mathrm{t}| \leq n$. Thus x^{t}, and so a, is in $A_n = (\mathrm{g} \oplus \mathrm{k}1)^n$. Thus $H_n = A_n$. In particular $H_1 = \mathrm{g} \oplus \mathrm{k}1 = P(H) \oplus \mathrm{k}1$, and so $\mathrm{g} = P(H)$.

3) Assume char $\mathrm{k} = p \neq 0$. First note that $\hat{\mathrm{g}} \subseteq P(H)$, and in fact $\hat{\mathrm{g}}$ is a restricted Lie subalgebra of $P(H)$. It is not difficult to see that $U(\mathrm{g}) = u(\hat{\mathrm{g}})$; we will work with H in this second form. Thus let $\{x_\lambda \mid \lambda \in \Lambda\}$ denote an ordered basis of $\hat{\mathrm{g}}$ and use multi-index notation for the restricted PBW basis monomials of $u(\hat{\mathrm{g}})$; the only difference from the characteristic 0 situation is that now $\mathrm{n} : \Lambda \to \{0, 1, \ldots, p-1\}$. Now let $A_1 = \hat{\mathrm{g}} \oplus \mathrm{k}1$ and $A^n = (A_1)^n$, and proceed as in 2). Since $\mathrm{n}(\lambda) < p$ for each $\lambda \in \Lambda$, each $\binom{\mathrm{n}(\lambda)}{\mathrm{m}(\lambda)} \neq 0$ and so $\binom{\mathrm{n}}{\mathrm{m}} \neq 0$ for any n. Thus as before $\alpha_{\mathrm{t}} \binom{\mathrm{t}}{\mathrm{s}} x^{\mathrm{s}} \otimes x^{\mathrm{t}-\mathrm{s}}$ is a non-trivial term in Δa, which gives $|\mathrm{t}| \leq n$. Thus x^t, and so a, is in $A_n = (\hat{\mathrm{g}} \oplus \mathrm{k}1)^n$. Thus $H_n = A_n$. In particular we have shown $H_1 = \hat{\mathrm{g}} \oplus \mathrm{k}1$ and so $P(H) = \hat{\mathrm{g}}$.
□

When $\mathfrak{g}$ is finite-dimensional, we can see explicitly the relationship between H_n and $\mathrm{Jac}(H^*)$ as shown in 5.2.9.

5.5.4 LEMMA. *Let $\mathfrak{g}$ be a finite-dimensional Lie algebra of characteristic 0, and fix a basis $\{x_1, \dots, x_d\}$ of $\mathfrak{g}$. Choose $f_i \in U(\mathfrak{g})^*$, all $1 \le i \le d$, by setting $f_i(x_j) = \delta_{ij}$ and $f_i = 0$ on all monomials in the $\{x_j\}$ of degree $\neq 1$. Then $U(\mathfrak{g})^* \cong \mathrm{k}[[f_1, \dots, f_d]]$, the algebra of formal power series in the $\{f_i\}$, as algebras.*

PROOF. Consider the PBW basis monomials x^{n} as before, and define $f^{(\mathrm{n})} \in U(\mathfrak{g})^*$ by $\langle f^{(\mathrm{n})}, x^{\mathrm{m}} \rangle = \delta_{\mathrm{n,m}}$. Then for any n, m, using 5.5.2

$$\begin{aligned} f^{(\mathrm{n})} f^{(\mathrm{m})}(x^{\mathrm{t}}) &= \sum_{0 \le \mathrm{s} \le \mathrm{t}} \binom{\mathrm{t}}{\mathrm{s}} \langle f^{(\mathrm{n})}, x^{\mathrm{s}} \rangle \langle f^{(\mathrm{m})}, x^{\mathrm{t-s}} \rangle \\ &= \sum_{0 \le \mathrm{s} \le \mathrm{t}} \binom{\mathrm{t}}{\mathrm{s}} \delta_{\mathrm{n,s}} \delta_{\mathrm{m,t-s}} = \binom{\mathrm{n+m}}{\mathrm{n}} f^{(\mathrm{n+m})}(x^{\mathrm{t}}). \end{aligned}$$

Setting $f^{\mathrm{n}} = \mathrm{n}! f^{(\mathrm{n})}$, we see that $f^{\mathrm{n}} f^{\mathrm{m}} = f^{\mathrm{n+m}}$. But for each i, note that $f_i = f^{(\mathrm{n}_i)}$, where $\mathrm{n}_i(j) = \delta_{ij}$. It now follows that as algebras, $U(\mathfrak{g})^* \cong \mathrm{k}[[f_1, \dots, f_d]]$. □

We now see the connection with 5.2.9. The Jacobson radical J of the power series algebra is just the ideal generated by all the $\{f_i\}$, and so J^n is generated by all monomials in the f_i of degree n. By construction of the f_i, and their product in 5.5.4, it is clear that $(J^{n+1})^{\perp}$ is the set of elements of $U(\mathfrak{g})$ of degree $\le n$. Thus this set must be H_n.

This fact can be used to give an alternate proof of 5.5.3 when k has char 0. For since $\{H_n\}$ is determined only by the coalgebra structure of $U(\mathfrak{g})$, we may assume that $\mathfrak{g}$ is abelian. It follows that any $a \in U(\mathfrak{g})$ lies in $U(\mathfrak{g}_1)$ for $\mathfrak{g}_1$ a finite-dimensional Lie subalgebra of $\mathfrak{g}$. The result is true for $U(\mathfrak{g}_1)$ by 5.5.4 and the above remarks. Now apply 5.2.12 to see that if $a \in U(\mathfrak{g})_n$, then $a \in U(\mathfrak{g}_1)_n$ and so has degree $\le n$.

We now consider the quantized enveloping algebras $U_q(\mathfrak{g})$, for $\mathfrak{g}$ a finite-dimensional semisimple Lie algebra of rank n over k of char 0. From the Appendix, we know that $U_q(\mathfrak{g}) = \mathbf{C}\langle E_i, F_i, K_i, K_i^{-1} \mid 1 \le i \le n \rangle$ where the K_i are all group-like and both E_i and F_i are K_i^{-1}, K_i-primitive.

The next lemma has been observed in both [M 93b] and [R 93b].

5.5.5 LEMMA. *$H = U_q(\mathbf{g})$ is pointed and $G(H)$ is the group generated by all the $\{K_i\}$. Thus $H_0 \cong \mathbf{k}\mathbf{Z}^n$.*

5.5.5 PROOF. Let G be the group generated by the $\{K_i\}$, let $A_0 = \mathbf{k}G$, and let $A_1 = \sum_i \mathbf{k}E_i + \mathbf{k}F_i + A_0$. Then $A_0 \subset A_1$ satisfies the hypotheses of 5.5.1, and $A_0 \subseteq H_0$. Thus $H_0 = A_0$ and so H is pointed with coradical $\mathbf{k}G$. □

In fact the lemma also shows that $A_n \subseteq H_n$ for all n, where $A_n = (A_1)^n$. By analogy with the situation for $U(\mathbf{g})$ in char 0, we ask:

5.5.6 QUESTION. If q is not a root of 1, is $H_n = A_n$ for $U_q(\mathbf{g})$?

The answer to the question is yes for $\mathbf{g} = s\ell(2)$ [M 93b]; the argument is a more careful version of 5.5.3 using q-binomial coefficients and the structure of C_n given in 5.4.1. For $\mathbf{g} = s\ell(n)$, Takeuchi has shown that $H_1 = A_1$, using arguments from [Tk 92a, Sec. 6].

Even if it is not possible to determine H_n for all n, it would still be very interesting to know that $H_1 = A_1$, for then we would have a complete description of the skew-primitive elements in $U_q(\mathbf{g})$.

§5.6 The structure of pointed cocommutative Hopf algebras

We prove here some basic facts about the structure of pointed cocommutative Hopf algebras. Although details may be found in both [S] and [A], the results are so important that we give a (somewhat sketchy) proof. Moreover our proof of 5.6.5 is somewhat more direct than those in [S] and [A].

First note that the "pointed" assumption is not really that special, since any cocommutative Hopf algebra becomes pointed by extending the base field. To see this fact, it suffices to show that if $\mathbf{k}$ is algebraically closed and H is cocommutative, then H is pointed. In that situation, let C be a simple subcoalgebra of H. By 5.1.4 C^* is a (finite-dimensional) commutative simple algebra over $\mathbf{k}$, and thus $C^* \cong \mathbf{k}$. Then also $C \cong \mathbf{k}$ and so is one-dimensional. Thus H is pointed.

We first consider the structure of cocommutative coalgebras.

5.6.1 DEFINITION. A coalgebra C is *irreducible* if any two non-zero subcoalgebras have non-zero intersection. A subcoalgebra D of a coalgebra C is an *irreducible component* of C if it is a maximal irreducible subcoalgebra.

It is easy to see that C is irreducible $\Leftrightarrow$ it contains a unique simple subcoalgebra C_0.

The lemma gives the basic properties of irreducible coalgebras.

5.6.2 LEMMA. *Let C be a coalgebra and $\{C_\alpha\}$ a family of subcoalgebras.*
1) *If $C = \sum_\alpha C_\alpha$, then any simple subcoalgebra of C lies in one of the C_α*
2) *if $\bigcap_\alpha C_\alpha \neq 0$ and each C_α is irreducible, then $\sum_\alpha C_\alpha$ is irreducible*
3) *any irreducible subcoalgebra of C is contained in a unique irreducible component*
4) *a sum of distinct irreducible components is direct.*

PROOF. 1) Let D be a simple subcoalgebra of C. Since D is finite-dimensional, it lies in some finite sum $\sum_{i=1}^n C_{\alpha_i}$. By induction on n, it suffices to prove that if $D \subset C_\alpha + C_\beta$, then either $D \subset C_\alpha$ or $D \subset C_\beta$. Suppose $D \not\subset C_\alpha$. Then $D \cap C_\alpha = 0$ since D is simple. Now choose $f \in C^*$ with $f|_D = \varepsilon_D$ and $f|_{C_\alpha} = 0$. Then for any $d \in D, f \rightharpoonup d = d$ but for $c \in C_\alpha, f \rightharpoonup c = 0$. Thus since $D \subseteq C_\alpha + C_\beta$,

$$d = f \rightharpoonup d \in (f \rightharpoonup C_\alpha) + (f \rightharpoonup C_\beta) = f \rightharpoonup C_\beta \subseteq C_\beta.$$

Thus $D \subseteq C_\beta$.
2) Since $\bigcap_\alpha C_\alpha \neq 0$, it must contain a simple subcoalgebra D, and D is the unique simple in each C_α by irreducibility. By 1), D is also the unique simple subcoalgebra of $\sum C_\alpha$ and thus $\sum C_\alpha$ is irreducible.
3) By 2) the sum E of all irreducible subcoalgebras containing the given subcoalgebra is irreducible. By construction E is maximal irreducible, and so is an irreducible component. It is clearly unique.
4) Suppose that $\{C_\alpha\}$ is some collection of distinct irreducible components and that $0 \neq C_\beta \cap \sum_{\alpha \neq \beta} C_\alpha$. Let D be the unique simple subcoalgebra of C_β; it follows that $D \subseteq C_\beta \cap \sum_{\alpha \neq \beta} C_\alpha$. Now by 1) applied to $\sum_{\alpha \neq \beta} C_\alpha$, we must have $D \subset C_\gamma$, for some $\gamma \neq \beta$. Thus $C_\beta \cap C_\gamma \neq 0$. By 2), $C_\beta + C_\gamma$ is irreducible. But both C_β and C_γ are maximal irreducible subcoalgebras, and so $C_\beta = C_\beta + C_\gamma = C_\gamma$, a contradiction. Thus the sum was direct. □

5.6.3 THEOREM. *If C is a cocommutative coalgebra, then C is the direct sum of its irreducible components.*

PROOF. By the Lemma, parts 3) and 4), it suffices to show that C is the sum of irreducible subcoalgebras. Thus, choose $x \in C$ and let $C(x)$ be the subcoalgebra generated by x; it suffices to show that $C(x)$ is such a sum. Replacing C by $C(x)$, we may assume C is finite-dimensional, by 5.1.1. Thus C^* is a finite-dimensional commutative algebra.

But now an old theorem about commutative algebras says that $C^* = A_1 \oplus \cdots \oplus A_n$, where the A_i are local rings. Thus $C \cong C^{**} \cong A_1^* \oplus \cdots \oplus A_n^*$ as coalgebras. It follows by 5.1.4, 3), that each A_i^* is irreducible since A_i has a unique maximal ideal. □

This result is certainly false if C is not cocommutative; for example, H_4 has irreducible components k1 and kg, whose sum is only two-dimensional.

Now for any H recall the set $G = G(H)$ of group-like elements. For each $x \in G$, let H_x denote the irreducible (connected) component containing x.

5.6.4 COROLLARY. *Let H be a Hopf algebra with $G = G(H)$. Then*
1) $H_x H_y = H_{xy}$ *and* $SH_x \subseteq H_{x^{-1}}$, *for all* $x, y \in G$; *in particular* H_1 *is a subHopfalgebra of* H
2) H_1 *is a* kG*-module algebra via* $x \cdot h = xhx^{-1}$, *for all* $x \in G, h \in H_1$
3) *if* H *is pointed cocommutative, then* $H_1 \# \mathrm{k}G \cong H$ *via* $h \# x \mapsto hx$.

PROOF. We show 1) and 2) together. The map $\phi_x : H \to H$ given by $h \mapsto xh$ is clearly a coalgebra automorphism of H, for each $x \in G$. Thus $\phi_x(H_1) = xH_1$ is the irreducible component of H containing x, and so $xH_1 = H_x$. Similarly $H_1 x = H_x$. Thus $xH_1 = H_1 x$, and $xH_1x^{-1} = H_1$.

We claim that $(H_1)^2 = H_1$. First, it follows by 5.1.10 that $H_1 \otimes H_1$ is connected, with $(H_1 \otimes H_1)_0 = \mathrm{k}(1 \otimes 1)$. The map $f : H_1 \otimes H_1 \to H$ given by $h \otimes k \mapsto hk$ is a coalgebra morphism, and thus 5.3.5 shows that $\mathrm{Im} f = (H_1)^2$ is connected. Since H_1 is the maximal irreducible component containing 1, it follows that $(H_1)^2 \subseteq H_1$; equality holds since $1 \in H_1$. Thus H_1 is a bialgebra. Moreover

$$H_x H_y = (xH_1)(yH_1) = xy(H_1)^2 = xyH_1 = H_{xy},$$

proving the first part of 1).

It remains only to show that $SH_x \subseteq H_{x^{-1}}$. Now $S : H^{cop} \to H$ is a coalgebra morphism, and $(H_x)^{cop}$ is connected since H_x is. Thus $S(H_x^{cop})$ is also connected by 5.3.5. Since $x^{-1} = Sx$, it follows that $SH_x \subseteq H_{x^{-1}}$, for any $x \in G$.

3) Since H is pointed, the H_x are the only irreducible components of H, and by 5.6.3 we have $H = \oplus_{x \in G} H_x$ as coalgebras. Since $H_x = H_1 x$ by 1), in fact $H \cong H_1 \otimes \mathrm{k}G$ as coalgebras. For, each $h \in H$ can be written uniquely as $\sum_x h_x x$, where $h_x \in H_1$; then the map $\phi : H \to H_1 \otimes \mathrm{k}G$ given by $h \mapsto \sum_x h_x \otimes x$ is a coalgebra isomorphism. Next we see that $H \cong H_1 \# \mathrm{k}G$ as algebras. For, if $h, k \in H_1$, then

$$\phi((hx)(ky)) = \phi(hxkx^{-1}xy) = \phi(h(x \cdot k)xy) = h(x \cdot k) \otimes xy.$$

But this is the usual multiplication in $H_1 \# \mathrm{k}G$. Finally, ϕ is a Hopf algebra map, using $S(h \otimes x) = (1 \otimes x^{-1})(Sh \otimes 1)$ in $H_1 \# \mathrm{k}G$. □

The decomposition in 5.6.4, 3) is attributed to Cartier and Gabriel in [Di] and to Kostant in [S]. Using this decomposition, it follows that understanding the structure of pointed cocommutative Hopf algebras reduces to understanding the structure of H_1, the irreducible (and so connected) component of 1. In characteristic 0, this description is due independently to Cartier [Ca], who stated it in terms of formal groups, and to Kostant (unpublished); see [Di, Ch. 11], [S, 13.0.1], [A, 2.5.3].

5.6.5 THEOREM. *Let H be a cocommutative connected Hopf algebra over a field* k *of characteristic* 0. *Then $H \cong U(\mathrm{g})$, for $\mathrm{g} = P(H)$.*

The proof we give of 5.6.5 follows notes of Blattner [B]. We first make some comments about an arbitrary vector space V and its dual V^*. As noted in the discussion after 5.1.2, it can happen that $U \neq U^{\perp\perp}$ for a subspace U of V^*. We say that U is *closed* if $U = U^{\perp\perp}$; more generally, $U^{\perp\perp}$ is the *closure* $\bar{U}$ of U. It is not difficult to show that finite-dimensional subspaces of V^* are closed.

Now let C be a connected (= pointed irreducible) coalgebra, with $C_0 = \mathrm{k}1$. Then $M = C_0^{\perp}$ is the unique maximal ideal of C^*. Writing $\overline{M^n} = (M^n)^{\perp\perp}$ for the closure of M^n, we see by 5.2.9 that $\cap_{n \geq 0} \overline{M^n} = 0$, since $\overline{M^n} = C_{n-1}^{\perp}$. Thus the $\{\overline{M^n}\}$ give a descending filtration of C^*. We construct a graded algebra from this filtration, as follows:

For each $n \geq 0$, define $gr_n C^* = \overline{M^n}/\overline{M^{n+1}}$ and $gr C^* = \oplus_{n \geq 0} gr_n C^*$. Then $gr\ C^*$ is a graded algebra. For, first note that $\overline{M^n}\,\overline{M^m} \subseteq \overline{M^{n+m}}$; this follows since $\overline{M^n} = C_{n-1}^{\perp}$, $\overline{M^m} = C_{m-1}^{\perp}$, $\overline{M^{m+n}} = (C_{m+n-1})^{\perp}$, and the fact that

$\Delta C_{m+n-1} \subseteq \sum_{i=0}^{m+n-1} C_i \otimes C_{m+n-1-i}$ by 5.2.2. We then get an induced map $gr_n C^* \times gr_m C^* \to gr_{n+m} C^*$.

5.6.6 LEMMA. *Let C be a connected coalgebra with $\mathbf{g} = P(C)$. Then*
1) $gr_1 C^* \cong \mathbf{g}^*$
2) *if also* $\mathbf{g}$ *is finite-dimensional, then* $\dim gr_n C^* < \infty$ *for all n, and $gr\ C^*$ is generated by* 1 *and* $gr_1 C^*$.

PROOF. 1) Note $\overline{M} = M$ since $M = C_0^\perp$. By 5.3.2, $\overline{M^2} = (M^2)^{\perp\perp} = C_1^\perp = (\mathbf{k}1 \oplus \mathbf{g})^\perp$. Thus if $\phi : C^* \to (\mathbf{k}1 \oplus \mathbf{g})^*$ denotes the restriction map, Ker $\phi = (\mathbf{k}1 \oplus \mathbf{g})^* = \overline{M^2}$. Thus ϕ induces an isomorphism ψ of $C^*/\overline{M^2}$ with $(\mathbf{k}1 \oplus \mathbf{g})^\perp$. But $M/\overline{M^2}$ is the (unique) maximal ideal of $C^*/\overline{M^2}$ and corresponds to the set $\{w \in (\mathbf{k}1 + \mathbf{g})^* \mid \langle w, 1\rangle = 0\} \cong \mathbf{g}^*$.

2) Since M^n is dense in $\overline{M^n}$, it follows that $(M^n + \overline{M^{n+1}})/\overline{M^{n+1}}$ is dense in $\overline{M^n}/\overline{M^{n+1}}$. Now multiplication in $gr\ C^*$ determines a linear map

$$\psi_n : (M/\overline{M^2})^{\otimes^n} \to \overline{M^n}/\overline{M^{n+1}}$$

with Im $\psi_n = (M^n + \overline{M^{n+1}})/\overline{M^{n+1}}$. Since $M/\overline{M^2} \cong \mathbf{g}^*$ by 1), and $\mathbf{g}^*$ is finite-dimensional, it follows that Im ψ_n is finite-dimensional, and so closed. By density, ψ_n is surjective. Thus $\overline{M^n}/\overline{M^{n+1}} = $ Im ψ_n is finite-dimensional, for all n. Also ψ_n surjective, for all n, gives that $gr\ C^*$ is generated by $gr_1 C^*$ and 1. □

5.6.7 LEMMA. *Assume char* $\mathbf{k} = 0$, *let* $\mathbf{g}$ *be a finite-dimensional Lie algebra, and let $C = U(\mathbf{g})$. Then the linear isomorphism $\psi : \mathbf{g}^* \to gr_1 U(\mathbf{g})^*$ extends to a graded algebra isomorphism*

$$\tilde{\psi} : S(\mathbf{g}^*) \to gr\ U(\mathbf{g})^*,$$

where here $S(\mathbf{g}^*)$ is the symmetric algebra on $\mathbf{g}^*$.

PROOF. Since char $\mathbf{k} = 0$, $\mathbf{g} = P(C)$ by 5.5.3, 2). By 5.5.4, $U(g)^* \cong \mathbf{k}[[f_1, \ldots, f_d]]$, where the f_i were dual to a basis $\{x_1, \ldots, x_d\}$ of $\mathbf{g}$. Now in this case M consists of all power series with 0 constant term, hence it is the ideal generated by $f_1, \ldots, f_d$. It follows that M^n is generated by all monomials in the $\{f_i\}$ of total degree n, so that M^n consists of all power series involving only those $f^{\mathbf{m}}$ with $|\mathbf{m}| \geq n$. But this is precisely the annihilator in $U(\mathbf{g})^*$ of the set of monomials $x^{\mathbf{m}}$ with $|\mathbf{m}| < n$. Thus M^n is already closed.

Moreover $gr_n C^* = M^n/M^{n+1}$ has as a basis the cosets of f^m with $|m| = r$. It follows that $gr\ C^* \cong \mathbf{k}[f_1, \ldots, f_d]$ as a graded algebra. Since $\{f_1, \ldots, f_d\}$ is a basis of $\mathbf{g}^*, \mathbf{k}[f_1, \ldots, f_d] = S(\mathbf{g}^*)$, the symmetric algebra over $\mathbf{k}$ on the vector space $\mathbf{g}^*$. □

PROOF of 5.6.5. By the universal property of enveloping algebras, the identity map $\mathbf{g} = P(H) \hookrightarrow H$ extends to a Hopf algebra map $\phi : U(\mathbf{g}) \to H$. Moreover ϕ is injective by 5.3.3. By the PBW theorem we know that $U(\mathbf{g})$ is linearly isomorphic to the symmetric algebra $S(\mathbf{g})$, and that the isomorphism preserves the coalgebra structure. Thus we may reduce to the following situation:

$$\phi : S(\mathbf{g}) \to H \text{ is a coalgebra injection and } \phi|_{\mathbf{g}} = id_{\mathbf{g}}.$$

Thus we may regard $S(\mathbf{g})$ as a subcoalgebra of the coalgebra $C = H$.

Let $\mathbf{h}$ be a non-zero finite-dimensional subspace of $\mathbf{g}$, and let $C(\mathbf{h})$ be the sum of all subcoalgebras D of C such that $D \cap \mathbf{g} \subseteq \mathbf{h}$. Now $S(\mathbf{h}) \cap \mathbf{g} = \mathbf{h}$; thus $S(\mathbf{h}) \subseteq C(\mathbf{h})$ and also $P(C(\mathbf{h})) = \mathbf{h}$.

The restriction map $\phi^* : C(\mathbf{h})^* \to S(\mathbf{h})^*$ is surjective and continuous, and $\phi^*(M) = N$, where M and N are the unique (closed) maximal ideals of $C(\mathbf{h})^*$ and $S(\mathbf{h})^*$, respectively. Hence $\phi^*(M^n) = N^n$ and so $\phi^*(\overline{M^n}) \subseteq \overline{N^n}$, for all n. Thus ϕ^* induces a map

$$gr\ \phi^* : gr\ C(\mathbf{h})^* \to gr\ S(\mathbf{h})^*.$$

Regarding $\mathbf{h}$ as an abelian Lie algebra, $U(\mathbf{h}) = S(\mathbf{h})$, and thus by 5.6.7 we have $gr\ S(\mathbf{h})^* \cong S(\mathbf{h}^*)$ as graded algebras. By 5.6.6, $gr\ C(\mathbf{h})^*$ is generated by $gr_1 C(\mathbf{h})^* \cong \mathbf{h}^*$; also $C(\mathbf{h})^*$, and hence $gr\ C(\mathbf{h})^*$, is commutative. Thus we have a canonical surjective morphism of graded algebras $\psi : S(\mathbf{h}^*) \to gr\ C(\mathbf{h})^*$. The composition $(gr\phi^*) \circ \psi : S(\mathbf{h})^* \to S(\mathbf{h}^*)$ is the identity on $\mathbf{h}^*$, and so is the identity. Thus $gr\ \phi^*$ is a bijection.

We claim that ϕ^* is injective (and so a bijection). Choose $0 \neq f \in \ker \phi^*$. Since $\cap_{n \geq 0} \overline{M^n} = 0$ by 5.2.9, we may find m such that $f \in \overline{M^m}$ but $f \notin \overline{M^{m+1}}$. Let $\bar{f}$ be the class of f in $gr_m C(\mathbf{h})^*; \bar{f} \neq 0$. But $\phi^*(f) = 0$ implies $(gr\ \phi^*)\bar{f} = 0$, a contradiction. Thus ϕ^* is injective, so $\phi : S(\mathbf{h}) \hookrightarrow C(\mathbf{h})$ is surjective, and so $S(\mathbf{h}) = C(\mathbf{h})$.

Finally, let D be any finite-dimensional subcoalgebra of H. Then $\mathbf{h} = D \cap \mathbf{g}$ is finite-dimensional and $D \subseteq C(\mathbf{h})$. Thus $D \subseteq S(\mathbf{h}) \subseteq S(\mathbf{g})$. But H

is the union of all such D; thus $H \subseteq S(\mathfrak{g})$ and so $H = S(\mathfrak{g})$. This proves the theorem. □

In characteristic $p \neq 0$, 5.6.5 is false, as is seen by the important example of divided power algebras.

5.6.8 EXAMPLE. Consider a set of symbols $\{x^{(0)} = 1, x^{(1)}, \ldots, x^{(n)}, \ldots\}$ and let these be a basis of H over k. H is an algebra via $x^{(k)} \cdot x^{(\ell)} = \binom{k+\ell}{k} x^{(k+\ell)}$, and a coalgebra via

$$\Delta x^{(n)} = \sum_{\ell=0}^{n} x^{(n-\ell)} \otimes x^{(\ell)} \text{ and } \varepsilon(x^{(n)}) = \delta_{n,0}.$$

Setting $S(x^{(n)}) = (-1)^n x^{(n)}$ makes H into a Hopf algebra. If k has characteristic 0, then H is a familiar object; namely, $H \cong \mathrm{k}[x]$, by sending $x^{(n)}$ to $\frac{1}{n!}x^n$. However if k has characteristic $p \neq 0, H$ is new. H is irreducible but is not generated by its primitive elements $P(H) = \mathrm{k}x^{(1)}$; thus it is not an enveloping algebra. In characteristic p it is possible that the sequence $x^{(n)}$ of divided powers is finite; in this case dim $H = p^m$ for some m, and H is spanned by $\{x^{(i)} \mid i = 0, \ldots, p^{m-1}\}$.

When the space of primitive elements of H is finite-dimensional, the coalgebra structure of H is well-understood. To see this, we describe the algebra structure of H^*; this was originally proved using formal groups. See [Ca, Thoerem 2], [DGa, p. 346], [W, 14.4]; also compare with 5.5.4.

5.6.9 THEOREM. *Let H be a cocommutative connected Hopf algebra over a perfect field* k *of characteristic $p \neq 0$ and assume that $P(H)$ is finite-dimensional. Then H^* is a truncated power series algebra, that is*

$$H^* \cong \mathrm{k}[[T_1, \ldots, T_{r+s}]]/(T_{r+1}^{p^{n_1}}, \ldots, T_{r+s}^{p^{n_s}})$$

for $r, s, n_i \geq 0$.

As a consequence of 5.6.9, it can be shown that as a coalgebra H is a finite tensor product of coalgebras, each of which is spanned by a sequence of divided powers. This fact can be interpreted as a kind of PBW theorem for H, in which ordinary powers of primitive elements are replaced by divided powers. This fact is shown in [S 67] by a purely Hopf algebraic argument; it is shown in [Di, p. 66] using 5.6.9.

If $P(H)$ is not finite-dimensional, things can go wrong; an example due to Newman shows that a primitive element can lie in arbitrarily long finite sequences of divided powers but in no infinite sequnece of divided powers. However some things can still be said in the general case. For example, it follows by 5.5.3 and the universal property of restricted enveloping algebras that $H \cong u(P(H)) \Leftrightarrow P(H)$ generates H as an algebra. More generally, see [Ca, Theorem 4].

§5.7 Semisimple cocommutative connected Hopf algebras

We now give an application of coradical filtrations to semisimplicity. Our main object is to prove the following:

5.7.1 THEOREM [S 71, DGa]. *Let H be a finite-dimensional semisimple cocommutative connected Hopf algebra of char $p \neq 0$. Then H is commutative. If also* k *is algebraically closed, then $H \cong (\mathrm{k}G)^*$, for G an abelian p-group.*

The proof we give is a recent simplification from [Ch 92] which avoids the use of divided powers. Recall that $Z(H)$ denotes the center of H.

5.7.2 LEMMA. *Consider the left adjoint action of H on itself.*
1) *If $a \in H$, then $a \in Z(H) \Leftrightarrow (ad\, h)(a) = \varepsilon(h)a$, for all $h \in H$*
2) *if H is cocommutative, then $\Delta(ad\, h)(k) = \sum (ad\, h_1)(k_1) \otimes (ad\, h_2)(k_2)$*
3) *if H is cocommutative, then $P(H)$ is ad H-stable.*

PROOF. 1) ($\Rightarrow$) is obvious. Conversely, assume $(ad\, h)(a) = \varepsilon(h)a$, for all $h \in H$. Then $ha = \sum h_1 a\varepsilon(h_2) = \sum h_1 a(Sh_2)h_3 = \sum (ad\, h_1)(a)h_2 = \sum \varepsilon(h_1)ah_2 = ah$, for all $h \in H$. Thus $a \in Z$.

2) Assuming H is cocommutative,

$$\begin{aligned}\Delta(ad\, h)(k) &= \sum (h_1 k S h_2)_1 \otimes (h_1 k S h_2)_2 \\ &= \sum h_1 k_1 S h_2 \otimes h_3 k_2 S h_4 \\ &= \sum (ad\, h_1)(k_1) \otimes (ad h_2)(k_2).\end{aligned}$$

3) This follows from 2). For, assume $x \in H$ with $\Delta x = x \otimes 1 + 1 \otimes x$.

Then

$$\begin{aligned}\Delta(ad\, h)(x) &= \sum (ad\, h_1)(x) \otimes (ad\, h_2)(1) + (ad\, h_1)(1) \otimes (ad\, h_2)(x) \\ &= \sum (ad\, h_1)(x) \otimes \varepsilon(h_2)1 + \varepsilon(h_1)1 \otimes (ad\, h_2)(x) \\ &= (ad\, h)(x) \otimes 1 + 1 \otimes (ad\, h)(x).\end{aligned}$$

□

5.7.3 LEMMA [S 71]. *Let H be a connected Hopf algebra, and A a finite-dimensional separable commutative subalgebra of H which is ad H-stable. Then $A \subseteq Z$, the center of H.*

PROOF. First, we may assume that A is spanned by primitive central idempotents $\{e_1, \ldots, e_m\}$. For, we may replace H by $H \otimes E$, where $E \supset \mathbf{k}$ is a splitting field for A; $H \otimes E$ is again connected, by a simpler version of the argument in 5.1.10. Let $\{H_n\}$ be the coradical filtration of H. We will show by induction on n that each H_n acts trivially on each $e = e_i$ via ad.

It is trivial that $H_0 = \mathbf{k}1$ acts trivially. Assume that H_{n-1} acts trivially, and choose $h \in H_m$. Since $H_m = \mathbf{k}1 + H_m^+$, we may assume that $\varepsilon(h) = 0$. By Lemma 5.3.2, $\Delta h = h \otimes 1 + 1 \otimes h + y$, where $y \in H_{n-1} \otimes H_{n-1}$; in fact we may check that $y \in H_{n-1}^+ \otimes H_{n-1}^+$. Since H is an H-module algebra under ad, it follows that

$$(*) \qquad (ad\, h)(e) = (ad\, h)(e^2) = (ad\, h)(e)e + e(ad\, h)(e) + 0,$$

since by induction on $n, ad\, H_{n-1}$ acts trivially on A. Since A is ad-stable and commutative, $(*)$ gives $(ad\, h)(e)e = 2(ad\, h)(e)e = 0$, and so $(ad\, h)(e) = 0 = \varepsilon(h)e$. Thus H_n acts trivially on e, and so by induction H acts trivially on A. Now 5.7.2, 1) implies that $A \subseteq Z$. □

PROOF of 5.7.1. We may assume H is non-trivial. Let $\mathbf{g} = P(H); \mathbf{g} \neq 0$ by 5.3.2. Moreover, $\mathbf{g}$ is a restricted Lie algebra, and by 5.3.3, $u(\mathbf{g}) \to H$ is injective. Thus we may assume $u(\mathbf{g}) \subset H$. Now $u(\mathbf{g})$ is semisimple by 3.2.3, and thus is commutative by Hochschild's theorem, 2.3.3. Moreover, $u(\mathbf{g})$ is $ad\, H$-stable by 5.7.2, 3). Thus H contains non-trivial commutative separable $ad\, H$-stable subHopfalgebras containing $P(H)$. Let A be a maximal such subHopfalgebra of H; by 5.7.3, $A \subseteq Z$.

We claim that $A = H$. To see this, we show by induction that $H_n \subseteq A$, where $\{H_n\}$ is the coradical filtration of H. Now $H_1 = \mathrm{k}1 + P(H) \subseteq A$ by construction. Assume that $H_{n-1} \subseteq A$ and choose $b \in H_n$. From 5.3.2, as before, we have

$$\Delta b = b \otimes 1 + 1 \otimes b + \sum_i c_i \otimes d_i,$$

where $c_i, d_i \in H_{n-1}$. Now choose any $h \in H_n^+$. By 5.7.2,

$$(*)\quad \Delta(ad\,h)(b) = (ad\,h)(b) \otimes 1 + 1 \otimes (ad\,h)(b) + \sum_i (ad\,h_1)(c_i) \otimes (ad\,h_2)(d_i).$$

Since $c_i, d_i \in A \subseteq Z, ad\,H$ is trivial on A, and $h \in H^+$, the last sum in $(*)$ is 0. Thus $(*)$ says that $(ad\,h)(b) \in P(H) \subseteq A$. It follows that $B = A + \mathrm{k}b$ is an *ad* H-stable subcoalgebra of H. Since $A \subseteq Z$, the Hopf subalgebra of H generated by B is commutative (and H-stable); it is also semisimple. By the maximality of A, we must have $B \subseteq A$, and so $b \in A$. That is, $H_n \subseteq A$ for all n, and so $A = H$.

Finally, assume that k is algebraically closed. Since H is commutative, $H \cong (\mathrm{k}G)^*$ for some abelian G by 2.3.1. If G were not a p-group, then G would have a non-trivial homomorphic image $\bar{G}$ such that p does not divide $|\bar{G}|$. Let $\pi : \mathrm{k}G \to \mathrm{k}\bar{G}$ be the quotient map; then $\pi^* : (\mathrm{k}\bar{G})^* \to (\mathrm{k}G)^*$ is injective. Since k is algebraically closed and p does not divide $|\bar{G}|$, $(\mathrm{k}\bar{G})^* \cong \mathrm{k}\bar{G}$. But then $(\mathrm{k}G)^*$ contains non-trivial group-like elements, so is not connected, a contradiction. □

A theorem of Nagata [Na] says that a fully reducible connected affine algebraic group G in positive characteristic is a torus (and thus that if k is algebraically closed and $H = \mathcal{O}(G)$, then $H \cong \mathrm{k}\mathbf{Z}^{(n)}$ for some n). Nagata's theorem was generalized independently by Demazure and Gabriel [DGa, p. 509] and by Sweedler [S 71]. The Hopf algebra version of their theorem, which is also called "Nagata's theorem" in [DGa], is the following:

5.7.4 THEOREM [S 71], [DGa]. *Suppose* k *has characteristic* $p \neq 0$ *and* H *is a commutative cosemisimple Hopf algebra such that* $\pi_0(H) = \mathrm{k}$. *Then* H *is cocommutative. If also* k *is algebraically closed, then* $H \cong \mathrm{k}G$, *for* G *an abelian group such that the torsion elements have* p*-power torsion.*

Here $\pi_0(H)$ means the maximal separable k-subalgebra of H; when k is algebraically closed, $\pi_0(H)$ is just the span of the idempotents in A. Sweedler's proof of 5.7.4 reduces to 5.7.1 by the following steps [S 71; A, Sec. 4.6]:

First, one may assume that H is affine. Since $\pi_0(H) = \mathbf{k}$, it can then be shown that $\cap_{n=1}^{\infty} I^n = 0$ for any proper ideal I of H. In particular, if $H^{(p^n)} = \{h^{p^n} \mid h \in H\}$, then $\cap_n HH^{+(p^n)} = 0$. Thus if $h \in H$, and C is the (finite-dimensional) subcoalgebra generated by h, then for some n we must have $C \cap HH^{+(p^n)} = 0$. Thus $C \hookrightarrow H/HH^{+(p^n)} = K$, a finite-dimensional Hopf algebra. Since K is commutative, cosemisimple, and $\pi_0(K) = \mathbf{k}$, it follows that K^* is cocommutative, semisimple, and connected. By 5.7.1, K^* is commutative. Thus K, and so C, is cocommutative.

Chapter 6

Inner Actions

The main topics in this Chapter are first, Skolem-Noether type theorems for actions of Hopf algebras, and second, the question as to when a largest inner subHopfalgebra or coalgebra exists for any action of a Hopf algebra. We also include a brief discussion of "X-inner" actions; that is, those actions which become inner when extended to a quotient ring.

§6.1 Definitions and examples

Since we sometimes wish to consider actions of coalgebras and not just of Hopf algebras, we give a more general definition of inner action.

6.1.1 DEFINITION. Let C be a coalgebra and $B \subseteq A$ be algebras. Consider an action $C \otimes B \to A$, given by $c \otimes b \mapsto c \cdot b$, which measures B to A (that is, 2) and 3) of 4.1.1 are satisfied). Then the measuring is *inner* if there exists a convolution invertible map $u \in \text{Hom}\,(C, A)$ such that for all $h \in C, b \in B$,

$$h \cdot b = \sum u(h_1) b\, u^{-1}(h_2). \tag{6.1.2}$$

Thus the definition applies when $C = H$ is a Hopf algebra and $A = B$ is an H-module algebra. In that case, we may always assume $u(1) = 1 = u^{-1}(1)$; for if not, simply replace u by $v(h) = u(1)^{-1}u(h)$, all $h \in H$.

We frequently use an equivalent form of 6.1.2, namely that $\sum (h_1 \cdot b) u(h_2) = u(h)b$, all $h \in C, b \in B$.

6.1.3 EXAMPLE. The trivial action of H on A is inner: set $u(h) = u^{-1}(h) = \varepsilon(h)$.

6.1.4 EXAMPLE. The (left) adjoint action of H on itself is inner (see 3.4.1). We set $u(h) = h$ and $u^{-1}(h) = Sh$.

6.1.5 EXAMPLE. If the action of H on A is inner and $g \in G(H)$, then g acts as an inner automorphism of A. For, $\Delta g = g \otimes g$ gives $u(g)u^{-1}(g) = 1 = u^{-1}(g)u(g)$, and thus $u^{-1}(g) = u(g)^{-1}$. Then $g \cdot a = u(g)au(g)^{-1}$, for all $a \in A$. Conversely if a group G acts as inner automorphisms of A, via

$g\cdot a = u_g a u_g^{-1}$ for each $g \in G$, then the action of $H = \mathbf{k}G$ is inner, by setting $u(\sum \alpha_g g) = \sum \alpha_g u_g$ and $u^{-1}(\sum \alpha_g g) = \sum \alpha_g u_g^{-1}$.

6.1.6 EXAMPLE. If the action of H is inner and $x \in P(H)$, then x acts as an inner derivation of A. For, $\Delta x = x \otimes 1 + 1 \otimes x$ gives $u(x)u^{-1}(1) + u(1)u^{-1}(x) = u(x) + u^{-1}(x) = \varepsilon(x) = 0$; thus $u^{-1}(x) = -u(x)$. Then $x \cdot a = u(x)a - au(x) = [u(x), a]$ for all $a \in A$. Conversely if a Lie algebra $\mathbf{g}$ acts as inner derivations of A, then the action of $H = U(\mathbf{g})$ is inner, as follows: given an appropriate map $u \in \mathrm{Hom}_{\mathbf{k}}(\mathbf{g}, A)$, one may extend u to $U(\mathbf{g})$ using a given basis $\{x_i\}$ of $\mathbf{g}$, via $u(1) = 1$, $u(x_1^{e_1}\cdots x_n^{e_n}) = u(x_1)^{e_1}\cdots u(x_n)^{e_n}$, and $u^{-1}(x_1^{e_1}\cdots x_n^{e_n}) = u^{-1}(x_n)^{e_n}\cdots u^{-1}(x_1)^{e_1}$.

6.1.7 EXAMPLE [BCM]. Let A be an algebra graded by the finite group G; then A is a $(\mathbf{k}G)^*$-module algebra (see 4.1.7). Let us assume that the action is inner, induced by $u \in \mathrm{Hom}((\mathbf{k}G)^*, A)$. Using the basis $\{p_x\}$ of $(\mathbf{k}G)^*$ as in 1.3.7, we see that

$$(*) \qquad p_x \cdot a = \sum_{y \in G} u(p_{xy^{-1}})au^{-1}(p_y).$$

As a more explicit example, let $A = M_n(B)$, the $n \times n$ matrices over some other algebra B, and assume that $|G| = n$. Fix an ordering $\{x_1, \ldots, x_n\}$ of the elements of G, and index the rows and columns of A by G. Then $\{e_{xy} \mid x, y \in G\}$ denote the usual matrix units. If we set $A_x = \sum_{yz^{-1}=x} Be_{yz}$, for all $x \in G$, then $A_xA_y \subseteq A_{xy}$ and $A = \oplus_{x\in G} A_x$. Thus A is a G-graded algebra. In fact this grading is inner: we may define $u(p_x) = e_{xx}$ and $u^{-1}(p_x) = e_{x^{-1}x^{-1}}$, all $x \in G$.

When $G = \mathbf{Z}_n$, it is easy to see what the graded components of A look like. Symbolically:

$$A = \begin{bmatrix} A_{\bar 0} & A_{\bar 1} & \ldots & A_{\overline{n-1}} \\ A_{\overline{n-1}} & A_{\bar 0} & \ldots & A_{\overline{n-2}} \\ \ldots & \ldots & \ldots & \ldots \\ A_{\bar 1} & A_{\bar 2} & \ldots & A_{\bar 0} \end{bmatrix}$$

§6.2 A Skolem-Noether theorem for Hopf algebras

The classical Skolem-Noether theorem asserts that if A is a simple Artinian ring with center Z, and $B \supseteq Z$ a simple subalgebra of A with $[B : Z] < \infty$, then any isomorphism of B into A which fixes Z extends to an inner automorphism of A. An analogous result for derivations was proved by Jacobson. In the case of derivations, the simplicity of B can be replaced by the weaker hypothesis of semisimplicity; however this is false for automorphisms. A natural question is to what extent these results can be generalized to actions of Hopf algebras.

The first such result was due to Sweedler [S 68], who proved a theorem for pointed cocommutative Hopf algebras (we note that the statement of his theorem is incorrect, since he misstated the classical result; however the rest of his argument is correct). More recently, [OQ] proved a Skolem-Noether theorem for strongly graded rings, which was extended to Hopf Galois extensions in [BM 89]. [Ma 90a] gave sufficient conditions for the Skolem-Noether theorem to hold for any coalgebra measuring of an Azumaya algebra A, which included the case of central simple algebras; the Azumaya algebra case was studied further in [BaU]. In a slightly different direction, [Ko 91] proved a result for A simple Artinian. Finally [Mi] has a new, short proof which includes many of the above results, though not the full strength of [Ko 91]. We follow here Koppinen's approach, because it is the best result for A simple Artinian; moreover the arguments follow the outline of the classical proof.

The first proposition follows Sweedler's argument very closely. It reduces the general problem to the case of simple coalgebras.

6.2.1 PROPOSITION. *Let A be a simple Artinian ring with center $Z(A) = \mathrm{k}1$ and let B be a finite-dimensional semisimple* k*-subalgebra of A. Let C be a* k*-coalgebra which measures B to A. If the measuring is inner for each simple subcoalgebra D of C, then it is inner on C.*

PROOF. Write $C_0 = \oplus_\alpha D_\alpha$, where the D_α are the simple subcoalgebras of C and C_0 is the coradical of C. If the measuring is inner on each D_α, say via $u_\alpha \in \mathrm{Hom}(D_\alpha, A)$, then $u_0 = \sum_\alpha u_\alpha$ induces the measuring on C_0. Moreover u_0 is convolution invertible since $\mathrm{Hom}(C_0, A) = \oplus_\alpha \mathrm{Hom}(D_\alpha, A)$. Thus the measuring is inner on C_0.

Now consider $A \otimes C$ as a right B-module via

$$(a \otimes c) \cdot b = \sum a(c_1 \cdot b) \otimes c_2. \tag{6.2.2}$$

Then $A \otimes C$ becomes a left $A \otimes B^{op}$-module via $(a \otimes b^o) \cdot (a' \otimes c) = (aa' \otimes c) \cdot b$, and it follows that $A \otimes C_0$ is an $A \otimes B^{op}$-submodule of $A \otimes C$. Since A is central simple and B finite-dimensional, $A \otimes B^{op}$ is semisimple Artinian. Thus there exists an $A \otimes B^{op}$-projection $\pi : A \otimes C \to A \otimes C_0$.

Next define $\tilde{u} : A \otimes C_0 \to A$ by $a \otimes c \mapsto au_0(c)$. It is straightforward to check that $\tilde{u}$ is an $A \otimes B^{op}$-map, using that $\sum (c_1 \cdot b)u_0(c_2) = u_0(c)b$, all $c \in C_0, b \in B$, by 6.1.2. We may then define the composite map $u : C \to A$ by $u(c) = \tilde{u}\pi(1 \otimes c)$.

We claim that $u(c)b = \sum (c_1 \cdot b)u(c_2)$, for all $c \in C, b \in B$. For, $u(c)b = (1 \otimes b^o) \cdot \tilde{u}\pi(1 \otimes c) = \tilde{u}\pi((1 \otimes b^o) \cdot (1 \otimes c)) = \sum \tilde{u}\pi(c_1 \cdot b \otimes c_2) = \sum ((c_1 \cdot b) \otimes 1^o) \cdot \tilde{u}\pi(1 \otimes c_2) = \sum (c_1 \cdot b)u(c_2)$. Thus the measuring will be inner on C, via u, provided we know u is invertible. But this follows by 5.2.10, since u is invertible on C_0. □

We can obtain Jacobson's theorem as a consequence of the Proposition. For, if A is simple Artinian and B is a finite-dimensional subalgebra as above, let $\delta : B \to A$ be a derivation of B to A. Let C be the coalgebra $\mathbf{k}i + \mathbf{k}\delta \subseteq \text{Hom}(B, A)$, where i is the inclusion $B \subset A$, $i \in G(C)$, and $\delta \in P_{ii}(C)$. Then $C_0 = \mathbf{k}i$, and 6.2.1 applies to show that δ is inner.

In fact the Proposition is true more generally: one can replace the hypotheses on A and B by the assumption that A is injective as a (B, A)-module [Ko 91, 3.2]. We now state Koppinen's theorem.

6.2.3 THEOREM [Ko 91]. *Let A be simple Artinian with center $Z(A) = \mathbf{k} \cdot 1$ and let B be a finite-dimensional simple subalgebra of A. Let C be a $\mathbf{k}$-coalgebra which measures B to A. Assume also that $B^{op} \otimes D^*$ is a simple algebra for each simple subcoalgebra D of C. Then the measuring is inner.*

PROOF. By the Proposition, we may assume that $C = D$ is simple and so finite-dimensional by 5.1.1. Then $A \otimes B^{op} \otimes C^*$ is simple Artinian, since A is central simple and $B^{op} \otimes C^*$ is simple and finite-dimensional by hypothesis. Thus it has a unique simple left module V, and all of its left modules are direct sums of copies of V.

We consider $A \otimes C$ as an $A \otimes B^{op} \otimes C^*$-module in two ways. In both cases the left A-action is just left multiplication on A, and the left C^*-action is the action on $1 \otimes C$ obtained by dualizing the right C-coaction via Δ, as in 1.6.4. The difference is in the two B^{op} actions. First, $M_1 = A \otimes C$ is a left B^{op}-module in the obvious way via $b^o \cdot (a \otimes c) = ab \otimes c$. Second, $M_2 = A \otimes C$ is a left B^{op}-module as in 6.2.2 via $b^o \cdot (a \otimes c) = \sum a(c_1 \cdot b) \otimes c_2$.

Since C is finite-dimensional, both M_1 and M_2 are finitely-generated $A \otimes B^{op} \otimes C^*$ -modules, and so $M_1 \cong V^{(n_1)}$, $M_2 \cong V^{(n_2)}$ for some $n_1, n_2 > 0$. Thus there exists a module embedding f between the two modules, in one direction or the other. Since M_1 and M_2 have the same underlying space and the same left A-action, $f \in \mathrm{End}_A(A \otimes C)$. Since $A \otimes C$ is an Artinian A-module and f is injective, f must also be surjective. Thus $M_1 \cong M_2$, and we may assume $f : M_1 \to M_2$. Note that f is a $B^{op} \otimes C^*$-module map and thus also a C- comodule map.

We interrupt the proof for a lemma:

LEMMA. *$End_A^C(A \otimes C) \cong Hom(C, A)^{op}$ as algebras.*

PROOF. Here $\mathrm{End}(A \otimes C)$ consists of the left A-module, right C-comodule maps on $A \otimes C$ and $\mathrm{Hom}(C, A)$ is an algebra under convolution. Define ϕ : $\mathrm{End}(A \otimes C) \to \mathrm{Hom}(C, A)$ by $\phi(f)(c) = (id \otimes \varepsilon) f(1 \otimes c)$ and $\psi : \mathrm{Hom}(C, A) \to \mathrm{End}(A \otimes C)$ by $\psi(g)(a \otimes c) = \sum ag(c_1) \otimes c_2$. Since f is a right C-comodule map, $(id \otimes \Delta) f = (f \otimes id)(id \otimes \Delta)$ and thus $f = (id \otimes \varepsilon \otimes id)(f \otimes id)(id \otimes \Delta)$. Hence for all $a \in A, c \in C$,

$$
\begin{aligned}
f(a \otimes c) &= a \cdot f(1 \otimes c) = a \cdot (id \otimes \varepsilon \otimes id)(f \otimes id)(id \otimes \Delta)(1 \otimes c) \\
&= \sum a \cdot (id \otimes \varepsilon \otimes id) f(1 \otimes c_1) \otimes c_2 \\
&= \sum a\phi(f)(c_1) \otimes c_2.
\end{aligned}
$$

It follows that $\psi \circ \phi = id$; similarly $\phi \circ \psi = id$, and it is easy to see ψ is an anti-algebra map. Thus $\mathrm{End}(A \otimes C)$ and $\mathrm{Hom}(C, A)$ are anti-isomorphic. □

We now finish the proof of 6.2.3. For $f : M_1 \to M_2$ as above, let $v = \phi(f) \in \mathrm{Hom}(C, A)$, as in the lemma. Since f is bijective in $\mathrm{End}(A \otimes C), v$ is invertible. Since f is a left B^{op}-map, $f(ab \otimes c) = b^o \cdot f(a \otimes c)$, for all

$b \in B, a \otimes c \in A \otimes C$. Using $f = \psi(v)$ as in the lemma, we see

$$\sum ab\ v(c_1) \otimes c_2 = b^o \cdot \sum a\ v(c_1) \otimes c_2 = \sum a\ v(c_1)(c_2 \cdot b) \otimes c_3.$$

Letting $a = 1$ and applying $id \otimes \varepsilon$ gives $b\ v(c) = \sum v(c_1)(c_2 \cdot b)$, or $c \cdot b = \sum v^{-1}(c_1) b\ v(c_2)$, all $c \in C, b \in B$. Thus the measuring is inner. □

We mention two major special cases of the theorem. The first was proved independently (and a little earlier) by Masuoka [Ma 90a].

6.2.4 COROLLARY. *Let $A = B$ be a simple algebra, finite-dimensional over its center* k. *Let C be a* k*-coalgebra. Then any measuring of A by C is inner.*

PROOF. $B^{op} = A^{op}$ is central simple, and D^* is simple by 5.1.4. Thus $B^{op} \otimes D^*$ is always simple. □

6.2.5 COROLLARY. *Let A be simple Artinian with center* k, *and let B be a finite-dimensional simple subalgebra of A. Let C be a pointed* k*-coalgebra. Then any measuring of B to A by C is inner.*

PROOF. $D = \mathbf{k}g$, for $g \in G(H)$, and so $B^{op} \otimes D^* \cong B^{op}$, which is simple. □

§6.3 Maximal inner subcoalgebras

If a group G acts as automorphisms of an algebra A, then the set $N = \{g \in G \mid g$ acts as an inner automorphism of $A\}$ is a normal subgroup of G. It follows that $\mathbf{k}N$ is the (unique) largest subHopfalgebra of $\mathbf{k}G$ which is inner on A in the sense of 6.1.1. This leads to the question, posed to us in 1988 by C. Sutherland, as to whether for any Hopf algebra action there is always a largest inner subHopfalgebra.

Although the answer to the question is no, in general, even when H is cocommutative [Sch 93a], nevertheless some nice positive results are known. We prove two of these results, due to Masuoka.

6.3.1 THEOREM [Ma 90b]. *For any coalgebra C which measures an algebra A, there is a largest inner subcoalgebra C_{inn}.*

We follow a simpler proof given by Schneider in [Sch 93a].

6.3.2 LEMMA. 1) *Assume $D \subseteq C$ are coalgebras, A is an algebra, and $u \in Hom(D, A)$ is invertible. Then there exists $v \in Hom(C, A)$ such that v is* *invertible and* $v|_D = u$.

2) *Assume* $E \subset D$ *are subcoalgebras of* C, *that* C *measures* A, *and that* E *and* D *are inner on* A *via* $v \in Hom(E, A)$ *and* $u \in Hom(D, A)$ *respectively. Then* D *is inner via some* $\hat{u} \in Hom(D, A)$ *such that* $\hat{u}\ |_E = v$.

PROOF. 1) Write the coradical $C_0 = \oplus_{\alpha \in I} C_\alpha$, where the C_α are the simple subcoalgebras of C. Let $J = \{\alpha \in I \mid C_\alpha \subseteq D\}$; then

$$C_0 = (\oplus_{\alpha \in J} C_\alpha) \oplus (\oplus_{\beta \in I - J} C_\beta) = D_0 \oplus E.$$

Using that $D_0 = D \cap C_0$ by 5.1.9, it follows that $D \cap E = 0$. Let $\tilde{\varepsilon} : E \to A$ be the trivial map $e \mapsto \varepsilon(e)1$. Then $u \oplus \tilde{\varepsilon} : D \oplus E \to A$ is invertible on $D \oplus E$. Extend $u \oplus \tilde{\varepsilon}$ to any linear map $v : C \to A$. On $C_0 \subset D \oplus E, v$ is invertible. Thus v is invertible on C by 5.2.10.

2) Define $w = (u\ |_E)^{-1} * v \in \mathrm{Hom}(E, A)$; it is invertible. We claim that w and w^{-1} are in $\mathrm{Hom}(E, Z(A))$, where $Z(A)$ is the center of A. For, if $c \in E, a \in A$, then

$$c \cdot a = \sum u(c_1)au^{-1}(c_2) = \sum v(c_1)av^{-1}(c_2).$$

Thus

$$\begin{aligned} \sum u^{-1}(c_1)v(c_2)av^{-1}(c_3) &= au^{-1}(c) \\ \text{and so} \qquad (\sum u^{-1}(c_1)v(c_2))a &= a(\sum u^{-1}(c_1)v(c_2)) \\ w(c)a &= aw(c). \end{aligned}$$

Thus $w(c) \in Z(A)$. Now by 1), there exists an invertible $\hat{w} \in \mathrm{Hom}(D, Z(A))$ such that $\hat{w}\ |_E = w$.

Now define $\hat{u} = u * \hat{w} : D \to A$; $\hat{u}$ is invertible, and

$$\hat{u}\ |_E = u\ |_E * \hat{w}\ |_E = u\ |_E * w = u\ |_E * (u\ |_E)^{-1} * v = v.$$

Finally D is inner via $\hat{u}$, since it is inner via u and $\hat{w}$ has central values. □

PROOF of 6.3.1. First, if E and D are inner subcoalgebras of C, then $E + D$ is inner. For, if E is inner via u, 6.3.2, 2) applied to $E \cap D \subset D$ gives that D is inner via some v such that $v\ |_{E \cap D} = u\ |_{E \cap D}$. Then the map $w : E + D \to A$

given by $w(e+d) = u(e) + v(d) \in A$ is well defined, and $E + D$ is inner via w.

Now apply Zorn's lemma to the pairs (D, u) such that the subcoalgebra D of C is inner on A via $u : D \to A$. Let D' be maximal among such subcoalgebras. Then D' contains any other inner subcoalgebra E, since $D' \subset E + D'$ and $E + D'$ is inner by the above. Thus $D' = E + D'$ by maximality of D', and so $E \subset D$. □

6.3.3. THEOREM [Ma 90b]. *Let A be an H-module algebra for a pointed Hopf algebra H. Then there is always a largest inner subHopfalgebra H_{inn}.*

Again we follow Schneider's proof.

6.3.4 LEMMA. *Assume that the coalgebra C measures A. Let D be a pointed coalgebra and $\pi : D \to C$ a surjective coalgebra map. If the induced D-measuring of A is inner, then the C-measuring of A is inner.*

PROOF. We may assume that D is finite-dimensional. For, by 5.1.1, $D = \cup_\alpha D_\alpha$, where the D_α are finite-dimensional subcoalgebras, and thus $C = \cup_\alpha \pi(D_\alpha)$. If each $\pi(D_\alpha)$ is inner, then C is inner by 6.3.1. Thus we may assume $D = D_\alpha$, for some α. We proceed by induction on the dimension of D. The goal is to show that D is inner via $u : D \to A$ such that u and u^{-1} vanish on $I = ker(\pi)$; for then $v : C \to A$ given by $v(\pi(d)) = u(d)$, all $d \in D$ is well-defined and invertible, and thus C is inner via v.

Let E be a maximal subcoalgebra of D. Since D is pointed, it follows that dim $D/E = 1$. For, using 5.1.7, we know that $D^*/\text{Jac}(D^*) = D^*/D_0^\perp \cong (D_0)^* \cong \mathrm{k}^{(n)}$. Then $E^\perp$, as a minimal ideal of D^*, must be one-dimensional, and so E has codimension 1. By induction, $\pi(E)$ is inner via some invertible map $w : \pi(E) \to A$. If $I = \text{Ker }\pi$ is not contained in E, then $E + I = D$ (since E has codimension one) and $C = \pi(D)$ is inner. If $I \subseteq E$, consider the invertible map $E \xrightarrow{\pi} \pi(E) \xrightarrow{w} A$. E is inner via $w \circ \pi$ since π is a coalgebra map. By 6.3.2, 2), there is an invertible map $u : D \to A$ such that $u \mid_E = w\pi$ and D is inner via u. In particular $u(I) = 0$ and $u^{-1}(I) = 0$. Hence C is inner as noted above. □

PROOF of 6.3.3. First, $D = H_{inn}$ is a subcoalgebra by 6.3.1. We claim it is a bialgebra. Consider the map $\pi : D \otimes D \to D \cdot D \subset H$ given by $c \otimes d \mapsto c \cdot d$;

π is a surjective coalgebra map. Also $D \otimes D$ is pointed by 5.1.9 and 5.1.10. Now $D \otimes D$ is inner on A, for let $u : D \to A$ be the invertible map inducing D. Then for all $c, d \in D, a \in A$,

$$(c \otimes d) \cdot a \equiv (cd) \cdot a = c \cdot (d \cdot a) = \sum u(c_1)u(d_1)au^{-1}(d_2)u^{-1}(c_2).$$

By setting $v(c \otimes d) = u(c)u(d)$, we see that $D \otimes D$ is inner via v. Now 6.3.4 implies that $D \cdot D$ is inner; by the maximality of D, $D \cdot D \subset D$ and D is a subalgebra (and so a subbialgebra) of H.

To finish we will show that H_{inn} has an antipode. Since H_{inn} is pointed, $(H_{inn})_0 = \sum \mathrm{k}g$, for $g \in G(H_{inn})$. For each such g, g is inner, and thus $g \cdot a = uau^{-1}$ for all $a \in A$, for some unit $u = u(g)$ in A. But then $g^{-1} \cdot a = u^{-1}au$ and so g^{-1} is inner; thus the subcoalgebra $\mathrm{k}g^{-1} \in H_{inn}$ by the maximality of H_{inn}. That is, S stabilizes the coradical of H_{inn}. Thus id is $*$-invertible on $(H_{inn})_0$, and so is $*$-invertible on H_{inn} by 5.2.10. Thus H_{inn} has an antipode, which is necessarily the antipode on H. Thus H_{inn} is a Hopf algebra. □

Returning to group actions, recall also that if G acts on A and N is the subgroup of G of inner automorphisms, then N is normal in G and G/N acts on A^N. Schneider has shown that this fact can be extended to the case of any pointed cocommutative H. Recall the definition of normal subHopfalgebras from §3.4.

6.3.5 THEOREM [Sch 93a]. *Let H be pointed cocommutative and let A be an H-module algebra. Then H_{inn} is a normal subHopfalgebra of H, and $\bar{H} = H/HH_{inn}^{+}$ acts on $A^{H_{inn}}$.*

PROOF. $D = H_{inn}$ is a subHopfalgebra by 6.3.3. We claim that D is $ad_\ell H$-stable. Let $D_H = \{\sum(ad\ h)(d) \mid h \in H, d \in D\}$ and consider $\pi : H \otimes D \to D_H \subset H$ via $h \otimes d \mapsto (ad\ h)(d)$. Since H is cocommutative , 5.7.2, 2) implies that π is a coalgebra map, and that D_H is a subcoalgebra of H. Now D is inner via some $u : D \to A$. Then $v : H \otimes D \to A$ given by $v(h \otimes d) = h \cdot u(d)$, $h \in H, u \in D$, is invertible with inverse $v^{-1}(h \otimes d) = h \cdot u^{-1}(d)$ since H measures A. Moreover $H \otimes D$ is inner via v since for all $h \in H, d \in D$, and $a \in A$,

$$\begin{aligned}\pi(h \otimes d) \cdot a &= (\sum h_1 d(Sh_2)) \cdot a \\ &= \sum h_1 \cdot (d \cdot ((Sh_2) \cdot a)) \\ &= \sum h_1 \cdot (\sum u(d_1)((Sh_2) \cdot a)u^{-1}(d_2)) \\ &= \sum (h_1 \cdot u(d_1))(h_2 \cdot (Sh_4 \cdot a))(h_3 \cdot u^{-1}(d_2)) \\ &= \sum (h_1 \cdot u(d_1))a(h_2 \cdot u^{-1}(d_2)) \\ &= \sum v(h_1 \otimes d_1)av^{-1}(h_2 \otimes d_2),\end{aligned}$$

where cocommutativity was used to see $\sum h_2 Sh_4 = \varepsilon(h_2)$.

Now apply 6.3.4 to see that D_H is inner on A. Thus $D_H \subset H_{inn} = D$ by 6.3.1, and so $(ad_\ell H)(D) \subseteq D$. A similar argument works for $ad_r H$, and thus D is normal in H.

By 3.4.2, it follows that $HD^+ = D^+H = I$ is a Hopf ideal of H and so $\bar{H} = H/HD^+$ is a Hopf algebra. It is then straightforward to see that the induced action of $\bar{H}$ on A^D is well-defined, since $D^+ \cdot A^D = 0$. Thus A^D becomes an $\bar{H}$-module algebra. □

We do not know whether or not the cocommutativity hypothesis can be removed from 6.3.5.

§6.4 X-inner actions and extending to quotients

A fundamental technique in the work on group actions in the 70's and 80's was to consider the *X-inner automorphisms* of a ring R; when R is prime, these are the automorphisms of R which become inner when extended to the Martindale quotient ring of R. They were introduced into ring theory about 1975 by V.K. Kharchenko and at about the same time into operator algebras by G. Pedersen and G. Elliot. Many times the X-inner automorphisms were the "obstruction" in a problem, and better results could be obtained by looking at X-outer groups (that is, those whose only X-inner element was the identity). Thus it would be nice to be able to extend this method to Hopf algebra actions.

We first need to define the various Martindale quotient rings; for a reference, see [M 80, Ch.3] or [P, Ch.3]. Let R be any ring, and let $\mathcal{F}$ denote the filter of ideals of R which have zero left and right annihilator. We first consider the left quotient ring $Q^\ell(R)$. Let $\mathcal{S}$ be the set of all pairs (I, f), where $I \in \mathcal{F}$ and $f : I \to R$ is a left R-module map. Define $(I, f) \sim (J, g)$ if $f = g$ on some $K \in \mathcal{F}$, $K \subseteq I \cap J$. Then $Q^\ell(R) = \mathcal{S}/\sim$. More compactly:

$$Q^\ell(R) = \varinjlim_{I \in \mathcal{F}} \operatorname{Hom}_R({}_R I, R). \tag{6.4.1}$$

$Q^\ell(R)$ becomes a ring as follows: for (I, f) and (J, g), $(I \cap J, f + g)$ determines addition and $(IJ, f \circ g)$ the multiplication. R embeds into $Q^\ell(R)$ as right multiplications on $I = R$, and any $q \in Q^\ell(R)$ has the property that there exists $I \in \mathcal{F}$ such that $Iq \subseteq R$.

The right Martindale quotient ring $Q^r(R)$ is defined similarly, using right R-module maps. Given both $Q^\ell(R)$ and $Q^r(R)$, one may then define the symmetric Martindale quotient ring $Q(R)$ as

$$\begin{aligned} Q(R) &= \{q \in Q^\ell(R) \mid qI \subseteq R, \text{ some } I \in \mathcal{F}\} \\ &= \{q \in Q^r(R) \mid Iq \subseteq R, \text{ some } I \in \mathcal{F}\}. \end{aligned}$$

A direct limit definition is given in [P], which is in fact closer to the definition used in C^*-algebras:

6.4.2 DEFINITION. $Q(R) = \varinjlim_{I \in \mathcal{F}} \{(f, g) \mid f \in \operatorname{Hom}_R({}_R I, R), g \in \operatorname{Hom}_R(I_R, R)$, and $(af)b = a(gb)$ all $a, b \in I$, for some $I \in \mathcal{F}\}$.

Here g is written on the left and f on the right. In this formulation, $R \hookrightarrow Q$ via $a \mapsto (r_a, \ell_a)$, where r_a (resp. ℓ_a) denotes right (left) multiplication by a.

For the reader unfamiliar with these constructions, one may check that if R is a commutative domain with quotient field F, then $Q^\ell(R) = Q^r(R) = Q(R) = F$, so in the classical case these quotients are nothing new. Also, if R is simple, clearly $Q^\ell(R) = Q^r(R) = Q(R) = R$. For more general rings, however, they are more interesting; see [M 80] and [P] for more examples. In particular if $M_\infty(\mathbf{k})$ is the set of all (countably) infinite matrices over $\mathbf{k}$, F the subset of finite matrices, and $\mathbf{k}1$ the scalar matrices, let $R = \mathbf{k}1 + F$. Then

$Q^\ell(R)$ is the set of all row-finite matrices, $Q^r(R)$ the column-finite matrices and $Q(R)$ the matrices which are both row- and column-finite.

Now we return to actions of Hopf algebras. Let A be an H-module algebra. It will be useful to replace $\mathcal{F}$ by $\mathcal{F}_H$, the filter of H-stable ideals of A with zero annihilator. One may repeat the above constructions, obtaining $Q^\ell_H(A)$, $Q^r_H(A)$, and finally $Q_H(A)$, the *H-symmetric ring of quotients of* A. Note that $Q_H \subset Q$, and similarly $Q^\ell_H \subset Q^\ell$ and $Q^r_H \subset Q^r$.

In general Q_H can be genuinely smaller than Q. For example, let $A = \mathbf{k}[x]$, the polynomials over a field $\mathbf{k}$ of characteristic 0, and let $H = U(\mathbf{g})$, where $\mathbf{g}$ is the one-dimensional Lie algebra generated by $\frac{d}{dx}$. Then A is H-simple (that is, there are no proper H-stable ideals) and thus $Q_H(A) = A$. However $Q(A) = \mathbf{k}(x)$, the rational functions in x.

One may use the same A for a group algebra example. Namely, let $H = \mathbf{k}G$, where $G = \langle g \rangle$ is the cyclic group generated by the automorphism $x \mapsto x+1$ of A. Since $\mathbf{k}$ has characteristic $0, A$ is G-simple, and thus $Q_H(A) = Q_G(A) = A$, although $Q(A) \neq A$. However if G is finite this cannot happen, since if $I \in \mathcal{F}$ then $\cap_{g\in G} I^g \in \mathcal{F}_G$, and consequently $Q_G(A) = Q(A)$.

Finally we get to the point of this section: trying to extend the H-action on A to these various quotient rings.

By generalizing the earlier definition for automorphisms, Cohen showed that the H-action extends to the H-quotient rings.

6.4.3 PROPOSITION [C 86]. *Let H have a bijective antipode, let A be an H-module algebra, and let $\mathcal{F}_H$ be the filter of H-stable ideals with zero annihilator. Then*
1) $Q^\ell_H(A)$ is an H-module algebra, with an H-action extending the action on A, as follows:

Let $f : {}_AI \to A$, for $I \in \mathcal{F}_H$, determine an element of $Q^\ell_H(A)$. Then define $h \cdot f : I \to A$ by

$$a(h \cdot f) = \sum h_2 \cdot [(S^{-1}h_1 \cdot a)f]$$

for all $a \in I, h \in H$.
2) $Q^r_H(A)$ is an H-module algebra, with an H-action extending the action on A, as follows:

Let $g : I_A \to A$, for $I \in \mathcal{F}_H$, determine an element of $Q^r_H(A)$. Then *define* $h \cdot g : I \to A$ *by*

$$(h \cdot g)a = \sum h_1 \cdot [g(Sh_2 \cdot a)]$$

for all $a \in I, h \in H$.

3) $Q_H(A)$ *is an* H*-module algebra by using the actions in 1) and 2). That is, if* $f : {}_AI \to A$ *and* $g : I_A \to A$*, then define*

$$h \cdot (f, g) = (h \cdot f, h \cdot g).$$

In fact the extension of the H-action to each of these quotients is unique; this is shown in [M 93a].

Given 6.4.3, we would then be tempted to define an H-action to be X-inner if it is inner considered as an action on $Q_H(A)$ (or on $Q^\ell_H(A)$ or $Q^r_H(A)$). However in the group action case, it was the full quotient $Q(A)$ and not the G-quotient $Q_G(A)$ which was used. Thus we should consider the question of when an H-action extends to $Q(A)$. This seems much more difficult; however some partial results are obtained in [M 93a]. We first need a definition.

6.4.4 DEFINITION. Let A be an H-module algebra. Then the H-action on A is *$\mathcal{F}$-continuous* if given any $I \in \mathcal{F}$ and $h \in H$, there exists $J \in \mathcal{F}$ such that $h \cdot J \subseteq I$.

Note that being $\mathcal{F}$-continuous is equivalent to saying that given I, the inverse image of h on I contains some $J \in \mathcal{F}$; this is just the usual definition of continuity in a topological group, where A is a group under addition and $\mathcal{F}$ is a basis of neighborhoods of 0 in A.

6.4.5 PROPOSITION [M 93a]. *Let* H *have bijective antipode and let* A *be an* H*-module algebra. If the* H*-action on* A *is* $\mathcal{F}$*-continuous, then it extends to an action on* $Q(A)$.

This can be considered a generalization of 6.4.3 since any H-action is trivially $\mathcal{F}_H$-continuous, using $J = I$.

6.4.5 reduces the problem to determining which actions are $\mathcal{F}$-continuous. We have

6.4.6 THEOREM [M 93a]. *Let* H *be a pointed Hopf algebra, and let* A *be an* H*-module algebra. Then the* H*-action on* A *is* $\mathcal{F}$*-continuous.*

The proof uses an inductive argument on the coradical filtration $\{H_n\}$ of H, and the description of the elements in H_n given by the Taft-Wilson theorem, 5.4.1. In fact the results in [M 93a] were proved when A was only a twisted H-module (see 7.1.3), although with the assumption that the H-action was "biinvertible". In [MSch], 6.4.5 is generalized to crossed product actions (7.1.2).

We close by offering our definition of X-inner for Hopf actions.

6.4.7 DEFINITION. Let A be an H-module algebra. Given that the action extends to $Q(A)$ or $Q_H(A)$ where appropriate,
1) the H-action is *X-inner* if it becomes inner when extended to $Q(A)$
2) the H-action is *X_H-inner* if it becomes inner when extended to $Q_H(A)$.

Of course if the action is X_H-inner, it is also X-inner since $Q_H \subset Q$.

6.4.8 COROLLARY. *Let H be a pointed Hopf algebra and let A be an H-module algebra. Then there exists a largest X-inner subHopfalgebra of H.*

PROOF. Note the antipode of H is bijective by 5.2.11. Now use 6.4.6, 6.4.5, and 6.3.3. □

One would then like to define an X-outer action of H to be one in which $H_{inn} = \mathrm{k}1$, where H_{inn} is the largest inner subcoalgebra of H as in 6.3.1. However, such actions are not very well behaved even for $H = U(\mathfrak{g})$, and a stronger definition is probably needed. See, for example, [Sch 93a].

Chapter 7

Crossed Products

The crossed products we consider here are generalizations of the smash products studied in Chapter 4, in which the action of the Hopf algebra is twisted by a cocycle. This construction has been extensively studied for group actions. In this chapter we first characterize a Hopf crossed product as a certain kind of extension, and discuss the question as to when a Hopf algebra is a crossed product over a normal subHopfalgebra. We then consider when two crossed products are isomorphic, with particular interest in inner actions. Finally we look at the "Maschke-type" problem of when a crossed product is semiprime.

§7.1 Definitions and examples

Recall from 4.1.1 that a Hopf algebra H *measures* an algebra A if there is a $\mathbf{k}$-linear map $H \otimes A \to A$, given by $h \otimes a \mapsto h \cdot a$, such that $h \cdot 1 = \varepsilon(h)1$ and $h \cdot (ab) = \sum (h_1 \cdot a)(h_2 \cdot b)$, for all $h \in H, a, b \in A$.

7.1.1 DEFINITION. Let H be a Hopf algebra and A an algebra. Assume that H measures A and that σ is an invertible map in $\mathrm{Hom}_{\mathbf{k}}(H \otimes H, A)$. The *crossed product* $A \#_\sigma H$ of A with H is the set $A \otimes H$ as a vector space, with multiplication

$$(a\#h)(b\#k) = \sum a(h_1 \cdot b)\sigma(h_2, k_1)\#h_3 k_2$$

all $h, k \in H, a, b \in A$. Here we have written $a\#h$ for the tensor $a \otimes h$.

7.1.2 LEMMA [DT 86, BCM]. *$A\#_\sigma H$ is an associative algebra with identity element $1\#1$ $\Leftrightarrow$ the following two conditions are satisfied:*
1) A is a twisted H-module; that is, $1 \cdot a = a$, all $a \in A$, and

$$(7.1.3) \qquad h \cdot (k \cdot a) = \sum \sigma(h_1, k_1)(h_2 k_2 \cdot a)\sigma^{-1}(h_3, k_3),$$

all $h, k \in H, a \in A$.

2) σ is a cocycle. That is, $\sigma(h,1) = \sigma(1,h) = \varepsilon(h)1$, all $h \in H$, and

$$\sum [h_1 \cdot \sigma(k_1, m_1)]\sigma(h_2, k_2 m_2) = \sum \sigma(h_1, k_1)\sigma(h_2 k_2, m) \tag{7.1.4}$$

all $h, k, m \in H$.

Note that A need not be an H-module and that σ does not necessarily have values in the center of A.

The proof of the lemma is fairly straightforward. In the rest of these notes, we assume that all crossed products are associative with identity $1\#1$, and thus that the conditions in the lemma hold.

We note that crossed products in this generality were introduced independently in [DT 86] and in [BCM]. They were studied much earlier in [S 68] in the case A was commutative and H cocommutative; in that case A is always an H-module and 7.1.3 is not needed.

7.1.5 EXAMPLE. Consider the case when σ is trivial; that is, $\sigma(h,k) = \varepsilon(h)\varepsilon(k)1$, for all $h, k \in H$. Then 7.1.3 simply says that A is an H-module, and 7.1.4 is trivial. Thus A is an H-module algebra. Moreover, the definition of multiplication in 7.1.1 reduces to the multiplication in a smash product, 4.1.4, and so $A\#_\sigma H = A\#H$.

If the action is trivial, then we write $A\#_\sigma H = A_\sigma[H]$, and call the crossed product a *twisted product*. This generalizes the notion of a twisted group algebra.

7.1.6 EXAMPLE. Let $H = \mathrm{k}G$ be a group algebra. Then we obtain the familiar conditions for a group crossed product:

$$g \cdot (h \cdot a) = \sigma(g,h)(gh \cdot a)\sigma(g,h)^{-1}$$

$$[g \cdot \sigma(h,k)]\sigma(g,hk) = \sigma(g,h)\sigma(gh,k)$$

for all $g, h, k \in G$. Moreover multiplication in $A\#_\sigma \mathrm{k}G$ becomes

$$(a\#g)(b\#h) = a(g \cdot b)\sigma(g,h)\#gh,$$

the usual multiplication as in [M 80] and [P].

As a special case of this example, we note that for any group G and normal subgroup N of G, we may write $\mathrm{k}G = \mathrm{k}N\#_\sigma \mathrm{k}[G/N]$, a crossed product of $\mathrm{k}N$ with the quotient group $\bar{G} = G/N$. For each coset $\bar{x} \in \bar{G}$, choose a coset representative $\gamma(\bar{x}) \in \bar{x}$; for simplicity assume $\gamma(\bar{1}) = 1$. Since $G = \bigcup_{\bar{x} \in \bar{G}} N\gamma(\bar{x})$, $kG = (kN)\gamma(\bar{G})$ and we may multiply in kG by

$$(n\gamma(\bar{x}))(m\gamma(\bar{y})) = n[\gamma(\bar{x})m\gamma(\bar{x})^{-1}] \, [\gamma(\bar{x})\gamma(\bar{y})\gamma(\bar{x}\bar{y})^{-1}]\gamma(\bar{x}\bar{y})$$

$$= n(\bar{x} \cdot m)\sigma(\bar{x}, \bar{y})\gamma(\bar{x}\bar{y})$$

for all $n, m \in N, \bar{x}, \bar{y} \in G$, where $\bar{x} \cdot m = (ad\gamma(\bar{x}))(m)$ and $\sigma(\bar{x}, \bar{y}) = \gamma(\bar{x})\gamma(\bar{y})\gamma(\bar{x}\bar{y})^{-1} \in N$.

Since N is normal in G, kN is stable under this action of $\bar{G}$ and σ has values in N. Thus we have an action, a cocycle, and a multiplication as in 7.1.1; one may also check directly that 7.1.3 and 7.1.4 hold (alternatively, since we know kG is associative, these properties follow from 7.1.2). Thus $kG = kN \#_\sigma k[G/N]$. This fact has proved useful in studying group algebras themselves; for by proving results about crossed products, one can then use induction on a chain of normal subgroups. See [P].

By a similar argument, one can check that crossed products for group actions are "transitive" in the following sense: given a crossed product $A \#_\sigma kG$ and a normal subgroup N of G, then we can find a cocycle $\tau : G/N \times G/N \to A \#_\sigma kN$ such that $A \#_\sigma kG \cong (A \#_\sigma kN) \#_\tau k[G/N]$. In Chapter 8 we consider to what extent this result can be extended to arbitrary Hopf algebras.

7.1.7 EXAMPLE. Let $H = U(\mathbf{g})$ for a Lie algebra $\mathbf{g}$. Several definitions of "differential crossed products " with $U(\mathbf{g})$ have appeared in the literature; we consider two of them here, First, they were defined in [McR, 1.7.12] and [Ch 87] as follows: a k-algebra $B \supset A$ is a crossed product of A by $U(\mathbf{g})$, written $B = A * U(\mathbf{g})$, provided there is a vector space embedding of $\mathbf{g}$ into B, via $x \mapsto \bar{x}$, such that for all $x, y \in \mathbf{g}, a \in A$,

i) $\bar{x}a - a\bar{x} = \delta_x(a) \in A$, where $\delta_x \in \mathrm{Der}_k(A)$

ii) $\bar{x}\bar{y} - \bar{y}\bar{x} = \overline{[xy]} + \tau(x, y)$, where $\tau : \mathbf{g} \times \mathbf{g} \to A$

iii) for a given basis $\{x_\alpha\}$ of $\mathbf{g}$, B is a free right (and left) A-module with the standard monomials in $\{\bar{x}_\alpha\}$ as a basis.

One may verify that τ and δ_x satisfy

$$(7.1.8) \quad \tau([xy], z) + \tau([zx], y) + \tau([yz], x) = \delta_x(\tau(y, z)) + \delta_z(\tau(x, y)) + \delta_y(\tau(z, x))$$

$$(7.1.9) \qquad \delta_x\delta_y(a) - \delta_y\delta_x(a) = \delta_{[xy]}(a) + [\tau(x, y), a]$$

for all $a \in A, x, y, z \in \mathfrak{g}$. Also, τ is clearly alternating and bilinear. That is, τ is a Lie cocycle from $\mathfrak{g}$ to A.

This leads us to the second definition: one begins with a linear map $\mathfrak{g} \to \mathrm{Der}_{\mathrm{k}} A$ given by $x \mapsto \delta_x$ and a Lie cocycle $\tau : \mathfrak{g} \times \mathfrak{g} \to A$ satisfying 7.1.8 and 7.1.9. Then one may form the Lie algebra $\mathcal{L} = A^- \times_\tau \mathfrak{g}$, an extension of A^- by $\mathfrak{g}$, and its enveloping algebra $U(\mathcal{L})$. In $U(\mathcal{L})$, let I be the ideal generated by the set $\{1_{U(A^-)} - 1_A, a \cdot b - ab \mid a, b \in A\}$, where ab is the usual multiplication in A and $a \cdot b$ denotes multiplication in $U(A^-)$. Then define the crossed product to be $A \times_\tau U(\mathfrak{g}) = U(\mathcal{L})/I$.

This construction is a common generalization of work of [BGR, Theorem 4.2] in which no cocycle was present, and work of [Mc, §2.4] in which a cocycle appeared but A was a commutative $U(\mathfrak{g})$-module. In fact both of these constructions are isomorphic to a Hopf crossed product:

7.1.10 THEOREM [M 88]. *Let A be an algebra and $\mathfrak{g}$ a Lie algebra. Then the following are equivalent, for an algebra B:*
1) $B \cong A * U(\mathfrak{g})$ as in [McR], [Ch 87],
2) $B \cong A \times_\tau U(\mathfrak{g})$ as above,
3) $B \cong A \#_\sigma U(\mathfrak{g})$ as in 7.1.1, where $\sigma : U(\mathfrak{g}) \times U(\mathfrak{g}) \to A$ is a Hopf cocycle as in 7.1.4.

The proof depends on Corollary 7.2.8.

7.1.11 EXAMPLE. Let $F \subset E$ be fields containing k such that E/F is a finite Galois extension with Galois group G. We will see in Example 8.2.6 that $E \cong F\#_\sigma(\mathrm{k}G)^*$, a crossed product of F with $H = (\mathrm{k}G)^*$. In fact the action of H on F is trivial, so the product in 7.1.1 depends only on the cocycle σ :

$$(a\#p_x)(b\#p_y) = \sum_{\substack{st=x \\ ur=y}} ab\sigma(p_s, p_u)\#p_t p_r = \sum_{st=x} ab\sigma(p_s, p_{yt^{-1}})\#p_t$$

We close this section by proving a necessary and sufficient condition for A to be an H-module when H is cocommutative.

7.1.12 PROPOSITION [BCM]. *Let H be cocommutative and A a twisted H-module which is measured by H. Then A is an H-module $\Leftrightarrow \sigma(H \otimes H) \subseteq Z(A)$, the center of A.*

PROOF. ($\Leftarrow$) If $\sigma(H \otimes H) \subseteq Z(A)$, then 7.1.3 becomes $h \cdot (k \cdot a) = hk \cdot a$ since H is cocommutative. Thus A is an H-module.
($\Rightarrow$) We may assume that H is pointed, since otherwise we may pass to $H \otimes \bar{\mathbf{k}}$, where $\bar{\mathbf{k}}$ is the algebraic closure of $\mathbf{k}$ (see §5.6). Then $H \otimes \bar{\mathbf{k}}$ is pointed and $Z(A \otimes \bar{\mathbf{k}}) = Z(A) \otimes \bar{\mathbf{k}}$. Now proceed by induction on the coradical filtration $\{H_n\}$ of H. We will show that for all pairs $(n, m), \sigma(H_n \otimes H_m) \subseteq Z(A)$.

If $n = m = 0$, recall $H_0 = \mathbf{k}G$ since H is pointed. For $g, h \in G$, 7.1.3 becomes $(gh \cdot a) = \sigma(g, h)(gh \cdot a)\sigma(g, h)^{-1}$, since $g \cdot (h \cdot a) = gh \cdot a$ by assumption. But $A = gh \cdot A$, and thus $\sigma(g, h) \in Z(A)$.

Now assume it is true for (n', m') if either $n' < n$ or if $m' < m$, and consider H_n and H_m. Since H is pointed, it suffices to show the result for $h \in H_n$ such that $\Delta h = x \otimes h + \sum_i h'_i \otimes h''_i$ for some $x \in G(H)$, where $h''_i \in H_{n-1}$; since H is cocommutative, also $\Delta h = h \otimes x + \sum h''_i \otimes h'_i$. Similarly we may assume that $k \in H_m$ satisfies $\Delta k = y \otimes k + \sum_j k'_j \otimes k''_j = k \otimes y + \sum k''_j \otimes k'_j$, for some $y \in G(H)$ and $k''_j \in H_{m-1}$. Rewriting (7.1.2) as

$$\sum h_1 \cdot (k_1 \cdot a)\sigma(h_2, k_2) = \sum \sigma(h_1, k_1)(h_2 k_2 \cdot a)$$

and using the inductive assumption that all $\sigma(h, k''_j), \sigma(h''_i, k)$, and $\sigma(h''_i, k''_j)$ are in $Z(A)$, we obtain after simplifying that

$$(xy \cdot a)\sigma(h, k) = \sigma(h, k)(xy \cdot a).$$

Since $xy \cdot A = A$, it follows that $\sigma(h, k) \in Z(A)$. □

§7.2 Cleft extensions and existence of crossed products

We wish to characterize crossed products $B = A \#_\sigma H$ as special kinds of extensions $A \subset B$. To do this, we first need a definition. Recall the notion of comodule algebra from 4.1.2 and of coinvariants from 1.7.1.

7.2.1 DEFINITION. Let $A \subset B$ be $\mathbf{k}$-algebras, and H a Hopf algebra.

1) $A \subset B$ is a (*right*) H-*extension* if B is a right H-comodule algebra with $B^{\mathrm{co}H} = A$.
2) The H-extension $A \subset B$ is H-*cleft* if there exists a right H-comodule map $\gamma : H \to B$ which is (convolution) invertible.

We note that we may always assume that $\gamma(1) = 1$ in 2), for if not, we may replace γ by $\gamma' = \gamma(1)^{-1}\gamma$.

The main result of this section is the following:

7.2.2 THEOREM. *An H-extension $A \subset B$ is H-cleft $\Leftrightarrow B \cong A\#_\sigma H$.*

The theorem follows from the next two propositions. The first is due to Doi and Takeuchi, although the same construction was given in [BCM] under the assumption that γ was a coalgebra splitting. It extends the construction for $N \subset G$ given in 7.1.6.

7.2.3 PROPOSITION [DT 86]. *Let $A \subset B$ be a right H-extension, which is H-cleft via $\gamma : H \to B$ such that $\gamma(1) = 1$. Then there is a crossed product action of H on A, given by*

$$h \cdot a = \sum \gamma(h_1) a \gamma^{-1}(h_2), \text{ for all } a \in A, h \in H, \tag{7.2.4}$$

and a convolution invertible map $\sigma : H \otimes H \to A$ given by

$$\sigma(h,k) = \sum \gamma(h_1)\gamma(k_1)\gamma^{-1}(h_2 k_2), \text{ all } h, k \in H. \tag{7.2.5}$$

This action gives B the structure of an H-crossed product over A. Moreover the algebra isomorphism $\Phi : A\#_\sigma H \to B$ given by $a\#h \mapsto a\gamma(h)$ is both a left A-module and right H-comodule map, where $A\#_\sigma H$ is a right H-comodule via $a\#h \mapsto \sum a\#h_1 \otimes h_2$.

We require a technical lemma.

7.2.6 LEMMA. *Assume that $A \subset B$ is a right H-extension, via $\rho : B \to B \otimes H$, and that $A \subset B$ is H-cleft via γ with $\gamma(1) = 1$. Then*
1) $\rho \circ \gamma^{-1} = (\gamma^{-1} \otimes S) \circ \tau \circ \Delta$
2) for any $b \in B$, $\sum b_0 \gamma^{-1}(b_1) \in A = B^{coH}$.

PROOF. 1) First observe that since ρ is an algebra map, $\rho\circ\gamma^{-1}$ is the inverse of $\rho \circ \gamma = (\gamma \otimes id) \circ \Delta$. Let $\theta = (\gamma^{-1} \otimes S) \circ \tau \circ \Delta$. Then

$$\begin{aligned}
[(\rho\circ\gamma) * \theta](h) &= \sum [(\gamma \otimes id) \circ \Delta(h_1)][(\gamma^{-1} \otimes S) \circ \tau \circ \Delta(h_2)] \\
&= \sum [\gamma(h_1) \otimes h_2][\gamma^{-1}(h_4) \otimes S h_3] \\
&= \sum \gamma(h_1)\gamma^{-1}(h_4) \otimes h_2 S h_3 = \varepsilon(h) 1 \otimes 1.
\end{aligned}$$

Thus θ is a right inverse of $\rho\circ\gamma$, and so $\theta = \rho\circ\gamma^{-1}$ by uniqueness of inverses.

2)
$$\begin{aligned}\rho(\sum b_0\gamma^{-1}(b_1)) &= \sum \rho(b_0)\rho(\gamma^{-1}(b_1))\\ &= \sum b_0\gamma^{-1}(b_3)\otimes b_1(Sb_2) \qquad \text{using 1)}\\ &= \sum b_0\gamma^{-1}(b_1)\otimes 1\end{aligned}$$

Thus 2) follows by the definition of B^{coH}. □

The lemma enables us to define an inverse to Φ. Namely, define

$$\Psi : B \to A\#_\sigma H \quad \text{by} \quad b \mapsto \sum b_0\gamma^{-1}(b_1)\#b_2.$$

PROOF of 7.2.3. We first show that $h\cdot a \in A$, for $a \in A, h \in H$. Now

$$\begin{aligned}\rho(h\cdot a) &= \rho\left(\sum \gamma(h_1)a\gamma^{-1}(h_2)\right) = \sum \rho\circ\gamma(h_1)\rho(a)\rho\circ\gamma^{-1}(h_2)\\ &= \sum(\gamma(h_1)\otimes h_2)(a\otimes 1)(\gamma^{-1}(h_4)\otimes Sh_3)\\ &\qquad \text{using } \rho\circ\gamma = (\gamma\otimes id)\circ\Delta,\ A = B^{coH}, \text{ and 7.2.6, 1)}\\ &= \sum \gamma(h_1)a\gamma^{-1}(h_4)\otimes h_2Sh_3 = h\cdot a\otimes 1.\end{aligned}$$

Thus $h\cdot a \in B^{coH} = A$. Moreover it is easy to see that H measures A, since the action behaves like the adjoint action.

Similarly we see that σ has values in A. For if $h, k \in H$, then by 7.2.6

$$\begin{aligned}\rho(\sigma(h,k)) &= \sum \rho\gamma(h_1)\rho\gamma(k_1)\rho\gamma^{-1}(h_2k_2)\\ &= \sum(\gamma(h_{11})\otimes h_{12})(\gamma(k_{11})\otimes k_{12})(\gamma^{-1}(h_{22}k_{22})\otimes S(h_{21}k_{21}))\\ &= \sum \gamma(h_1)\gamma(k_1)\gamma^{-1}(h_4k_4)\otimes h_2k_2(Sk_3)(Sh_3)\\ &= \sum \gamma(h_1)\gamma(k_1)\gamma^{-1}(h_2k_2)\otimes 1 = \sigma(h,k)\otimes 1.\end{aligned}$$

Thus $\sigma(h,k) \in B^{coH} = A$.

Next we show that Φ and Ψ are mutual inverses. First, if $b \in B$, then $\Phi\Psi(b) = \sum \Phi(b_0\gamma^{-1}(b_1)\#b_2) = \sum b_0\gamma^{-1}(b_1)\gamma(b_2) = b$. Next, choose $a\#h \in$

$A\#_\sigma H$. Then

$$\begin{aligned}\Psi\Phi(a\#h) &= \Psi(a\gamma(h)) = \sum a_0\gamma(h)_0\ \gamma^{-1}(a_1\gamma(h)_1)\#a_2\gamma(h)_2 \\ &= \sum a\gamma(h)_0\ \gamma^{-1}(\gamma(h)_1)\#\gamma(h)_2 \quad \text{since } \rho(a) = a\otimes 1 \\ &= a(\sum \gamma(h_1)\gamma^{-1}(h_2)\#h_3) \quad \text{since } \rho\circ\gamma = (\gamma\otimes id)\circ\Delta \\ &= a\#h.\end{aligned}$$

Thus $\Psi = \Phi^{-1}$. Moreover, Φ is an algebra map. For,

$$\begin{aligned}\Phi(a\#h)\Phi(b\#k) &= a\gamma(h)b\gamma(k) \\ &= \sum a\gamma(h_1)b\gamma^{-1}(h_2)\gamma(h_3)\gamma(k_1)\gamma^{-1}(h_4k_2)\gamma(h_5k_3) \\ &= \sum a(h_1\cdot b)\sigma(h_2,k_1)\gamma(h_3k_2) \\ &= \Phi((a\#h)(b\#k)).\end{aligned}$$

Thus $B \cong A\#_\sigma H$. Since we know B is an associative algebra, the conditions 7.1.3 and 7.1.4 follow from 7.1.2.

Also, Φ is clearly a left A-module map. It is also a right H-comodule map, since

$$\begin{aligned}\rho(\Phi(a\#h)) &= \rho(a\gamma(h)) = \rho(a)\rho(\gamma(h)) \\ &= (a\otimes 1)\sum \gamma(h_1)\otimes h_2 \\ &= \sum \Phi(a\#h_1)\otimes h_2 \\ &= (\Phi\otimes id)\circ(id\otimes\Delta)(a\#h)\end{aligned}$$

and $id\otimes\Delta$ gives the H-comodule structure on $A\#_\sigma H$. This proves the Proposition. □

The second proposition used in 7.2.2 is due to [BM 89], improving a partial result in [BCM].

7.2.7 PROPOSITION [BM 89]. *Let $A\#_\sigma H$ be a crossed product, and define the map $\gamma : H \to A\#_\sigma H$ by $\gamma(h) = 1\#h$. Then γ is convolution-invertible, with inverse*

$$\gamma^{-1}(h) = \sum \sigma^{-1}(Sh_2, h_3)\#Sh_1.$$

In particular $A \hookrightarrow A\#_\sigma H$ is H-cleft.

PROOF. Set $\mu(h) = \sum \sigma^{-1}(Sh_2, h_3)\#Sh_1$. Then it is straightforward to verify that μ is a left inverse for γ. For,

$$\begin{aligned}(\mu * \gamma)(h) &= \sum (\sigma^{-1}(Sh_2, h_3)\#Sh_1)(1\#h_4) \\ &= \sum \sigma^{-1}(Sh_3, h_4)\sigma(Sh_2, h_5)\#(Sh_1)h_6 \quad \text{by 7.1.1} \\ &= \sum \varepsilon(Sh_2)\varepsilon(h_3)1\#(Sh_1)h_4 \\ &= \varepsilon(h)1\#1.\end{aligned}$$

To check that μ is a right inverse for γ is more complicated. First, by a computation similar to the above, one may check that

$$(\gamma * \mu)(h) = \sum [h_1 \cdot \sigma^{-1}(Sh_4, h_5)]\sigma(h_2, Sh_3)\#1.$$

Thus it suffices to prove that $\sum [h_1 \cdot \sigma^{-1}(Sh_4, h_5)]\sigma(h_2, Sh_3) = \varepsilon(h)1$. We claim that

$$h \cdot \sigma^{-1}(k, m) = \sum \sigma(h_1, k_1 m_1)\sigma^{-1}(h_2 k_2, m_2)\sigma^{-1}(h_3, k_3).$$

This follows from the fact that , from 7.1.4,

$$h \cdot \sigma(k, m) = \sum \sigma(h_1, k_1)\sigma(h_2 k_2, m_1)\sigma^{-1}(h_3, k_3 m_2)$$

and that the map $h \otimes k \otimes m \mapsto h \cdot \sigma^{-1}(k, m)$ is the convolution inverse of $h \otimes k \otimes m \mapsto h \cdot \sigma(k, m)$ in $\mathrm{Hom}(H \otimes H \otimes H, A)$. We now use the claim to see that

$$\begin{aligned}&\sum [h_1 \cdot \sigma^{-1}(Sh_4, h_5)]\sigma(h_2, Sh_3) \\ &\quad = \sum \sigma(h_1, (Sh_8)h_9)\sigma^{-1}(h_2 Sh_7, h_{10})\sigma^{-1}(h_3, Sh_6)\sigma(h_4, Sh_5) \\ &\quad = \sum \sigma(h_1, (Sh_6)h_7)\sigma^{-1}(h_2 Sh_5, h_8)\varepsilon(h_3)\varepsilon(Sh_4) \\ &\quad = \sum \sigma(h_1, (Sh_4)h_5)\sigma^{-1}(h_2 Sh_3, h_6) \\ &\quad = \sum \varepsilon(h_3)\sigma(h_1, 1)\sigma^{-1}(1, h_4)\varepsilon(h_2) = \varepsilon(h)1.\end{aligned}$$

This proves the proposition.   □

Note that for smash products $A \# H$, it is trivial that γ is invertible: simply let $\gamma^{-1}(h) = 1 \# Sh$.

We have also proved 7.2.2, by combining 7.2.3 and 7.2.7.

As a consequence, we may show the analog for Lie algebras of the fact shown for groups in 7.1.6. It is also the basic ingredient in showing 7.1.10 mentioned earlier.

7.2.8 COROLLARY. *Let* $\mathbf{g}$ *be a Lie algebra and* $\mathbf{h}$ *a Lie ideal, and let* π : $\mathbf{g} \to \bar{\mathbf{g}} = \mathbf{g}/\mathbf{h}$ *be the canonical map. Then* $U(\mathbf{g}) \cong U(\mathbf{h}) \#_\sigma U(\bar{\mathbf{g}})$, *a Hopf crossed product of* $U(\mathbf{h})$ *with* $U(\bar{\mathbf{g}})$.

PROOF. By 7.2.3, it suffices to show that $U(\mathbf{g})$ is a $U(\bar{\mathbf{g}})$-comodule algebra, with coinvariants $U(\mathbf{h})$, such that the extension is cleft. First, $U(\mathbf{g})$ is a right $U(\bar{\mathbf{g}})$-comodule algebra via $\rho = (id \otimes \pi) \circ \Delta$. Now if $w \in U(\mathbf{h})$, then $\sum w_1 \otimes \pi(w_2) = \sum w_1 \otimes \varepsilon(w_2)\bar{1} = w \otimes \bar{1}$, and thus $U(\mathbf{h}) \subseteq U(\mathbf{g})^{coU(\bar{\mathbf{g}})} = A$. Before showing the other containment, we show that the extension is cleft.

Let $\{\bar{x}_\alpha\}$ be an ordered basis of $\bar{\mathbf{g}}$, and for each α choose $x_\alpha \in \mathbf{g}$ such that $\pi x_\alpha = \bar{x}_\alpha$. Define $\gamma \in \mathrm{Hom}_k(U(\bar{\mathbf{g}}), U(\mathbf{g}))$ by $\gamma(\bar{x}_{\alpha_1} \ldots \bar{x}_{\alpha_n}) = x_{\alpha_1} \ldots x_{\alpha_n}$ for any standard monomial in $U(\bar{\mathbf{g}})$; then γ is invertible, with $\gamma^{-1} = S\gamma$. Moreover γ is a comodule map; that is, $(\gamma \otimes id) \circ \bar{\Delta} = (id \otimes \pi) \circ \Delta \circ \gamma = \rho \circ \gamma$. Thus $A \subset U(\mathbf{g})$ is cleft.

Now the proof of 7.2.3 shows that the map $A \otimes U(\bar{\mathbf{g}}) \to U(\mathbf{g})$ given by $a \otimes \bar{h} \mapsto a\gamma(\bar{h})$ is a linear isomorphism. By the PBW theorem, the restriction of this map to $U(\mathbf{h}) \otimes U(\bar{\mathbf{g}})$ maps onto $U(\mathbf{g})$. Thus $A = U(\mathbf{h})$.   □

Since for groups and Lie algebras, homomorphisms give crossed products, we ask:

7.2.9 QUESTION. Let $\pi : H \to \bar{H}$ be a Hopf algebra surjection and consider H as a right $\bar{H}$-comodule algebra via $\rho = (id \otimes \pi) \circ \Delta$. For $A = H^{co\bar{H}}$, when is it true that $H \cong A \#_\sigma \bar{H}$?

The question is false in general, even when $I = \mathrm{Ker}\ \pi$ is of the form HK^+ for K a normal subHopfalgebra; this has been known for a long time. For, it is clear that a necessary condition is that H is free over A. Thus Oberst and Schneider's non-free example gives a counterexample:

7.2.10 EXAMPLE. Recall from 3.5.2 that H is commutative but that H is not free over K. Now K is normal in H and $I = HK^+$ is a Hopf ideal of H; thus $\bar{H} = H/HK^+$ is a Hopf algebra and $\pi : H \rightarrow \bar{H}$ a Hopf algebra surjection. Moreover, $K = H^{co\bar{H}}$ (this follows by 3.4.4 since H is faithfully flat over K). Thus we cannot have $H \cong A\#_\sigma \bar{H}$, for $A = H^{co\bar{H}}$.

Some positive results are known, however; in [OSch 74] it is shown that it is true if H is pointed cocommutative. This gives an alternate proof of 7.2.8. More recently some progress has been made on the question by Schneider [Sch 92]: for example, it is true whenever H is finite-dimensional, or when H is pointed provided Ker π is of the form HK^+ for some normal subHopfalgebra K. We consider this question further in §8.4.

We give another application of 7.2.7 and 7.2.3. Now for any crossed product $A\#_\sigma H$, it is trivially true that $A\#_\sigma H \cong A \otimes H$ as a left A-module; thus $A\#_\sigma H$ is free as a left A-module. In fact this is also true on the right:

7.2.11 COROLLARY [BM 89]. *Let $A\#_\sigma H$ be a crossed product. Then $A\#_\sigma H \cong H \otimes A$ as right A-modules, provided the antipode S of H is bijective.*

PROOF. We give a simplification by Schneider of our original proof. By 7.2.7 we know that the map $\gamma : H \rightarrow A\#_\sigma H$ via $h \mapsto 1\#h$ is an invertible right H-comodule map, where the comodule structure maps for H and $A\#_\sigma H$ are given by Δ and by $\rho = id \otimes \Delta$, respectively. Let $\bar{S}$ denote the (composition) inverse of S and set $\mu = \gamma^{-1} \circ \bar{S}$. Note that μ is inverible under twist convolution (see 1.4.1) with inverse $\hat{\mu} = \gamma \circ \bar{S}$. It follows that if $B = A\#_\sigma H$, then B^{op} is a right H^{op}-comodule algebra which is cleft via

$$\mu : H^{op} \rightarrow B^{op}.$$

Thus by 7.2.3, $A^{op} \otimes H^{op} \cong B^{op}$ as left A^{op}-modules, where the isomorphism is given by

$$a^{op} \otimes h^{op} \mapsto a^{op}\mu(h)^{op} = (\gamma^{-1}(\bar{S}(h))a)^{op}.$$

It follows that the map $\alpha : H \otimes A \rightarrow B$, given by $h \otimes a \mapsto \gamma^{-1}(\bar{S}(h))a$, is an isomorphism of right A-modules. □

The special case of 7.2.11 when σ is trivial (that is, we have a smash product) is very easy. For then $\gamma^{-1}(h) = 1\#Sh$ and so

$$\alpha : H \otimes A \to B \quad \text{by} \quad h \otimes a \mapsto (1\#h)(a\#1)$$

is an A-module isomorphism with inverse $\beta : B \to H \otimes A$ given by $\beta(a\#h) = \sum h_2 \otimes (\bar{S}h_1 \cdot a)$. This fact has already been observed in 4.4.3, 1); in that case we actually have $A\#H = (1\#H)(A\#1)$. We do not know if this is true in general:

7.2.12 QUESTION. Let $A\#_\sigma H$ be a crossed product. Is it always true that $A\#_\sigma H = (1\#H)(A\#1)$?

Knowing this fact would be very useful in studying ideals in crossed products. A somewhat weaker fact is known: if the antipode S of H is bijective, then for any H-stable ideal I of A, $I\#_\sigma H = (A\#_\sigma H)(I\#1)$ [MSch].

§7.3 Inner actions and equivalence of crossed products

Before looking at the general case of isomorphism of crossed products, we consider the special case of inner actions. In this situation the crossed product can be replaced by another one in which the action becomes trivial but the cocycle has been changed.

7.3.1 PROPOSITION [BCM]. *Let $A\#_\sigma H$ be a crossed product such that the action of H on A is inner, via some invertible $u \in Hom(H, A)$. Define $\tau \in Hom(H \otimes H, A)$ by*

$$\tau(h,k) = \sum u^{-1}(k_1)u^{-1}(h_1)\sigma(h_2,k_2)u(h_3k_3) \tag{7.3.2}$$

Then τ is a cocycle and $A\#_\sigma H \cong A_\tau[H]$, a twisted product with trivial action, via an algebra isomorphism which is also a left A-module, right H-comodule map.

PROOF. Define $\phi : A\#_\sigma H \to A_\tau[H]$ by $a\#h \mapsto \sum au(h_1) \otimes h_2$. Then ϕ has an inverse $\psi : A_\tau[H] \to A\#_\sigma H$ given by $a \otimes h \mapsto \sum au^{-1}(h_1)\#h_2$. It is straightforward to check that ϕ and ψ are inverses. We check that ϕ is an algebra map: for all $a, b \in A$, $h, k \in H$,

$$\begin{aligned}
\phi((a\#h)(b\#k)) &= \phi(\sum a(h_1 \cdot b)\sigma(h_2, k_1)\#h_3k_2) \\
&= \sum a(h_1 \cdot b)\sigma(h_2, k_1)u(h_3k_2) \otimes h_4k_3 \\
&= \sum a\ u(h_1)b\ u^{-1}(h_2)\sigma(h_3, k_1)u(h_4k_2) \otimes h_5k_3 \\
&= \sum a\ u(h_1)b\ u(k_1)\tau(h_2, k_2) \otimes h_3k_3 \\
&= (\sum a\ u(h_1) \otimes h_2)(\sum b\ u(k_1) \otimes k_2) \\
&= \phi(a\#h)\phi(b\#k).
\end{aligned}$$

Since $A_\tau[H] \cong A\#_\sigma H$ as algebras, $A_\tau[H]$ is associative, and thus τ is a cocycle. It is clear that ϕ is a left A-module, right H-comodule map. □

7.3.3 EXAMPLE. Let $A\#H$ be a smash product such that the H-action is inner via some $u \in \text{Alg}(H, A)$; in this case we say that the action is *strongly inner*. Then $A\#H \cong A \otimes H$, for the cocycle τ in 7.3.2 becomes trivial.

As a particular case of the example, let H act on itself via the (left) adjoint action. Then $u(h) = h$ is certainly an algebra map, and thus $H\#H \cong H \otimes H$.

The converse of 7.3.1 is also true: that is, if $A\#_\sigma H \cong A_\tau[H]$ for some twisted product, by an isomorphism ϕ which is a left A-module, right H-comodule map, then the original action must have been inner, via some $u \in \text{Hom}(H, A)$ such that ϕ is given as in 7.3.1.

More generally, one can give necessary and sufficient conditions for two crossed products to be isomorphic. This is shown by Doi in [D 89]; it was also done independently by R.J. Blattner in 1985 (unpublished). We follow Blattner's formulation.

7.3.4 THEOREM. *Let A be an algebra and H be a Hopf algebra, with two crossed product actions $h \otimes a \mapsto h \cdot a$, $h \otimes a \mapsto h \cdot' a$ with respect to two cocycles $\sigma, \sigma' : H \otimes H \to A$, respectively. Assume that*

$$\phi : A\#_\sigma H \to A\#'_{\sigma'} H$$

is an algebra isomorphism, which is also a left A-module, right H-comodule *map. Then there exists an invertible map* $u \in Hom(H, A)$ *such that for all* $a \in A, h, k \in H$,

1) $\phi(a\#h) = \sum au(h_1)\#'h_2$

2) $h \cdot' a = \sum u^{-1}(h_1)(h_2 \cdot a)u(h_3)$

3) $\sigma'(h, k) = \sum u^{-1}(h_1)(h_2 \cdot u^{-1}(k_1))\sigma(h_3, k_2)u(h_4k_3)$.

Conversely given a map $u \in Hom(H, A)$ *such that 2) and 3) hold, then the map* ϕ *in 1) is an isomorphism.*

PROOF. Define $u \in \mathrm{Hom}(H, A)$ by $u(h) = (id \otimes \varepsilon)\phi(1\#h)$, all $h \in H$. Then $(id \otimes \varepsilon)\phi(a\#h) = (id \otimes \varepsilon)[(a \otimes 1)\phi(1\#h)] = au(h)$, as ϕ is a left A-module map. Since ϕ is a right H-comodule map, we have

$$(id \otimes \Delta) \circ \phi = (\phi \otimes id) \circ (id \otimes \Delta).$$

Apply $id \otimes \varepsilon \otimes id$ to both sides of the equation. The left side becomes ϕ, and the right side becomes $([(id \otimes \varepsilon) \circ \phi] \otimes id) \circ (id \otimes \Delta)$, which evaluated at $a\#h$ is

$$\sum (id \otimes \varepsilon) \circ \phi(a\#h_1) \otimes h_2 = \sum au(h_1) \otimes h_2$$

by the above. This proves 1).

Similarly, as $\phi^{-1} : A\#'_{\sigma'}H \to A\#_\sigma H$ is an isomorphism satisfying the same hypotheses as ϕ, we may set $v(h) = (id \otimes \varepsilon)\phi^{-1}(1\#h)$ and conclude as above that $\phi^{-1}(a\#'h) = \sum av(h_1)\#h_2$. We claim that $v = u^{-1}$. For,

$$1\#h = \phi^{-1}\phi(1\#h) = \phi^{-1}(\sum u(h_1)\#'h_2) = \sum u(h_1)v(h_2)\#h_3.$$

Applying $id \otimes \varepsilon$ to both sides, we see that $\sum u(h_1)v(h_2) = \varepsilon(h)1$. Similarly we see $\sum v(h_1)u(h_2) = \varepsilon(h)1$, and thus $v = u^{-1}$.

Now the equation $\phi^{-1}((a\#'h)(b\#'k)) = \phi^{-1}(a\#'h)\phi^{-1}(b\#'k)$ becomes

$$\sum a(h_1 \cdot' b)\sigma'(h_2, k_1)v(h_3k_2)\#h_4k_3 = \sum av(h_1)(h_2 \cdot bv(k_1))\sigma(h_3, k_2)\#h_4k_3.$$

Set $a = b = 1$ and apply $id \otimes \varepsilon$ to both sides, obtaining

$$\sum \sigma'(h_1, k_1)v(h_2k_2) = \sum v(h_1)(h_2 \cdot v(k_1))\sigma(h_3, k_2).$$

This proves 3), after inverting $v(hk)$ and using $v = u^{-1}$.

Again use the above equation with $a = 1$ and $k = 1$, and apply $id \otimes \varepsilon$ to both sides to see

$$\sum (h_1 \cdot' b)v(h_2) = \sum v(h_1)(h_2 \cdot b).$$

Inverting v gives 2).

The converse follows as in the proof of 7.3.1. □

As an application, we return to 7.2.3 and see how changing the map $\gamma : H \to B$ will change the action 7.2.4 and the cocycle 7.2.5.

7.3.5 COROLLARY. *Let $A \subset B$ be a right H-extension which is H-cleft via $\gamma, \gamma' : H \to B$, with $\gamma(1) = \gamma'(1) = 1$. Let $A\#_\sigma H$ and $A\#'_{\sigma'}H$ be the two representations of B as a crossed product over A with H, with the two actions and cocycles σ, σ'as described in 7.2.3 and define $u = \gamma * (\gamma')^{-1}$ in $Hom(H, B)$. Then the actions and cocycles are related as in 7.3.4, 2) and 3).*

PROOF. Let $\Phi : A\#_\sigma H \to B$ and $\Phi' : A\#'_{\sigma'}H \to B$ be the corresponding isomorphisms from 7.2.3; thus $\Phi(a\#h) = a\gamma(h)$ and $\Phi'(a\#'h) = a\gamma'(h)$, all $a \in A, h \in H$. Since Φ and Φ' are both left A-module, right H-comodule maps, so is $\Theta = (\Phi')^{-1}\Phi : A\#_\sigma H \to A\#'_{\sigma'}H$. Apply 7.3.4 to get that $\Theta(a\#h) = \sum au(h_1)\#'h_2$. Applying Φ' to both sides, we see that

$$a\gamma(h) = \sum au(h_1)\gamma'(h_2).$$

Setting $a = 1$ gives $\gamma = u * \gamma'$. The result follows. □

7.3.4 suggests the following definition:

7.3.6 DEFINITION. Let H be a Hopf algebra and A an algebra. Two crossed products $A\#_\sigma H$ and $A\#'_{\sigma'}H$ are *equivalent* if there exists an algebra isomorphism $\phi : A\#_\sigma H \to A\#'_{\sigma'}H$ which is a left A-module, right H-comodule morphism.

One would hope to extend the equivalence criteria in 7.3.4 to get a general cohomology theory for algebras over Hopf algebras; however this looks very difficult. As a beginning, it was shown by Sweedler that for H cocommutative and A commutative, there is a bijective correspondence between the second cohomology group $\mathcal{H}^2(H, A)$ and the equivalence classes of H-cleft extensions B of A [S 68]. Note that in this case A is an H-module, and in addition all the crossed products in a given equivalence class have the same H-action, by 7.3.4, 2); only the cocycle may be different.

This result has been extended in [D 89] to the following: assume H is cocommutative and consider a given crossed product action of H on A; since H is cocommutative the center $Z(A)$ of A is a (left) H-module algebra under this action. Then there is a bijective correspondence between $\mathcal{H}^2(H, Z(A))$ and the equivalence classes of H-cleft extensions of A which have the same H-action as in the given crossed product action.

§7.4 Generalized Maschke theorems and semiprime crossed products

We first consider when crossed products are semisimple Artinian. This will generalize 2.2.1 as well as the [CF 86] result for smash products mentioned after 4.4.7; the general form was finished in [BM 89]. For a crossed product $A\#_\sigma H$, recall that the map $\gamma : H \to A\#_\sigma H$, given by $h \mapsto 1\#h$, is invertible in $\mathrm{Hom}(H, A\#_\sigma H)$ by 7.2.7, although we will not need the actual formula for γ^{-1}. It then follows from the fact that $(1\#h)(a\#1) = \sum(h_1 \cdot a)\#h_2$, writing $a = a\#1$, that

$$h \cdot a = \sum \gamma(h_1) a \gamma^{-1}(h_2) \tag{7.4.1}$$

for all $h \in H, a \in A$.

7.4.2 THEOREM [BM 89]. *Let $A\#_\sigma H$ be a crossed product for a finite-dimensional, semisimple Hopf algebra H.*
1) Let $V \in {}_{A\#_\sigma H}\mathcal{M}$. If $W \subseteq V$ is a submodule which has a complement in ${}_A\mathcal{M}$, then W has a complement in ${}_{A\#_\sigma H}\mathcal{M}$.
2) If A is semisimple Artinian, then so is $A\#_\sigma H$.

PROOF. Clearly 2) follows from 1). For 1), we begin as in 2.2.1. Let $\pi : V \to W$ be an A-projection and choose $t \in \int_H^r$ with $\varepsilon(t) = 1$. However our averaging function is slightly different than the map in 2.2.1; for any $v \in V$, we define

$$\tilde{\pi}(v) = \sum \gamma^{-1}(t_1)\pi(\gamma(t_2)v).$$

This is closer to the map in [CF 86] with S replaced by γ^{-1}; it corresponds to the right adjoint action, whereas the map in 2.2.1 corresponds to the left adjoint action.

We first check, as in [CF 86], that $\tilde{\pi}$ is a left A-map. For $a \in A, v \in V$,

$$\begin{aligned}
\tilde{\pi}(av) &= \sum \gamma^{-1}(t_1)\pi(\gamma(t_2)av) \\
&= \sum \gamma^{-1}(t_1)\pi((t_2 \cdot a)\gamma(t_3)v) \quad \text{by 7.4.1} \\
&= \sum \gamma^{-1}(t_1)(t_2 \cdot a)\pi(\gamma(t_3)v) \quad \text{since } \pi \text{ is an } A\text{-map} \\
&= \sum a\gamma^{-1}(t_1)\pi(\gamma(t_2)v) \quad \text{by 7.4.1} \\
&= a\tilde{\pi}(v).
\end{aligned}$$

Thus $\tilde{\pi}$ is an A-map. To show $\tilde{\pi}$ is also an H-map, we need a different argument than in 2.2.1 or in [CF 86]. First, we observe

$$(*) \qquad h \otimes \Delta t = \sum h_1 \otimes t_1 h_2 \otimes t_2 h_3$$

similarly to the equation $(*)$ in 2.2.1. Now for any $h \in H, v \in V$,

$$\begin{aligned}
\tilde{\pi}(\gamma(h)v) &= \sum \gamma^{-1}(t_1)\pi(\gamma(t_2)\gamma(h)v) \\
&= \sum \gamma^{-1}(t_1)\pi(\sigma(t_2, h_1)\gamma(t_3 h_2)v) \quad \text{by 7.1.1} \\
&= \sum \gamma^{-1}(t_1)\sigma(t_2, h_1)\pi(\gamma(t_3 h_2)v) \quad \text{since } \pi \text{ is an } A\text{-map} \\
&= \sum \gamma(h_1)\gamma^{-1}(t_1 h_2)\pi(\gamma(t_2 h_3)v) \quad \text{by 7.1.1} \\
&= \sum \gamma(h)\gamma^{-1}(t_1)\pi(\gamma(t_2)v) \quad \text{by } (*) \\
&= \gamma(h)\tilde{\pi}(v).
\end{aligned}$$

Thus $\tilde{\pi}$ is a left $A\#_\sigma H$-map.

Finally if $w \in W$, then $\tilde{\pi}(w) = \sum \gamma^{-1}(t_1)\gamma(t_2)w = w$ since $\pi|_W = id$. Thus $\tilde{\pi}^2 = \tilde{\pi}$, and so Ker $\tilde{\pi}$ is an $A\#_\sigma H$-complement for W in V. □

We next consider the more general situation when A is not Artinian. Recall that A is semiprimitive if $\text{Jac}(A) = (0)$; equivalently, the intersection of the primitive ideals is 0. For inner actions, the standard method of induced modules works to show that $A\#H$ is semiprimitive.

7.4.3 COROLLARY. *Let H be finite-dimensional and semisimple, and $A\#_\sigma H$ a crossed product with A semiprimitive. Then $A\#_\sigma H$ is semiprimitive if the H-action on A is inner.*

PROOF. By 7.3.1, $A\#_\sigma H \cong A_\tau[H]$, a twisted product with trivial action. For any (left) A-module V, we form the induced module $\bar{V} = A_\tau[H] \otimes_A V$. Since the elements of H commute with the elements of A, $\bar{V} \cong V^{(n)}$ as an A-module, where $n = \dim_{\mathbf{k}} H$. Thus if V is a completely reducible A-module, so is $\bar{V}$.

Now an algebra A is semiprimitive $\Leftrightarrow$ there exists a faithful completely reducible A-module, say V. Constructing $\bar{V}$ as above, $\bar{V}$ is certainly faithful as an $A_\tau[H]$-module. It is also completely reducible as an $A_\tau[H]$-module by 7.4.2. Thus $A_\tau[H]$ is semiprimitive since it has a faithful completely reducible module. □

The above argument does not extend to $A\#_\sigma H$ when the action of H on A is not inner, since it is false in general that if V is a completely reducible A-module, then $\bar{V} = (A\#_\sigma H) \otimes_A V$ is completely reducible as an A-module even when σ is trivial, as we show in our next example.

7.4.4 EXAMPLE. Let G be a finite group. Let $A = \mathbf{k}G, H = (\mathbf{k}G)^*, V$ the trivial one dimensional G-module $\mathbf{k}$, and form $\bar{V} = (A\#H) \otimes_A V$, where H acts on A by $\rightharpoonup$. Let $\{p_x \mid x \in G\}$ be the basis of $(\mathbf{k}G)^*$ dual to the basis G of $\mathbf{k}G$. We have by 4.1.7

$$(1\#p_y)(x\#1) = x\#p_{x^{-1}y}$$

for $x, y \in G$, which shows that $A\#H$ is a free right A-module with basis $\{1\#p_x \mid x \in G\}$. Thus we may identify $\bar{V}$ with H via $p_x \leftrightarrow (1\#p_x) \otimes_A 1$. Moreover, the left action of A on $\bar{V}$ corresponds to the action of A on H given by $x \cdot p_y = p_{xy}$. In fact,

$$(x\#1)(1\#p_y) \otimes_A 1 = (1\#p_{xy})(x\#1) \otimes_A 1 = (1\#p_{xy}) \otimes_A 1.$$

This shows that as an A module, $\bar{V}$ is isomorphic to A under the left regular representation of A on itself. Now if the characteristic of $\mathbf{k}$ divides the order of $G, \mathbf{k}G$ is not completely reducible as a left G-module. □

We can extend 7.4.3 to show that A semiprime implies that $A\#_\sigma H$ is semiprime when H is inner, by using the "primitivity machine" of Lorenz and Passman [LoP]. The method was used in [BCM] for the case when H was also cocommutative.

If R is any ring with 1, let $\prod R$ denote the complete direct product of copies of R indexed over the natural numbers, and let $\sum R$ denote the direct sum. Now define

$$\hat{R} = \prod R / \sum R$$

and denote the elements of $\prod R$ by (a_n).

7.4.5 LEMMA [LoP]. *The following are equivalent:*
1) R is semiprime,
2) $\hat{R}$ has no nonzero nil ideals,
3) $\hat{R}$ is semiprime.

The next lemma was proved in [LoP] for $H = \mathrm{k}G$.

7.4.6 LEMMA. *Let $A\#_\sigma H$ be a crossed product, with H finite-dimensional. Then*

$$\widehat{A\#_\sigma H} \cong \hat{A}\#_\sigma H.$$

PROOF. The action of H on A can be extended to $\prod A$ by letting it act on each component; since $\sum A$ is stable under this action, the induced action of H on $\hat{A}$ is well defined. Similarly $\sigma : H \otimes H \to A$ may be extended to $\bar{\sigma} : H \otimes H \to \hat{A}$ by defining

$$\bar{\sigma}(h,k) = (\sigma(h,k), \ldots, \sigma(h,k), \ldots) + \sum A.$$

Clearly $\hat{A}\#_\sigma H$ is a crossed product.

The fact that $\widehat{A\#_\sigma H} \cong \hat{A}\#_\sigma H$ follows by exactly the same argument used by Lorenz and Passman [LoP], except that a fixed basis $\{h_1, \ldots, h_n\}$ of H is used instead of the group elements $\{g \mid g \in G\}$. □

7.4.7 THEOREM [BM 89]. *Let H be finite-dimensional and semisimple, and $A\#_\sigma H$ a crossed product with A a semiprime algebra. Then $A\#_\sigma H$ is semiprime if the action of H is inner.*

PROOF. We form $\widehat{A\#_\sigma H}$; by Lemma 7.4.5, it suffices to show $\widehat{A\#_\sigma H}$ is semiprime; by Lemma 7.4.6, this is equivalent to showing $\hat{A}\#_\sigma H$ is semiprime.

Now since A is semiprime, $\hat{A}$ has no nil ideals by Lemma 7.4.5. Thus the polynomial ring $\hat{A}[x]$ is semiprimitive by Amitsur's theorem. By Corollary

7.4.3, $\hat{A}[x]\#_\sigma H$ is also semiprimitive, where $\hat{A}[x]\#_\sigma H$ makes sense by letting H act trivially on x. But then $\hat{A}[x]\#_\sigma H = (\hat{A}\#_\sigma H)[x]$ is semiprimitive. Thus $\hat{A}\#_\sigma H$ has no nilpotent ideals, so is semiprime. □

7.4.7 has recently been improved in [MSch] to the case when the action of H is X_H-inner; recall from Chapter 6 that this means that the action of H becomes inner when extended to the H-symmetric quotient ring Q_H of A. A crucial ingredient there is the fact mentioned at the end of §7.2 about H-stable ideals I of A.

One might hope here to consider the X-inner and X-outer cases separately, and then put them together. A major difficulty is that it is not clear what the appropriate definition of X-outer should be, and if so if it would imply that $A\#_\sigma H$ is semiprime. For some partial results, particularly when H is pointed, see [BeM 92] and [Sch 93a].

In 7.4.7 we obtained a semiprimeness result by restricting the action of H. The other known positive results are proved by restricting H itself. As noted before 4.4.7, it is known when $H = \mathrm{k}G$ [FM] or when $H = (\mathrm{k}G)^*$ [CM]; the result for $(\mathrm{k}G)^*$ will be proved in 9.5.3. As a consequence of these facts and the results of §5.6, we obtain:

7.4.8 PROPOSITION. *Let H be finite-dimensional and semisimple, and let $A\#_\sigma H$ be a crossed product with A semiprime. Then $A\#_\sigma H$ is semiprime in the following two cases:*
1) H is commutative
2) H is pointed cocommutative [Ch 92].

PROOF. 1) This follows from the result for $(\mathrm{k}G)^*$ mentioned above and Harrison's Theorem 2.3.1, by extending the base field.
2) By 5.6.4 $H \cong K\#\mathrm{k}G$ where K is connected and $G = G(H)$. If char $\mathrm{k} = 0$, then $K = \mathrm{k}1$ by 5.6.5 and so $H = \mathrm{k}G$; we are now done by [FM]. Thus, assume char $\mathrm{k} = p \neq 0$. Since H is semisimple, also K is semisimple by 2.2.2. Thus by 5.7.1 and 2.3.1, we may assume $K \cong (\mathrm{k}L)^*$ for some group L. By an argument similar to 7.1.6, we see that

$$A\#_\sigma H \cong (A\#_\sigma K)\#_\tau \mathrm{k}G.$$

But now $A\#_\sigma K$ is semiprime by [CM], and then $(A\#_\sigma K)\#_\tau \mathrm{k}G$ is semiprime by [FM]. □

In fact the result in [Ch 92] is a bit more general; A is only required to be H-semiprime.

We close by restating 4.4.7 for crossed products, as it remains an important open question.

7.4.9 QUESTION. If H is finite-dimensional and semisimple and A is semiprime, is any crossed product $A\#_\sigma H$ semiprime?

§7.5 Twisted H-comodule algebras

In this last section we consider a variation on the crossed product, in which H is replaced by an H-comodule algebra. In this case the cocycle has values in $\mathbf{k}$, on which H acts trivially; thus we assume $\sigma : H \otimes H \to \mathrm{k}$ satisfies

$$\sum \sigma(k_1, m_1)\sigma(h_2, k_2m_2) = \sum \sigma(h_1, k_1)\sigma(h_2k_2, m)$$

for all $h, k, m \in H$, and that $\sigma(h, 1) = \sigma(1, h) = \varepsilon(h)$, for all $h \in H$.

7.5.1 DEFINITION. Let A be a left H-comodule algebra and let σ be as above. The *σ-twisted comodule algebra* ${}_\sigma A$ is the set A as a vector space, with elements written as $\bar{a}$, for each $a \in A$, and with multiplication

$$\bar{a} \cdot \bar{b} = \sum \sigma(a_{-1}, b_{-1})\overline{a_0 b_0},$$

where $\rho : A \to H \otimes A$, via $a \mapsto \sum a_{-1} \otimes a_0$, gives the H-comodule structure of A.

Strictly speaking, we should call ${}_\sigma A$ a *left* twisted comodule algebra. If A is a right H-comodule algebra, we can construct a *right* twisted comodule algebra A_σ, although now we need σ to be a "right cocycle", that is

$$\sum \sigma(k_2, m_2)\sigma(h, k_1m_1) = \sum \sigma(h_2, k_2)\sigma(h_1k_1, m).$$

7.5.2 EXAMPLE. Set $A = H$ and use $\Delta : H \to H \otimes H$; then we may construct ${}_\sigma H$ and H_σ. This is done in [Lu], and applied to the dual of the Drinfeld double. See §10.3. These twisted Hopf algebras generalize the notion of twisted group algebras.

7.5.3 EXAMPLE. Let $H = \mathrm{k}G$, so that A is a G-graded algebra. For homogeneous elements $a \in A_x$ and $b \in A_y$, where $x, y \in G$, the multiplication in

${}_\sigma A$ is given by

$$\bar{a} \cdot \bar{b} = \sigma(x, y)\overline{ab}.$$

This is an example of the "cocycle twist" used in [AST]. In fact they use a bigraded algebra A and form the double twist ${}_\sigma A_{\sigma^{-1}}$. That is, A is graded by G on the left and is also graded by G on the right, so that both of these twists make sense separately.

For more on twisted graded algebras, see [Z].

Chapter 8

Galois Extensions

The definition of Hopf Galois extension has its roots in the Chase-Harrison-Rosenberg approach to Galois theory for groups acting on commutative rings [CHR]. In 1969 Chase and Sweedler extended these ideas to coactions of Hopf algebras acting on a commutative **k**-algebra, for **k** a commutative ring [CS]; the general definition appears in [KT] in 1980.

In this chapter we first give a number of examples of Galois extensions. Next we characterize Galois extensions with the normal basis property, a result of Doi and Takeuchi. In Section 3 we study Galois extensions when H is finite-dimensional, and give some characterizations which are analogs of those in [CHR]. In Section 4 we apply the machinery of Galois extensions, normal bases, and crossed products to answer questions we considered earlier in Chapters 3 and 7. Finally we consider relative Hopf modules and faithfully flat Galois extensions.

§8.1 Definition and examples

The definition of Galois extension is given in terms of coactions, as this is more useful for arbitrary H.

8.1.1 DEFINITION. Let A be a right H-comodule algebra with structure map $\rho : A \to A \otimes H$. Then the extension $A^{coH} \subset A$ is *right H-Galois* if the map $\beta : A \otimes_{A^{coH}} A \to A \otimes_{\mathbf{k}} H$, given by $a \otimes b \mapsto (a \otimes 1)\rho(b)$, is bijective.

Some remarks are in order about the definition:

1) In the case considered by Chase and Sweedler, **k** was a commutative ring, A was assumed to be a (commutative) faithfully flat **k**-algebra, H was a finitely-generated projective **k**-algebra, and β was an isomorphism between $A \otimes_{\mathbf{k}} A$ and $A \otimes_{\mathbf{k}} H$. Among other things, this had the effect of forcing $A^{coH} = \mathbf{k}$.

2) [KT] only required β to be surjective. However, they were also assuming H was finite over **k**, in which case it follows that β is also injective; see 8.3.1.

Recent experience indicates that for infinite-dimensional Hopf algebras, the "right" definition of Galois is to require that β be bijective.

3) There seems to be an asymmetry in the definition of β; why not use $\beta'(a \otimes b) = \rho(a)(b \otimes 1)$? In fact if the antipode S is bijective, then β is surjective, injective, or bijective, respectively, $\Leftrightarrow$ β' is surjective, injective, or bijective.

This can be seen as follows: let $\phi \in \mathrm{End}(A \otimes H)$ be given by $\phi(a \otimes h) = \rho(a)(1 \otimes Sh)$; then ϕ is invertible with inverse $\phi^{-1}(a \otimes h) = (1 \otimes \bar{S}h)\rho(a)$, where $\bar{S}$ is the inverse of S. Now $\phi \circ \beta = \beta'$. This remark is [KT, 1.2]. Thus either β or β' may be used when S is bijective.

We next give some elementary examples to illustrate the definition. First, surely any definition of Galois should include the classical case of automorphisms of fields.

8.1.2 CLASSICAL GALOIS FIELD EXTENSIONS. Let G be a finite group acting as automorphisms on a field $E \supset \mathbf{k}$, and let $F = E^G$. Here the group algebra $\mathbf{k}G$ acts on E, and so its dual $H = (\mathbf{k}G)^*$ coacts, which is what we shall need for our new definition 8.1.1.

We know that E/F is classically Galois with group $G \Leftrightarrow G$ acts faithfully on $E \Leftrightarrow [E:F] = |G|$. To see that this is equivalent to 8.1.1, first assume E/F is classically Galois. Set $n = |G|$, write $G = \{x_1, \dots, x_n\}$ and let $\{b_1, \dots, b_n\}$ be a basis of E/F. Let $\{p_1, \dots, p_n\} \subset (\mathbf{k}G)^*$ be the dual basis to the $\{x_i\} \subset \mathbf{k}G$. The action of G on E determines the corresponding coaction $\rho : E \to E \otimes_{\mathbf{k}} (\mathbf{k}G)^*$ via $\rho(a) = \sum_{i=1}^n (x_i \cdot a) \otimes p_i$, and thus the Galois map $\beta : E \otimes_F E \to E \otimes (\mathbf{k}G)^*$ is given by $\beta(a \otimes b) = \sum_i a(x_i \cdot b) \otimes p_i$. Thus if $w = \sum_j a_j \otimes b_j \in \mathrm{Ker}\,\beta$, then $\sum_j a_j(x_j \cdot b_j) = 0$ for all i, by the independence of the $\{p_i\}$. Since G acts faithfully, Dedekind's lemma on independence of automorphisms gives that the $n \times n$ matrix $C = [x_i \cdot b_j]$ is invertible. Thus $a_j = 0$ for all j, and so $w = 0$. Thus β is injective, and so bijective since both tensor products are finite-dimensional F-algebras. The converse (that 8.1.1 implies classical Galois) is easier.

8.1.3 DIFFERENTIAL GALOIS THEORY. Let $E \supset \mathbf{k}$ be a field of characteristic $p > 0$, and let $\mathbf{g} \subset Der_{\mathbf{k}}E$ be a restricted Lie algebra of $\mathbf{k}$-derivations of E which is finite-dimensional over $\mathbf{k}$; here restricted simply means that

$\mathfrak{g}$ is closed under p^{th} powers. The restricted enveloping algebra $u(\mathfrak{g})$ acts on E, and so we consider its dual $H = u(\mathfrak{g})^*$. The H-coinvariants are $E^{\mathfrak{g}} = \{a \in E \mid x \cdot a = 0, \text{ all } x \in \mathfrak{g}\}$. Note that we are already assuming that $\mathfrak{g}$ acts faithfully on E; however this does not suffice for $E^{\mathfrak{g}} \subset E$ to be an H-Galois extension, as the following example shows:

Let $E = \mathbf{Z}_2(z)$, rational functions over $\mathrm{k} = \mathbf{Z}_2$, and let $\mathfrak{g}$ be the k-span of $d_1 = \frac{d}{dz}$ and $d_2 = z\frac{d}{dz}$. Over $\mathbf{Z}_2, \mathfrak{g}$ is the 2-dim solvable Lie algebra with relation $[d_1, d_2] = d_1$; it is restricted since $d_1^2 = 0$ and $d_2^2 = d_2$. Now $E^{\mathfrak{g}} = \mathbf{Z}_2(z^2)$, and $E^{\mathfrak{g}} \subset E$ is not H-Galois. For, $E \otimes_{\mathrm{k}} u(\mathfrak{g})^* \cong E \otimes_{E^{\mathfrak{g}}} (E^{\mathfrak{g}} \otimes_{\mathrm{k}} u(\mathfrak{g})^*)$ is 8-dim over $E^{\mathfrak{g}}$ but $E \otimes_{E^{\mathfrak{g}}} E$ is 4-dim over $E^{\mathfrak{g}}$, so β is not a bijection.

The difficulty with this example is that d_1 and d_2 are dependent over E; when k-independent derivations remain independent over E, the extension will be Galois. The next result follows from 8.3.5 and 8.3.7:

8.1.4 THEOREM. *$E^{\mathfrak{g}} \subset E$ is $u(\mathfrak{g})^*$-Galois $\Leftrightarrow$ $E \otimes_{\mathrm{k}} \mathfrak{g} \to \operatorname{Der} E$ is injective.*

This result differs from Jacobson's classical result on Galois theory for inseparable field extensions; for he assumes that $\mathfrak{g}$ is an E-space, and then obtains a Galois correspondence theorem between intermediate fields and restricted Lie subalgebras (see [A, Ch. 5]).

8.1.5 SEPARABLE FIELD EXTENSIONS. Surprisingly, it is possible for a finite separable field extensions $F \subset E$ to be H-Galois for some H although it is not Galois in the usual sense. The following example is due to Greither and Pareigis [GPa]; an exposition also appears in [Pa 90].

For any k, let H_{k} denote the Hopf algebra with algebra structure given by $H_{\mathrm{k}} = \mathrm{k}[c,s]/(c^2 + s^2 - 1, cs)$ and with coalgebra structure given by $\Delta c = c \otimes c - s \otimes s, \Delta s = c \otimes s + s \otimes c, \varepsilon(c) = 1, \varepsilon(s) = 0, S(c) = c$, and $S(s) = -s$; H_{k} is called the *circle Hopf algebra*. Now let $F = \mathbf{Q}$ and $E = F(\omega)$, where ω is the real 4th root of 2; $F \subset E$ is not Galois for any group G. However, it is $(H_{\mathrm{k}})^*$-Galois for $\mathrm{k} = \mathbf{Q}$. In this case H_{k} acts on E as follows:

$\cdot$	1	ω	ω^2	ω^3
c	1	0	$-\omega^2$	0
s	0	$-\omega$	0	ω^3

It is shown in [GPa] that $\mathbf{Q} \subset E$ is H_{k}^*-Galois; this will also be clear from some of the results in Section 3.

Note that when $k = \mathbf{Q}(i)$, $H_k \cong kZ_4$, the group algebra; that is, $\mathbf{Q}Z_4$ and $H_{\mathbf{Q}}$ are $\mathbf{Q}(i)$-forms of each other. In fact $\mathbf{Q} \subset E$ is also H^*-Galois for a second Hopf algebra H; this second Hopf algebra is a $\mathbf{Q}(\sqrt{-2})$-form of $\mathbf{Q}[\mathbf{Z}_2 \times \mathbf{Z}_2]$. Thus an extension can be Hopf Galois with two different Hopf algebras. On the other hand, there exist separable field extensions which are not Hopf Galois at all. Conditions for exactly when an extension is Hopf Galois are given in [Pa 90].

Another interesting recent paper on Hopf Galois field extensions is [Cl].

8.1.6 GRADED RINGS. Let G be any group and let A be a G-graded algebra; as in 4.1.7, this means that A is a kG-comodule algebra via $\rho : A \to A \otimes kG$ given by $a \mapsto \sum_{x \in G} a_x \otimes x$, where $a = \sum_x a_x \in \oplus_x A_x$. Also recall that for $H = kG$, $A^{coH} = A_1$, the identity component. A result of Ulbrich describes when these extensions are Galois:

8.1.7 THEOREM [U 81]. *$A_1 \subset A$ is kG-Galois $\Leftrightarrow$ A is strongly graded; that is, $A_x A_y = A_{xy}$, for all $x, y \in G$.*

PROOF. First note that A being strongly graded is equivalent to $A_x A_{x^{-1}} = A_1$, for all x. For if the more special condition holds, then

$$A_{xy} = A_{xy}A_1 = A_{xy}(A_{y^{-1}}A_y) \subseteq (A_{xy}A_{y^{-1}})A_y \subseteq A_x A_y \subseteq A_{xy}$$

and thus $A_x A_y = A_{xy}$, for all $x, y \in G$.

The Galois map $\beta : A \otimes_{A_1} A \to A \otimes kG$ is given by $a \otimes b \mapsto \sum_x ab_x \otimes x$. Thus β is surjective $\Leftrightarrow$ $1 \otimes y \in \operatorname{Im}\beta$, for all $y \in G$ $\Leftrightarrow$ there exist $a_i, b_i \in A$ such that $1 \otimes y = \sum_{x,i} a_i(b_i)_x \otimes x$, all $y \in G$ $\Leftrightarrow$ $\sum_i a_i(b_i)_y = 1$ and $\sum_i a_i(b_i)_x = 0$ for all $x \neq y$. This last condition is equivalent to $A_{y^{-1}}A_y = A_1$. Thus β is surjective $\Leftrightarrow$ A is strongly graded.

Finally, if A is strongly graded, then β is also injective (a fact not checked in [U 81] as the earlier definition of Galois was used; the following argument was shown to us by J. Osterburg). To see this, we construct an inverse for β. Since for each $x \in G, 1 \in A_1 = A_{x^{-1}}A_x$, we may write $1 = \sum_i c_{x^{-1},i}d_{x,i}$ where $c_{x^{-1},i} \in A_{x^{-1}}$ and $d_{x,i} \in A_x$. Now define $\alpha : A \otimes kG \to A \otimes_{A_1} A$ by $a \otimes x \mapsto \sum_i ac_{x^{-1},i} \otimes d_{x,i}$. Then $\alpha\beta = id$, for if $a, b \in A$,

$$\alpha\beta(a \otimes b) = \alpha(\sum_x ab_x \otimes x) = \sum_{x,i} ab_x c_{x^{-1},i} \otimes d_{x,i} = \sum_{x,i} a \otimes b_x c_{x^{-1},i}d_{x,i} = a \otimes b.$$

Thus β is injective, and so bijective. □

Strongly graded algebras are known to have many nice properties; in particular their module theory is very well behaved. See [Da], [P], [vO]. A major example is that of group crossed products $A = R * G$, where here $A_x = Rx$. Another example appeared in Chapter 6:

8.1.8 EXAMPLE. Consider the inner gradings of Example 6.1.7; we claim they are strongly graded. For, recall that $A_x = \sum_{yz^{-1}=x} Be_{yz}$, where the $\{e_{xy}\}$ are matrix units. Then for any $y \in G, e_{yy} = e_{y,x^{-1}y}e_{x^{-1}y,y} \in A_x A_{x^{-1}}$. Then $1 = \sum_y e_{yy} \in A_x A_{x^{-1}}$ and so $A_1 = A_x A_{x^{-1}}$; thus A is strongly graded as in 8.1.7, and so $A_1 \subset A$ is Galois.

A variation on this example shows that not all strongly graded rings are group crossed products; see 8.2.3.

8.1.9 GROUPS ACTING ON SETS. The Galois map β can be viewed as the dual of a natural map arising whenever a group G acts on a set X. For, write the action $\mu : X \times G \to X$ by $(x, g) \mapsto x \cdot g$, and consider the natural map $\alpha : X \times G \to X \times X$ given by $(x, g) \mapsto (x, x \cdot g)$. Note that the image of α is $X \times_Y X$, the fiber product of X with itself over Y, where $Y = X/G$ is the set of G-orbits on X. Moreover α is injective $\Leftrightarrow$ the action is free; that is, no $g \neq 1$ in G has any fixed points.

We now dualize this set-up; for simplicity assume that X and G are finite. Let $A = \mathrm{k}^X$, the algebra of functions from X to k under pointwise addition and multiplication, and let $H = \mathrm{k}^G \cong (\mathrm{k}G)^*$, a Hopf algebra. The G-action on X induces a left G-action on A, given by $g \cdot a(x) = a(x \cdot g)$, and thus a right H-coaction $\mu^* : A \to A \otimes H$. If we let $B = \mathrm{k}^Y$, the functions from X to k which are constant on G-orbits, then it follows that $B = A^G = A^{\mathrm{co}H}$. Since for any sets T and U, $\mathrm{k}^{T \times U} \cong \mathrm{k}^T \otimes \mathrm{k}^U$, the map α dualizes to

$$\alpha^* : A \otimes_B A \to A \otimes H.$$

It is now straightforward, using the definition of the transpose of a map, to verify that $\alpha^*(a \otimes b) = (a \otimes 1)\mu^*(b)$. Thus $\alpha^* = \beta$, our Galois map, and by the remarks in the previous paragraph, β is bijective $\Leftrightarrow \alpha : X \times G \to X \times_Y X$ is bijective $\Leftrightarrow$ the G-action is free.

This example is considered more generally at the end of §8.5, in the context of algebraic groups.

§8.2 The normal basis property and cleft extensions

A classical theorem in Galois theory says that if $F \subset E$ is a finite Galois extension of fields with Galois group G, then E/F has a normal basis: that is, there exists $a \in E$ such that the set $\{x \cdot a \mid x \in G\}$ is a basis for E over F. In this section we consider this property for H-extensions.

8.2.1 DEFINITION [KT]. Let $A \subset B$ be a right H-extension. The extension has the (right) *normal basis property* if $B \cong A \otimes H$ as left A-modules and right H-comodules.

Note that we have already seen this property in 3.3.1.

We first check that the definition is equivalent to the classical notion when H is finite-dimensional.

8.2.2 EXAMPLE. Let $\dim H = n < \infty$ and let $A \subset B$ be an H-extension. We may consider H^* acting on B with $B^{H^*} = A$. First assume that B has a normal basis over A in the usual sense; that is, for some $u \in B$ and $\{f_i\} \subset H^*, \{f_1 \cdot u, \cdots, f_n \cdot u\}$ is a basis for the free left A-module B. Recall from 2.1.3 that if $0 \neq t \in \int_H^\ell$, then $H = H^* \rightharpoonup t$; that is, the map from H^* to H given by $f \mapsto (f \rightharpoonup t)$ is a left H^*-module isomorphism. We may then define

$$\phi : A \otimes H \to B \quad \text{by} \quad a \otimes (f \rightharpoonup t) \mapsto a(f \cdot u).$$

$A \otimes H$ is a left H^*-module via $f \cdot (a \otimes h) = a \otimes (f \rightharpoonup h)$; this is the dual of the right comodule structure given by $id \otimes \Delta$, as in 1.6.5. Thus since ϕ is a left H^*-module map, it is a right H-comodule map. It is clearly a left A-module isomorphism, and thus $A \subset B$ has the normal basis property of 8.2.1. The converse is similar: if $\{f_1, \ldots, f_n\}$ is a k-basis for H^*, then $\{1 \otimes (f_1 \rightharpoonup t), \ldots, 1 \otimes (f_n \rightharpoonup, t)\}$ is an A-basis for $A \otimes H$. Given an isomorphism $\phi : A \otimes H \to B$, it follows that $\{f_1 \cdot u, \ldots, f_n \cdot u\}$ is an A-basis for B, where $u = \phi(1 \otimes t)$. Thus $A \subset B$ has a normal basis in the usual sense.

In the general case, not all Galois extensions have the normal basis property.

8.2.3 EXAMPLE. We consider a variation on Example 6.1.7. That is, let $A = M_3(\mathbf{k})$. A is $\mathbf{Z}_2$-graded by setting $A_{\bar{0}} = \begin{bmatrix} \mathbf{k} & \mathbf{k} & 0 \\ \mathbf{k} & \mathbf{k} & 0 \\ 0 & 0 & \mathbf{k} \end{bmatrix}$ and $A_{\bar{1}} = \begin{bmatrix} 0 & 0 & \mathbf{k} \\ 0 & 0 & \mathbf{k} \\ \mathbf{k} & \mathbf{k} & 0 \end{bmatrix}$.

A is strongly graded, and so $A_{\bar{0}} \subset A$ is Galois by 8.1.7. However $A \ncong A_{\bar{0}} \otimes \mathbf{k}\mathbf{Z}_2$ since $9 \neq 5 \cdot 2$.

The following theorem of Doi and Takeuchi characterizes Galois extensions with the normal basis property. Recall the notion of cleft extensions from 7.2.1.

8.2.4 THEOREM [DT 86]. *Let $A \subset B$ be an H-extension. Then the following are equivalent:*
1) $A \subset B$ is H-cleft
2) $A \subset B$ is H-Galois and has the normal basis property.

PROOF. 1) $\Rightarrow$ 2). By 7.2.3, cleft implies $B \cong A\#_\sigma H$ as left A-modules and right H-comodules; thus the normal basis property is satisfied. To see that $A \subset B$ is H-Galois, we construct an inverse α for the Galois map $\beta : B \otimes_A B \to B \otimes_{\mathbf{k}} H$ of 8.1.1; recall $\beta(a \otimes b) = (a \otimes 1)\rho(b)$, where $\rho : B \to B \otimes H$ is the comodule structure map.

Let $\gamma : H \to B$ be the $*$-invertible map such that $A \subset B$ is cleft. Define $\alpha : B \otimes_{\mathbf{k}} H \to B \otimes_A B$ by $\alpha(b \otimes h) = \sum b\gamma^{-1}(h_1) \otimes \gamma(h_2)$. Then

$$\beta\alpha(b \otimes h) = \sum (b\gamma^{-1}(h_1) \otimes 1)(\gamma(h_2) \otimes h_3) = \sum b\gamma^{-1}(h_1)\gamma(h_2) \otimes h_3 = b \otimes h,$$

where $\rho\gamma(h) = \sum \gamma(h_1) \otimes h_2$ since γ is an H-comodule map. Thus $\beta\alpha = id$ and β is surjective. To see that $\alpha\beta = id$ requires more work. First, if $b \in B$ and $\rho(b) = \sum b_0 \otimes b_1 \in B \otimes H$, then $\sum b_0\gamma^{-1}(b_1) \in A = B^{\mathrm{co}H}$ by 7.2.6, 2). Thus

$$\begin{aligned} \alpha\beta(a \otimes b) &= \alpha(\sum ab_0 \otimes b_1) = \sum ab_0\gamma^{-1}(b_1) \otimes \gamma(b_2) \\ &= \sum a \otimes b_0\gamma^{-1}(b_1)\gamma(b_2) = a \otimes b \end{aligned}$$

and so β is injective and so bijective. Thus $A \subset B$ is H-Galois.

2) $\Rightarrow$ 1). Assume $A \subset B$ is an H-extension such that the Galois map β is bijective and such that there exists a map $\phi : A \otimes H \to B$ which is a left A-module, right H-comodule isomorphism. Define $\gamma : H \to B$ by $\gamma(h) = \phi(1 \otimes h)$. Then γ is an H-comodule map since ϕ is. Thus we must show that γ is $*$-invertible.

To see this, define $g \in \mathrm{Hom}_A(B, A)$ by $g = (id \otimes \varepsilon) \circ \phi^{-1}$; clearly $g(\gamma(h)) = \varepsilon(h)1$. Then we define

$$\mu(h) = m(id \otimes g)\beta^{-1}(1 \otimes h).$$

We claim that $\mu = \gamma^{-1}$. First, for any $h \in H$,

$$\begin{aligned}
\gamma * \mu(h) &= \sum \gamma(h_1) m(id \otimes g)\beta^{-1}(1 \otimes h_2) \\
&= \sum m(id \otimes g)\beta^{-1}(\gamma(h_1) \otimes h_2) \\
&\qquad \text{since } \beta^{-1}(b \otimes h) = (b \otimes 1)\beta^{-1}(1 \otimes h), \text{ all } b \in B \\
&= m(id \otimes g)(1 \otimes \gamma(h)) \\
&\qquad \text{since } \beta(1 \otimes \gamma(h)) = \rho\gamma(h) = \sum \gamma(h_1) \otimes h_2, \text{ using} \\
&\qquad \text{that } \gamma \text{ is a comodule map} \\
&= \varepsilon(h)1.
\end{aligned}$$

For the other direction, note that $(id \otimes \Delta) \circ \beta = (\beta \otimes id) \circ (id \otimes \rho)$ and the fact that β is invertible gives $(id \otimes \rho) \circ \beta^{-1} = (\beta^{-1} \otimes id) \circ (id \otimes \Delta)$. Thus for any $h \in H$, writing $\beta^{-1}(1 \otimes h) = \sum_i a_i \otimes b_i$, it follows that

$$(*) \qquad \sum_i a_i \otimes b_{i_0} \otimes b_{i_1} = \sum \beta^{-1}(1 \otimes h_1) \otimes h_2.$$

Also note that for any $b \in B$, $b = \sum g(b_0)\gamma(b_1)$, since $\phi^{-1}(b) = (id \otimes \varepsilon \otimes id)(\sum \phi^{-1}(b_0) \otimes b_1) = \sum g(b_0) \otimes b_1$ and $\gamma(b_1) = \phi(1 \otimes b_1)$, using that ϕ is a comodule map. Then

$$\begin{aligned}
\mu * \gamma(h) &= \sum [m(id \otimes g)\beta^{-1}(1 \otimes h_1)]\gamma(h_2) \\
&= \sum_i [m(id \otimes g)(\sum a_i \otimes b_{i_0})]\gamma(b_{i_1}) \quad \text{by } (*) \\
&= \sum_i a_i (\sum g(b_{i_0})\gamma(b_{i_1})) = \sum_i a_i b_i \\
&= \varepsilon(h)1, \text{ since } (id \otimes \varepsilon) \circ \beta = m.
\end{aligned}$$

Thus γ is $*$-invertible and the extension is H-cleft. □

Combining the theorem with 7.2.2, we obtain:

8.2.5 COROLLARY. *Let $A \subset B$ be a right H-extension. Then $A \subset B$ is Galois with the normal basis property $\Leftrightarrow B \cong A \#_\sigma H$, a crossed product of A with H.*

8.2.6 REMARK. To see that $A \hookrightarrow A\#_\sigma H$ is Galois only requires the easy direction 1) $\Rightarrow$ 2) of 8.2.4, along with 7.2.7. In this case the inverse of the Galois map β is given explicitly by $\beta^{-1}((a\#h)\otimes k) = \sum (a\#h)\gamma^{-1}(k_1)\otimes\gamma(k_2)$, where $\gamma : H \to A\#_\sigma H$ via $h \mapsto 1\#h$ as in 7.2.7.

The Corollary gives a new view of classical Galois extensions:

8.2.7 EXAMPLE [BM 89, 1.2.6]. Let $F \subset E$ be a classical Galois field extension, with (finite) Galois group G. By 8.1.2 we know E/F is H-Galois for $H = (FG)^*$, and by 8.2.2 we know E/F has the normal basis property in the sense of 8.2.1. Thus E is a crossed product of F with H. Since E is commutative, the action of H is trivial by 7.2.4. Thus in fact $E \cong F_\sigma[H]$, a twisted product with H.

The "cleft" map $\gamma : H \to E$ can be given explicitly by a normal basis, as in 8.2.2 and the proof of 8.2.4. Assume that for some $u \in E$, $\{x \cdot u \mid x \in G\}$ is a basis of E over F, and recall that $\gamma(h) = \phi(1 \otimes h)$. Since $p_1 \in \int_H$ and $p_x = x^{-1} \rightharpoonup p_1$, we have $\gamma(p_x) = \phi(1 \otimes (x^{-1} \rightharpoonup p_1)) = x^{-1} \cdot u$. Knowing γ, the cocycle σ is then determined from the crossed-product multiplication 7.1.1:

$$(x^{-1} \cdot u)(y^{-1} \cdot u) = \gamma(p_x)\gamma(p_y) = \sum_{z \in G} \sigma(p_{xz^{-1}}, p_{yz^{-1}})(z^{-1} \cdot u).$$

These equations are solvable. For, the structure constants $a^z_{x,y} \in F$ of E/F with respect to the normal basis $\{x \cdot u\}$, that is the solutions of $(x \cdot u)(y \cdot u) = \sum_{z \in G} a^z_{x,y}(z \cdot u)$, satisfy the condition $a^z_{x,y} = a^1_{z^{-1}x, z^{-1}y}$ since the elements of G are automorphisms of E over F. Thus we can set $\sigma(p_x, p_y) = a^1_{x^{-1}, y^{-1}}$.

(One might wish to use 7.2.5 to compute σ, but this involves knowing γ^{-1}, an unpleasant formula.)

§8.3 Galois extensions for finite-dimensional H

In this section we assume that H is finite-dimensional and that A is an H-module algebra (and thus an H^*-comodule algebra). We first show that a number of other conditions are equivalent to $A^H \subset A$ being H^*-Galois; most of these conditions are the Hopf versions of ones considered by Chase-Harrison-Rosenberg in their Galois theory of groups acting on commutative rings. We then give several applications, including one to the Galois theory of division rings.

We first prove the result of Kreimer and Takeuchi mentioned in the second remark after 8.1.1. We give a simpler proof due to Schneider [Sch 92].

8.3.1 THEOREM[KT]. *Let H be finite-dimensional and A an H-module algera such that the Galois map $\beta : A \otimes_{A^H} A \to A \otimes H^*$ of 8.1.1 is surjective. Then*

1) *there exists a "dual basis" of A as a right A^H-module*
2) *β is injective, and so bijective.*

PROOF. 1) By 2.1.3, we know that $\theta : H^* \to H$ given by $f \mapsto (t \leftharpoonup f)$ is a right H^*-module isomorphism, for a given $0 \neq t \in \int_H^\ell$. Choose $T \in H^*$ such that $t \leftharpoonup T = 1$; then $T \in \int_{H^*}^\ell$, as was shown after 4.3.8. Now since β is surjective, we may find a_i, b_i in A such that

$$(*) \qquad 1 \otimes T = \beta(\sum_{i=1}^{n} a_i \otimes b_i) = \sum a_i b_{i_0} \otimes b_{i_1}.$$

For each $i = 1, \ldots, n$, let $\phi_i(a) = t \cdot (b_i a) \in A^H$, all $a \in A$. Then

$$\begin{aligned}
\sum_i a_i \phi_i(a) &= \sum_i a_i (t \cdot b_i a) = \sum_i a_i (t_1 \cdot b_i)(t_2 \cdot a) \\
&= \sum a_i \sum b_{i_0} \langle b_{i_1}, t_1 \rangle (t_2 \cdot a) \quad \text{by 1.6.4} \\
&= \sum \langle T, t_1 \rangle (t_2 \cdot a) \quad \text{by } (*) \\
&= (t \leftharpoonup T) \cdot a = a.
\end{aligned}$$

Thus $a_1, \ldots, a_n$ and $\phi_1, \ldots, \phi_n$ are a dual basis for A over A^H.

2) By the third remark after 8.1.1, it suffices to show β' is injective. Choose $\sum_j u_j \otimes v_j \in \operatorname{Ker} \beta'$, so that $\sum_j u_{j_0} v_j \otimes u_{j_1} = 0$. Now, using 1) and 1.6.4,

$$\begin{aligned}
\sum_j u_j \otimes v_j &= \sum_{j,i} a_i \phi_i(u_j) \otimes v_j = \sum_{i,j} a_i \otimes \phi_i(u_j) v_j \\
&= \sum_i a_i \otimes \sum_j (t \cdot b_i u_j) v_j \\
&= \sum_i a_i \otimes \sum b_{i_0} u_{j_0} \langle b_{i_1} u_{j_1}, t \rangle v_j = 0.
\end{aligned}$$

Thus Ker $\beta' = 0$ and so β' is injective. □

Next we recall from 4.5.1 that A is a left $A\#H$-module via $(a\#h)\cdot b = a(h\cdot b)$. This action determines an algebra map

$$\pi : A\#H \to \mathrm{End}(A_{A^H})$$

where A is a right A^H-module via right multiplication. We also have

8.3.2 LEMMA. *Let H be finite-dimensional and A an H-module algebra. Then $A^H \cong \mathrm{End}({}_{A\#H}A)^{op}$ as algebras.*

PROOF [BeM 86]. These arguments extend the known arguments for group actions. Define $\psi : A^H \to \mathrm{End}({}_{A\#H}A)$ by $a \mapsto a_r$, right multiplication by $a \in A^H$. Clearly ψ is injective. Now given any $\sigma \in \mathrm{End}({}_{A\#H}A)$ and $a \in A, \sigma(a\cdot 1) = a\sigma(1)$, and so $\sigma = \sigma(1)_r$. Moreover $\sigma(1) \in A^H$. For if $h \in H, h\cdot\sigma(1) = \sigma(h\cdot 1) = \varepsilon(h)\sigma(1)$, and so $\sigma(1) \in A^H$. Thus ψ is surjective. It is clearly an anti-homomorphism. □

We can now state the main theorem of this section:

8.3.3 THEOREM: *Let H be a finite-dimensional Hopf algebra and A a left H-module algebra. Then the following are equivalent:*

1) $A^H \subset A$ is right H^**-Galois*

2) a) *the map $\pi : A\#H \to End(A_{A^H})$ is an algebra isomorphism, and*

b) *A is a finitely-generated projective right A^H-module*

3) *A is a generator for ${}_{A\#H}\mathcal{M}$*

4) *if $0 \neq t \in \int_H^\ell$, then the map $[\ ,\ ] : A \otimes_{A^H} A \to A\#H$ given by $[a,b] = atb$ is surjective*

5) *for any $M \in {}_{A\#H}\mathcal{M}$, consider $A \otimes_{A^H} M^H$ as a left $A\#H$-module by letting $A\#H$ act on A via π. Then the map $\phi : A \otimes_{A^H} M^H \to M$ given by $a \otimes m_0 \mapsto a\cdot m_0$ is a left $A\#H$-module isomorphism.*

Part 1) $\Rightarrow$ 2) is [KT, 1.7]; 2) $\Rightarrow$ 1) is [U 82, 1.1]; 1) $\Leftrightarrow$ 4) appears implicitly in the proof of [KT, 1.7], although it is not explicitly stated; and 5) is a dual version of the "weak structure theorem" of [DT 89] (this will be considered further in §8.5). The proof of the theorem we give here comes from [CFM].

PROOF of 8.3.3. First, since we always have $(A^H)^{op} \cong \mathrm{End}({}_{A\#H}A)$ by 8.3.2, 2) $\Leftrightarrow$ 3) follows by a well-known theorem of Morita.

To see 1) $\Leftrightarrow$ 4), we essentially follow [KT, proof of 1.7]. As in 8.3.1, it follows from 2.1.3 that for $0 \neq t \in \int_H^\ell$, the map $\theta : H^* \to H$ given by $\theta(f) = t \leftharpoonup f$ is a bijection. It follows that $[\ ,\ ] = (id \otimes \theta) \circ \beta$, where $\beta : A \otimes_{A^H} A \to A \otimes H^*$ is the Galois map. For, if $a, b \in A$,

$$(id \otimes \theta) \circ \beta(a \otimes b) = \sum ab_0 \otimes (t \leftharpoonup b_1) = \sum a\langle b_1, t_1\rangle b_0 \otimes t_2$$

$$= \sum a(t_1 \cdot b) \otimes t_2 = (a\#t)(b\#1)$$

where in the last equality we have identified $A \otimes H$ with $A\#H$. Since θ is a bijection, it follows that $[\ ,\ ]$ is surjective $\Leftrightarrow$ β is surjective. Thus 1) $\Rightarrow$ 4). It also follows that 4) $\Rightarrow$ 1), since when H is finite-dimensional, β is injective if it is surjective by 8.3.1.

Before proceeding, we make a remark which will be used several times. Consider $M = A\#H$ as a left $A\#H$-module via left multiplication; then $(A\#H)^H = (1\#t)(A\#1) = tA$, for $0 \neq t \in \int_H^\ell$. This follows from the fact that $A\#H = (1\#H)(A\#1)$ by 4.4.3 and the definition of left integrals.

4) $\Rightarrow$ 3). Assuming $[\ ,\ ]$ is surjective, $(t) = AtA = A\#H$. Thus there exists $\{b_i\}, \{c_i\} \subset A$ such that $1\#1 = \sum_{i=1}^n b_i t c_i$. As in [BeM 86], define

$$\psi : A^{(n)} \to A\#H \quad \text{by} \quad (a_1, \dots, a_n) \mapsto \sum_i a_i t c_i.$$

Clearly ψ is additive, and surjective since $1\#1 \in \operatorname{Im} \psi$. Thus A will be a generator for ${}_{A\#H}\mathcal{M}$ provided ψ is a left $A\#H$-map. To see this, it suffices to show that $\psi_1 : A \to A\#H$ given by $a \mapsto atc$ is a left $A\#H$-map. This follows from 4.4.3, 2).

5) $\Rightarrow$ 4). Assume 5) and choose $M = A\#H$, an $A\#H$-module by left multiplication. By the above remark, $M^H = tA$ for $0 \neq t \in \int_H^\ell$. Thus 5) gives that $\phi : A \otimes_{A^H} tA \to A\#H$, via $a \otimes tb \mapsto atb$, is an isomorphism. Since t centralizes A^H, $A \otimes_{A^H} tA \cong A \otimes_{A^H} A$ via $a \otimes tb \mapsto a \otimes b$. Combining these two isomorphisms gives 4).

2) $\Rightarrow$ 4). Since A is A^H-projective by 2b), it has a "dual basis" over A^H as in 8.3.1; that is, there exist $b_1, \dots, b_s \in A$ and $\phi_1, \dots, \phi_s \in \operatorname{Hom}(A_{A^H}, A^H_{A^H})$ such that for all $a \in A$, $a = \sum_i b_i \phi_i(a)$. Thus $\sum b_i \phi_i = id$ in $\operatorname{End}(A_{A^H})$. Now $\operatorname{Hom}(A_{A^H}, A^H_{A^H}) \hookrightarrow \operatorname{End}(A_{A^H}) \cong A\#H$ by 2a). Thus for each $i = 1, \dots, s$, there exists $d_i \in A\#H$ with $\phi_i = \pi(d_i)$ and $\sum b_i d_i = 1\#1$ in $A\#H$. We

claim that $d_i \in (A\#H)^H$, where H acts as usual by left multiplication. For, if $h \in H$ and $a \in A$,

$$\pi(hd_i)(a) = h \cdot (d_i \cdot a) = h \cdot \phi_i(a) = \varepsilon(h)\phi_i(a) = \pi(\varepsilon(h)d_i)(a),$$

since $\phi_i(a) \in A^H$. Since π is an isomorphism, $hd_i = \varepsilon(h)d_i$, all h, and so $d_i \in (A\#H)^H$. By the above remark, $d_i = tc_i$, some $c_i \in A$. Thus $1\#1 = \sum b_i d_i = \sum b_i t c_i$ is in the image of $[\ ,\ ]$.

4) $\Rightarrow$ 5). Assuming 4), write $1\#1 = \sum b_i t c_i$ as before and let $d_i = tc_i$. Since $t \in \int_H^\ell$, $d_i M \subseteq M^H$ for any $M \in {}_{A\#H}\mathcal{M}$. Thus we may define $\psi : M \to A \otimes_{A^H} M^H$ by $m \mapsto \sum_i b_i \otimes d_i m$. We claim that ϕ and the given map ψ are inverses. For, $\phi\psi(m) = \phi(\sum b_i \otimes d_i m) = (\sum b_i d_i)m = m$ since $\sum b_i d_i = 1\#1$, and $\psi\phi(a \otimes m_0) = \psi(a \cdot m_0) = \sum b_i \otimes d_i \cdot a m_0 = \sum b_i \otimes (t \cdot (c_i a))m_0$ since $m_0 \in M^H$. Now $t \cdot (c_i a) \in A^H$ and $\sum_i b_i(t \cdot (c_i a)) = (1\#1) \cdot a = a$ implies $\psi\phi(a \otimes m_0) = a \otimes m_0$. Finally, ϕ is an $A\#H$-map. For, choose $a, b \in A, h \in H$, and $m_0 \in M^H$. Then

$$(b\#h)\phi(a \otimes m_0) = (b\#h)am_0 = b(1\#h)(a\#1) \cdot m_0 = b(\sum (h_1 \cdot a)\#h_2) \cdot m_0$$

$$= b(\sum (h_1 \cdot a)\varepsilon(h_2)m_0) = b(h \cdot a)m_0 = \phi[(b\#h) \cdot (a \otimes m_0)].$$

Thus ϕ is an isomorphism. □

8.3.4 REMARK. One can prove an alternate version of 8.3.3 in which A is considered as a left A^H, right $A\#H$-module. Here the right action is given by $a \cdot (b\#h) = \bar{S}h \cdot (ab)$. This determines an algebra map

$$\pi' : A\#H \to \operatorname{End}({}_{A^H}A).$$

The alternate version of 8.3.3 also uses β' instead of β.

Note that the map in 8.3.3, 4) has already appeared in §4.4 and in the Morita context in §4.5. Thus we can restate 4.5.4 as saying that A^H is Morita-equivalent to $A\#H$ if $A^H \subset A$ is Galois with a surjective trace map.

We can apply this fact, as well as 8.3.3, to the situation when $A\#H$ is simple.

8.3.5 COROLLARY. *Let H be finite-dimensional and A an H-module algebra such that $A\#H$ is a simple ring. Then*

1) $A^H \subset A$ is right H^*-Galois.

2) If also $A\#H$ is Artinian, then A^H is simple Artinian, A is a free right A^H-module of rank $n = \dim H$, and $A\#H \cong M_n(A^H)$.

PROOF. 1) $[A, A] = AtA$ is an ideal of $A\#H$ by 4.4.3, 3). Now use 8.3.3, 4).
2) Since $A\#H$ is simple Artinian, it is von Neumann regular, and thus there exists $x = \sum_i u_i\#h_i \in A\#H$ such that $txt = t$. Setting $c = \sum_i \varepsilon(h_i)u_i$, it follows that $tct = \hat{t}(c)t$, and so $t = txt = \hat{t}(c)t$. Thus $\hat{t}(c) = 1$ and $\hat{t}$ is surjective. Since $A^H \subset A$ is Galois with surjective trace, 4.5.4 or 4.5.5 implies A^H is Morita equivalent to $A\#H$, as noted above. Thus A^H is simple Artinian (alternatively, we could have used 4.3.4).

It then follows by an old lemma of Artin and Whaples that A is a free right A^H-module, say of rank m. Now by 8.3.3, 2), $A\#H \cong \mathrm{End}(A_{A^H}) \cong M_m(A^H)$. But then $m^2 = [A\#H : A^H] = [A\#H : A][A : A^H] = nm$. Thus $m = n$. □

8.3.6 PROPOSITION [CFM]. *Let H be finite-dimensional and A an H-module algebra such that $A\#H$ is simple Artinian. Then $A^H \subset A$ has the normal basis property, and consequently $A \cong A^H\#_\sigma H^*$, a crossed product of H^* with A^H.*

PROOF. It suffices to show the normal basis property, as the second statement follows from 8.2.5 and 8.3.5, 1). The argument for this which we give here follows an old argument of Nakayama for group actions.

Now A is Artinian by 8.3.5, 2) and so has finite length as a module over itself. Also $A \cong M^{(k)}$ for some k as a left $A\#H$-module, where M is the unique simple $A\#H$-module. Now $A\#H \cong M^{(\ell)}$ for some ℓ, and has length $n = \dim H$ as a left A-module. Thus $nk = \ell$, and so $A\#H \cong A^{(n)}$ as left $A\#H$-modules.

Now as left $A^H \otimes H$-modules, $A\#H \cong A \otimes H$ using 4.4.1, 1). Since A is free over A^H of rank n, it follows that $A\#H$ is free over $A^H \otimes H$ of rank n. Thus as left $A^H \otimes H$-modules,

$$A^{(n)} \cong A\#H \cong (A^H \otimes H)^{(n)}.$$

The Krull-Schmidt theorem now implies $A \cong A^H \otimes H$ as left $A^H \otimes H$-modules. We are now done, since $H \cong H^*$ as left H-modules by 2.1.3. □

We can now give an application to the Galois theory of division rings. It generalizes analogous results for fields, both classically for group actions and also for Hopf algebras [S, 10.1.1].

8.3.7 THEOREM [CFM]. *Let D be a left H-module algebra, where D is a division ring and H is finite-dimensional. Then the following are equivalent:*
1) $D^H \subset D$ *is right* H^**-Galois*
2) $[D : D^H]_r = dim\ H$ *or* $[D : D^H]_\ell = dim\ H$
3) $D^H \subset D$ *has the normal basis property.*
4) $D \cong D^H \#_\sigma H^*$
5) $D\#H$ *is simple*
6) D *is a faithful left or right* $D\#H$*-module.*

PROOF. First note that $D\#H$ is always Artinian since it is a finite D-module. Thus 5) $\Rightarrow$ 4) by 8.3.6. It is also clear that 4) $\Rightarrow$ 3) $\Rightarrow$ 2). Assuming 1), $\pi : D\#H \to \mathrm{End}(D_{D^H})$ is an isomorphism by 8.3.3, 2a), and so D is a faithful left $D\#H$-module; this proves 6). Now assume 6). Since D is always a simple left and right $D\#H$-module, 6) implies that $D\#H$ is either left or right primitive. Since $D\#H$ is Artinian, it is simple, proving 5).

It remains to show that 2) $\Rightarrow$ 1). Assume first that $[D : D^H]_r = n = \dim H$. Let $K = \mathrm{Ker}\ \pi$ and set $B = D\#H/K$. Then as in 8.3.2, $D^H \cong \mathrm{End}(\,_B D)^{op}$. Since D is a faithful simple B-module, the Density Theorem implies $B \cong \mathrm{End}(\ D_{D^H}) \cong M_n(D^H)$; in particular π is surjective. But now

$$[B : D^H]_r = n^2 = [D\#H : D]_r[D : D^H]_r = [D\#H : D^H]_r.$$

Since D^H is a division ring, this can only happen if $K = 0$. Thus π is injective, and so an isomorphism. Thus $D^H \subset D$ is Galois by 8.3.3, 2). Note we have used that $[D\#H : D]_r = n$; see the remark after 7.2.11. If we assume instead that $[D : D^H]_\ell = \dim H$, we proceed as before using the alternate version of 8.3.3 mentioned in 8.3.4. □

We next give an example to show that the conditions in 8.3.7 do not always hold, even for group actions. The example in 8.1.3 already shows the conditions fail when $H = u(\mathfrak{g})^*$.

8.3.8 EXAMPLE [CFM]. Let D be a division algebra of characteristic 2, of dimension 4 over its center Z, with an element $x \notin Z, x^2 \in Z$. Let g and h be inner automorphisms of D given by conjugation by x and $x+1$, respectively. Let $G = \langle g, h\rangle \cong \mathbf{Z}_2 \times \mathbf{Z}_2$; then the Hopf algebra $H = ZG$ acts on D.

Note that $D^H = \mathrm{Cent}_D(x) = Z[x]$, so $[D : D^H] = 2$ although $\dim H = 4$. Thus by 8.3.7 we know that $D^H \subset D$ is not Galois and so $D\#H$ is not simple. In fact it is not even semiprime: $D\#H = D * G$, the usual skew group ring, and in this ring $w = 1 + xg + (1+x)h$ is a central element with $w^2 = 0$. □

Given 8.3.6, one may ask when $A\#H$ is simple. The following is known:

8.3.9 THEOREM [BeM 92]. *Let A be simple with center $Z \supset \mathbf{k}$ and let H be of the form $H = K\#\mathbf{k}G$, where K is a connected Hopf algebra and $G = G(H)$. Assume A is a left H-module algebra such that*
1) $G \to Aut\, A/Inn\, Aut\, A$ is injective (that is, G is outer) and
2) $Z \otimes P(H) \to Der\, A/Inn\, Der\, A$ is injective.
Then $A\#H$ is simple.

The theorem is false if H is not of the form $K\#\mathbf{k}G$, as is seen by Example 4.4.8: for, $P(H_4) = 0$ and g is outer on $\mathbf{C}$, but $\mathbf{C}\#H$ is not simple.

We close with another corollary of 8.3.3, pointed out by Y. Doi. It shows that ideals in Galois extensions are very well-behaved.

8.3.10 COROLLARY. *Let H be finite-dimensional and let $A^H \subset A$ be right H^*-Galois. If I is any left (respectively right) H-stable ideal of A, then $I = A(I \cap A^H)$ (resp. $I = (I \cap A^H)A$).*

PROOF. First assume I is a left ideal. Apply 8.3.3, 5) with $M = I$; I is a left $A\#H$-module since it is H-stable. Thus $I \cong A \otimes_{A^H} (I \cap A^H)$ and so $I = A(I \cap A^H)$. If I is a right ideal, we use the alternate version of 8.3.3 mentioned in 8.3.4 in which A is considered as a right $A\#H$-module. □

We may apply 8.3.10 to crossed products, as follows: first, for any crossed product $A\#_\sigma H$, the extension $A \hookrightarrow A\#_\sigma H$ is right H-Galois by 8.2.5. Next, the standard action of H^* on H as in 1.6.5 extends naturally to $A\#_\sigma H$ by letting H^* act trivially on A. Thus 8.3.10 specializes to:

8.3.11 COROLLARY. *Let H be finite-dimensional and let $A\#_\sigma H$ be a crossed product. If I is any left (respectively right) ideal of $A\#_\sigma H$ which is H^*-stable, then $I = (A\#_\sigma H)(I \cap A)$ (resp. $I = (I \cap A)(A\#_\sigma H)$).*

The Corollary provides a little help on Question 7.4.9. For, it implies that if A is H-semiprime, then $A\#_\sigma H$ has no non-zero nilpotent ideals which are H^*-stable. However in general it seems quite difficult to construct stable ideals.

§8.4 Normal bases and Hopf algebra quotients

In this section we answer, for finite-dimensional Hopf algebras, two questions we have discussed earlier. The first is Question 7.2.9, about writing a Hopf algebra as a crossed product with a homomorphic image. As a consequence of this, we can prove Theorem 3.3.1, which says that any finite-dimensional Hopf algebra has a "normal basis" over a subHopfalgebra. These results are due to H.-J. Schneider.

We first restate 7.2.9 in a somewhat more general form. Let $A\#_\sigma H$ be any crossed product and $\pi : H \to \bar{H}$ a Hopf algebra surjection. Then $A\#_\sigma H$ is a right $\bar{H}$-comodule algebra via $\rho = (id_{A\#_\sigma H} \otimes \pi)(id_A \otimes \Delta)$; that is, $a\#h \mapsto \sum (a\#h_1) \otimes \pi(h_2)$.

8.4.1 QUESTION. Let $A\#_\sigma H$ and π be as above. When is it true that $A\#_\sigma H \cong (A\#_\sigma H)^{co\bar{H}} \#_\tau \bar{H}$ for some new cocycle τ?

As noted in Example 7.1.6, it is true when $H = \mathbf{k}G$, a group algebra. We know it is false in general, even if A is trivial; this is Example 7.2.10. However it is true in some important special cases, as we shall see.

First we need a definition. Recall the definition of the equalizer of a pair of maps from §3.4.

8.4.2 DEFINITION[MiMo]. Let C be a coalgebra and let V (respectively W) be right (left) C-comodules with structure maps ρ_V, ρ_W respectively. Then the *cotensor product* $V\square_C W$ of V and W over C is the equalizer of $id_V \otimes \rho_W,\ \rho_V \otimes id_W : V \otimes W \to V \otimes C \otimes W$. That is,

$$V\square_C W \subset V \otimes W \underset{\rho_V \otimes id}{\overset{id\otimes \rho_W}{\rightrightarrows}} V \otimes C \otimes W$$

is exact.

8.4.3 EXAMPLE. Let C be a coalgebra and V a right C-comodule. Then $V\square_C C \cong V$. Similarly $C\square_C W \cong W$ if W is a left C-comodule.

To see this, let $\rho : V \to V \otimes C$ be the structure map. Since ρ is a comodule map, it follows that $V \cong \rho(V) \subseteq V \square_C C$. Conversely if $u \in V \square_C C$, let $v = (id \otimes \varepsilon)u$; then $(\rho \otimes id)(u) = (id \otimes \Delta)(u)$ gives that $u = \rho(v)$. Thus $V \square_C C = \rho(V) \cong V$. A similar argument works for W. □

We also recall that for any right H-extension $B \subset A$, the coinvariants $B = A^{coH}$ can be defined by the exactness of the equalizer diagram

$$B \subset A \underset{\otimes 1}{\overset{\rho}{\rightrightarrows}} A \otimes H$$

The next result is somewhat more general than we need, but is of independent interest.

8.4.4 PROPOSITION [Sch 92]. *Let $B \subset A$ be a right H-Galois extension such that A is flat as a left B-module. Let $\bar{H} = H/I$ be a quotient coalgebra and quotient right H-module of H, and let $A' = A^{co\bar{H}}$. Then $A' \otimes_B A \cong A \square_{\bar{H}} H$, via the canonical Galois map β for $B \subset A$.*

PROOF. Let $\rho : A \to A \otimes H$ be the given comodule structure map; then as before, A is a right $\bar{H}$-comodule algebra via $\bar{\rho} = (id \otimes \pi) \circ \rho$, for $\pi : H \to \bar{H}$.

The defining exact sequence of A', namely $A' \subset A \underset{\otimes \bar{1}}{\overset{\bar{\rho}}{\rightrightarrows}} A \otimes \bar{H}$, is right B-linear, where B acts on $A \otimes \bar{H}$ by acting on A. Since A is flat over B, tensoring with $- \otimes_B A$ gives another exact sequence. One may check that the following diagram commutes:

$$\begin{array}{ccccc}
A' \otimes_B A & \xrightarrow{i \otimes id} & A \otimes_B A & \underset{id \otimes \bar{1}}{\overset{(id \otimes \tau)(\bar{\rho} \otimes id_A)}{\rightrightarrows}} & A \otimes_B A \otimes \bar{H} \\
\downarrow \tilde{\beta} & & \downarrow \beta & & \downarrow \phi \\
A \square_{\bar{H}} H & \xrightarrow{i} & A \otimes H & \underset{id \otimes \rho_H}{\overset{\bar{\rho} \otimes id_H}{\rightrightarrows}} & A \otimes \bar{H} \otimes H
\end{array}$$

Here the top sequence is the one described above followed by $id \otimes \tau$, the bottom sequence is the definition of the cotensor product, $\tilde{\beta} = \beta$ restricted to $A' \otimes_B A, \rho_H = (\pi \otimes id) \circ \Delta$, and $\phi(a \otimes b \otimes \bar{h}) \cdot = \sum ab_0 \otimes \overline{hb_1} \otimes b_2$. For example, for the upper arrows in the right hand square, we see that

$$\phi \circ (id \otimes \tau)(\bar{\rho} \otimes id_A)(a \otimes b) = \phi(\sum a_0 \otimes b \otimes \bar{a}_1) = \sum a_0 b_0 \otimes \overline{a_1 b_1} \otimes b_2$$

$$= \sum (ab_0)_0 \otimes (\overline{ab_0})_1 \otimes b_1 = (\bar{\rho} \otimes id_H)(\sum ab_0 \otimes b_1)$$

$$= (\bar{\rho} \otimes id_H) \circ \beta(a \otimes b).$$

By assumption, β is bijective from $A \otimes_B A$ to $A \otimes H$. Also ϕ is bijective, since ϕ is the composition

$$A \otimes_B A \otimes \bar{H} \xrightarrow{\beta \otimes id_H} A \otimes H \otimes \bar{H} \xrightarrow{id_A \otimes \psi} A \otimes \bar{H} \otimes H \quad ,$$

where $\psi(h \otimes v) = \sum vh_1 \otimes h_2$, for $h \in H, v \in V$ defines an isomorphism $\psi : H \otimes V \to V \otimes H$ for any right H-module V (note that $\psi^{-1}(v \otimes h) = \sum h_2 \otimes v(Sh_1)$). Thus by the commutativity of the diagram, the first map $\tilde{\beta}$ is bijective. □

8.4.5 REMARK. In the situation of 8.4.4, assume the extension is actually a crossed product $A \hookrightarrow A\#_\sigma H$; it is H-Galois by 8.2.5. Let $\gamma : H \to A\#_\sigma H$ be as in 7.2.7. Then one obtains from 8.4.4 an isomorphism

$$(A\#_\sigma H)' \otimes H \to (A\#_\sigma H)\square_{\bar{H}} H$$

given by $d \otimes h \mapsto \sum d\gamma(h_1) \otimes h_2$. Here the inverse map can be described explicitly by $d \otimes h \mapsto \sum d\gamma^{-1}(h_1) \otimes h_2$.

8.4.6 THEOREM[Sch 92, 2.2]. *Let $B = A\#_\sigma H$ be a crossed product, $\pi : H \to \bar{H}$ a Hopf algebra surjection, and $B' = B^{co\bar{H}}$ as before. Assume that both H and A are finite-dimensional. Then $B \cong B'\#_\tau \bar{H}$, as left B'-modules and right $\bar{H}$-comodules.*

PROOF. 1) $\bar{H}^*$ is a subHopfalgebra of H^*; thus by the Nichols-Zoeller theorem 3.1.5, H^* is free over $\bar{H}^*$, say $H^* \cong (\bar{H}^*)^{(t)}$ as left $\bar{H}^*$-modules. Then $H \cong \bar{H}^{(t)}$ as left $\bar{H}$-comodules. By Remark 8.4.5, $B' \otimes H \cong B\square_{\bar{H}} H$ as left B'-modules. Let $n = \dim H$. Then

$$B'^{(n)} \cong B' \otimes H \cong B\square_{\bar{H}} H \cong B\square_{\bar{H}} \bar{H}^{(t)} \cong B^{(t)}$$

as left B'-modules, where the last isomorphism uses Example 8.4.3.
2) We claim $B' \subset B$ is right $\bar{H}$-Galois. First, we know by 8.2.5 that $A \subset B$ is right H-Galois, and thus the map

$$\beta' : B \otimes_A B \to B \otimes H \text{ given by } b \otimes c \mapsto \sum b_0 c \otimes b_1$$

is a bijection by remark 3) after 8.1.1. It follows that the induced map

$$\bar{\beta}' : B \otimes_{B'} B \longrightarrow B \otimes \bar{H} \text{ given by } b \otimes c \mapsto \sum b_0 c \otimes \pi(b_1)$$

is surjective. Since H is finite-dimensional, $\bar{\beta}'$ is also bijective by 8.3.1. Thus $B' \subset B$ is Galois.

3) Combining 1) and 2), we have that

$$\begin{aligned} B^{(n)} &\cong B \otimes_{B'} (B')^{(n)} \cong B \otimes_{B'} B^{(t)} \cong (B \otimes_{B'} B)^{(t)} \\ &\cong (B \otimes \bar{H})^{(t)} \cong B^{(t)} \otimes \bar{H} \cong B'^{(n)} \otimes \bar{H} \cong (B' \otimes \bar{H})^{(n)} \end{aligned}$$

as left B'-modules and right $\bar{H}$-comodules. That is, the $\bar{H}$-extension $B' \subset B$ has the (right) normal basis property. Thus by Corollary 8.2.5, $B \cong B' \#_\tau \bar{H}$. □

We recall the notation $H^{co\bar{H}}$ and ${}^{co\bar{H}}H$ of §3.4. Note that 8.4.7, 2) is Theorem 3.3.1, the "normal basis" theorem for subHopfalgebras.

8.4.7 COROLLARY. *Let H be a finite-dimensional Hopf algebra.*

1) *Assume $\pi : H \to \bar{H}$ is a Hopf algebra surjection. Then*
   a) *$H \cong H^{co\bar{H}} \otimes \bar{H}$ as left $H^{co\bar{H}}$-modules and right $\bar{H}$-comodules*
   b) *$H \cong \bar{H} \otimes {}^{co\bar{H}}H$ as right ${}^{co\bar{H}}H$-modules and left $\bar{H}$-comodules.*

2) *Assume $K \subset H$ is a subHopfalgebra. Then*
   a) *$H \cong K \otimes H/K^+H$ as left K-modules and right H/K^+H-comodules*
   b) *$H \cong H/HK^+ \otimes K$ as right K-modules and left H/HK^+-comodules.*

PROOF. 1a) is the special case of the theorem with $B = H$. Then 1b) follows from a). For, suppose $\phi : H \to H^{co\bar{H}} \otimes \bar{H}$ is the isomorphism in a). Then the composition

$$H \xrightarrow{\bar{S}} H \xrightarrow{\phi} H^{co\bar{H}} \otimes \bar{H} \to \bar{H} \otimes^{co\bar{H}} H,$$

where $\bar{S}$ is the inverse of the antipode S of H, and the last map is given by $b \otimes h \mapsto \overline{S(h)} \otimes S(b)$, is the isomorphism giving b).

Similarly 1b) $\Rightarrow$ 1a).

2) We claim that 1a) implies 2b) for H^*. First, since $\pi : H \to \bar{H}$ is a surjection, $\pi^* : \bar{H}^* \to H^*$ is an injection, so we may assume $\bar{H}^* \subset H^*$. Next, it has already been noted that $B = H^{co\bar{H}}$ can be defined by the exactness of

the diagram $B \xrightarrow{i} H \underset{g}{\overset{f}{\rightrightarrows}} H \otimes \bar{H}$, where $f(h) = h \otimes \bar{1}$ and $g = (id \otimes \pi) \circ \Delta$. Dualizing this diagram gives the coequalizer diagram

$$H^* \otimes \bar{H}^* \underset{g^*}{\overset{f^*}{\rightrightarrows}} H^* \xrightarrow{i^*} B^* \quad .$$

Thus B^* can be defined by this exact sequence. Now Ker $i^* = \text{Im}(f^* - g^*)$ $= H^*(\bar{H}^*)^+$, and so $B^* \cong H^*/H^*(\bar{H}^*)^+$. It now follows from 1a) that

$$H^* \cong B^* \otimes \bar{H}^* \cong H^*/H^*(\bar{H}^*)^+ \otimes \bar{H}^*$$

as left B^*-comodules and right H^*-modules. Replacing H^* by H and $\bar{H}^*$ by K gives 2b).

Similarly 1b) for H implies 2a) for H^*, and so for $K \subset H$. □

We state another situation in which 8.4.1 is true:

8.4.8 THEOREM[Sch 92]. *Let $\pi : H \to \bar{H}$ be a Hopf algebra surjection. Assume*
a) H is injective as a right $\bar{H}$-comodule, and
b) the coradical of $\bar{H}$ is liftable along π.
If $B = A\#_\sigma H$ is a crossed product and $B' = B^{co\bar{H}}$, then $B \cong B'\#_\tau \bar{H}$.

Here the hypothesis b) means that on $\bar{H}_0$, there exists a coalgebra map $g : \bar{H}_0 \to H$ such that πg is the inclusion map $\bar{H}_0 \subset \bar{H}$.

We mention some situations in which the hypotheses of 8.4.8 hold: Assume that $\bar{H} = H/K^+H$ for some normal subHopfalgebra K of H. Then 8.4.8 applies if:
1) H is pointed, or
2) H_0 is cocommutative and $\bar{H}$ is connected, or
3) H_0 is cocommutative and B' is finite-dimensional.

The proof of 8.4.8 (and some of its consequences) depend on work in [Sch 90a]; some of these results are discussed in §8.5.

Schneider constructs a rather interesting example in [Sch 92] which shows that crossed products are not "transitive", in general. That is, there exists a Hopf algebra H and Hopf ideals $I \subset J$ of H such that: H is an H/I-crossed product, and H/I is an H/J-crossed product, but H is not an H/J-crossed product.

§8.5 Relative Hopf modules

Relative Hopf modules are a generalization of the (H,K)-Hopf modules considered in Chapter 3, in which the subHopfalgebra K of H is replaced by an algebra A.

8.5.1 DEFINITION [D 83]. Let A be a right H-comodule algebra, with structure map $\rho : A \to A \otimes H$. Then M is a *right* (H,A)-*Hopf module* if
1) M is a right H-comodule, via $\tau : M \to M \otimes H$
2) M is a right A-module
3) $\tau(m \cdot a) = \tau(m) \cdot \rho(a)$; that is, τ is a right A-module map via ρ.

Let $\mathcal{M}_A^H$ denote the category of right (H,A)-Hopf modules. Analogously we may define ${}_A\mathcal{M}^H$, ${}^H\mathcal{M}_A$, and ${}^H_A\mathcal{M}$ as we did for (H,K)-Hopf modules in §1.9.

8.5.2 EXAMPLE. When H is finite-dimensional, relative Hopf modules are simply $A\#H^*$-modules. As Doi points out, it is straightforward to see that ${}_A\mathcal{M}^H = {}_{A\#H^*}\mathcal{M}$, using the usual duality as in 1.6.4 between right H-comodules and left H^*-modules. We can also identify $\mathcal{M}_A^H$ with $\mathcal{M}_{A\#H^*}$, though this takes more work. As shown by Fischman, if $M \in \mathcal{M}_A^H$, then M is a left H^*-module by the usual dual correspondence, and A is a left H^*-module algebra. Now define a right action of $A\#H^*$ on M via $m \cdot (a\#f) = \bar{S}f \cdot (m \cdot a)$, all $m \in M, a \in A, f \in H^*$; one must check this is a module action. Conversely, right $A\#H^*$-modules give right (H,A)-Hopf modules. See [CFM, 1.6].

8.5.3 EXAMPLE. Let $H = \mathrm{k}G$ be a group algebra. Then we know that $A = \oplus_{g\in G} A_g$ is a G-graded algebra. The category $\mathcal{M}_A^H$ is precisely the category of G-graded right A-modules; see [NvO].

8.5.4 EXAMPLE. Let G be an affine algebraic group over an algebraically closed field $\mathbf{k}$; $G = \operatorname{Spec} H$ for H a commutative affine k-Hopf algebra. The category $\mathcal{M}_A^H$ consists of those right A-modules M with a left G-action such that

$$g \cdot (m \cdot a) = (g \cdot m)(g \cdot a),$$

for all $g \in G, m \in M, a \in A$. This category is usually written as ${}_G\mathcal{M}_A$, the *category of* (G,A)-*modules*.

If G is finite, we may use $H = (\mathbf{k}G)^*$ for any field $\mathbf{k}$.

The condition 1) $\Leftrightarrow$ 5) of 8.3.3 can be generalized to arbitrary H; this is work of Doi and Takeuchi.

8.5.5 PROPOSITION. *Let $B \subset A$ be any right H-extension, and consider the condition*

(*) for any $M \in \mathcal{M}_A^H$, the map

$$\phi_M : M^{coH} \otimes_B A \to M \text{ given by } \quad m_0 \otimes a \mapsto m_0 \cdot a$$

is a right (H, A)-Hopf module isomorphism.

1) [D 85] *If (*) holds, then $B \subset A$ is H-Galois.*

2) [DT 89] *If $B \subset A$ is H-Galois and ${}_BA$ is flat, then (*) holds.*

PROOF(sketch). 1) Assuming (*), let $M = A \otimes H$; M is a right H-comodule via $\tau = id \otimes \Delta$ and a right A-module using $\rho : A \to A \otimes H$ followed by right multiplication in $A \otimes H$. Then $M^{coH} = A$ and we get the isomorphism $A \otimes_B A \to A \otimes H$.

2) Assume the Galois map $\beta : A \otimes_B A \to A \otimes H$ is a bijection; note $\beta = \phi_{A \otimes H}$. To show any ϕ_M is an isomorphism, we define an inverse $\psi : M \to M \otimes_B A$ by $m \mapsto \sum (m_0 \otimes 1)\beta^{-1}(1 \otimes m_1)$, where M is a right H-comodule via $m \mapsto \sum m_0 \otimes m_1$. It is relatively straightforward to check that $\psi \circ \phi = id$ on $M^{coH} \otimes_B A$, and also that $\phi \circ \psi = id$ on M. However one needs to check that $\operatorname{Im} \psi \subseteq M^{coH} \otimes_B A$. To see this, one checks that $\operatorname{Im} \psi \subseteq (M \otimes_B A)^{coH}$, where $M \otimes_B A$ is a right H-comodule via $m \otimes a \mapsto \sum (m_0 \otimes a) \otimes m_1$. By flatness, $(M \otimes_B A)^{coH} = M^{coH} \otimes_B A$. □

As a consequence of 8.5.5, we can improve 8.3.11 to the following: let $A \#_\sigma H$ be any crossed product, and let I be any left (right) ideal of $A \#_\sigma H$ which is also a right H-subcomodule. Then $I = (A \#_\sigma H)(I \cap A)$ (respectively, $I = (I \cap A)(A \#_\sigma H)$).

In Chapter 4 we saw that there is a close relationship between A^H-modules and $A \# H$-modules, when H is finite-dimensional acting on A; in particular A^H and $A \# H$ are Morita-equivalent when $A^H \subset A$ is H^*-Galois and the trace is surjective. In that case the category equivalence between $\mathcal{M}_{A^H}$ and $\mathcal{M}_{A \# H}$ is given by $N \mapsto N \otimes_{A^H} A$. A recent theorem of Schneider shows that an analogous result is true for arbitrary H; his result also generalizes a theorem for algebraic groups on induction of modules and affine quotients.

8.5.6 THEOREM [Sch 90a]. *Let H be a Hopf algebra with bijective antipode and A a right H-comodule algebra with $B = A^{coH}$. Then the following are equivalent:*

1) a) *$B \subset A$ is right H-Galois and*
   b) *A is a faithfully flat left (or right) B-module;*
2) a) *the Galois map β is surjective and*
   b) *A is an injective H-comodule;*
3) *the map $\mathcal{M}_B \to \mathcal{M}_A^H$ given by $N \mapsto N \otimes_B A$ is an equivalence;*
4) *the map ${}_B\mathcal{M} \to {}_A\mathcal{M}^H$ given by $N \mapsto A \otimes_B N$ is an equivalence.*

8.5.6 extends a number of results in the literature. If $H = \mathrm{k}G$ is a group algebra, then 1a) means A is strongly graded, and the equivalence of 1) and 3) is due to Dade [Da, Theorem 2.8]. [DT 89] prove a result similar to 1) $\Rightarrow$ 3): they show that Galois plus the existence of a total integral implies 3). Recall from 4.3.9 that if H is finite-dimensional, then the extension $B \subset A$ has a total integral $\Leftrightarrow$ the trace is surjective. If also $B \subset A$ is Galois, then these conditions are equivalent to ${}_BA$ or A_B being faithfully flat [KT].

The difficult part of Schneider's theorem, however, is the equivalence of 1) and 2). This part has as a special case the following theorem on algebraic groups, due independently to Cline, Parshall, and Scott [CPS] and to Oberst [O].

8.5.7 THEOREM. *Let* k *be algebraically closed, and let $X \supset G$ be affine algebraic groups with G closed in X. Then the quotient X/G is affine $\Leftrightarrow$ induction of G-modules to X-modules is exact.*

The original proofs used the theory of affine groups; in particular [CPS] used Haboush's theorem that reductive groups are geometrically reductive. Later Doi [D 85] gave a purely Hopf algebraic proof; however his proof required that the Hopf algebra was commutative. In order to see that 8.5.7 is in fact a special case of 8.5.6, we first translate it (actually a more general version due to Oberst) into the language of schemes and Hopf algebras.

Thus, let X be an affine scheme and G an affine algebraic group scheme; that is, $X = \operatorname{Spec} A$ and $G = \operatorname{Spec} H$ for A a commutative affine k-algebra and H a commutative affine k-Hopf algebra. An action $\mu : X \times G \to X$ is determined by a coaction $\rho = \mu^* : A \to A \otimes_{\mathbf{k}} H$. The action is free if $\bar{\beta} : A \otimes_{\mathbf{k}} A \to A \otimes_{\mathbf{k}} H$ given by $a \otimes b \mapsto (a \otimes 1)\rho(b)$ is surjective; dually, this corresponds to saying that the map $X \times G \to X \times X$ given by $(x, g) \mapsto (x, x \cdot g)$ is a (closed) embedding. Note that for X a set, G a group, this gives the usual definition of free action, namely G acts with no fixed points (see 8.1.9 where this set-up was discussed when G and X were finite). In the situation of 8.5.7, where $G \subset X$ are groups, G acts freely on X by translation.

Next, we consider quotients. The "affine quotient" of X by G is defined to be $Y = \operatorname{Spec} B$, where $B = A^{coH}$; that is, the sequence

$$B \subset A \underset{\otimes 1}{\overset{\rho}{\rightrightarrows}} A \otimes H$$

is exact. This corresponds to an exact sequence of affine schemes

$$X \times G \underset{\pi}{\overset{\mu}{\rightrightarrows}} X \to Y ,$$

where $\pi(x, g) = x$. How is this "affine quotient" related to the true quotient X/G? In general, Y may be much smaller than X/G; for example if $X = GL_2(\mathbf{k})$ and $G = \{$upper triangular matrices in $X\}$, then Y is just a point but $X/G \approx \mathbf{P}^1$. Given a free action as above, it will happen that $Y = X/G$ provided the map $X \times G \to X \times_Y X$ is an isomorphism and $X \to Y$ is faithfully flat; in this case one says that X is a principal fibre bundle over Y with group G [MuF, p.16]. Note that the algebraic version of these conditions says

a) $\beta : A \otimes_B A \to A \otimes H$ is bijective, and

b) A is a faithfully flat B-module.

That is, the "affine quotient" property (in the case of free actions) is precisely 1) of 8.5.6.

Oberst's version of 8.5.7 says that for a free action, X is a principal fibre bundle over Y with group $G \Leftrightarrow$ the functor

$${}^{G}(-) : {}_{G,A}\mathcal{M} \to {}_{B}\mathcal{M} \text{ given by } M \mapsto M^G$$

is exact. Here ${}_{G,A}\mathcal{M}$ is the category of left (G, A)-modules, as in example 8.5.4, although here the compatibility condition is $g \cdot (a \cdot m) = (g \cdot a)(g \cdot m)$.

But this is the same as the category ${}_A\mathcal{M}^H$, the right H, left A-Hopf modules, for $G = \operatorname{Spec} H$.

Now in 8.5.6, 2 a) means that the action is free, and b) says that induction of modules is exact. Thus in the case A and H are commutative, 8.5.7 follows from 8.5.6.

Further results on the connection between Galois extensions and representation theory are given in [Sch 90b], particularly on induction and restriction of simple and indecomposable modules. This work generalizes a number of classical results for groups and Lie algebras, with more natural proofs. It demonstrates that faithfully flat Hopf Galois extensions are objects of fundamental interest.

Chapter 9

Duality

In this chapter we first study the "finite dual" H° defined in Chapter 1, as well as its subHopfalgebra H', and give a number of examples of these dual Hopf algebras. We also review the classical duality between algebraic groups and commutative algebras. We then turn to the topic of "duality for actions", which has its roots in von Neumann algebras, and sketch the proof of the Blattner-Montgomery theorem. Finally we give some applications of this result to prime ideals in graded algebras.

§9.1 H°

We recall from 1.2.3 that for any k-algebra A, the finite dual A° of A was defined to be the set of $f \in A^*$ which vanish on an ideal I of A of finite codimension. We first give some other characterizations of A°, which in fact appear in [S] and [A].

9.1.1 LEMMA. *Let (A, m, u) be an algebra and $f \in A^*$. Then the following are equivalent:*

1) *f vanishes on a right ideal of A of finite codimension*
2) *f vanishes on a left ideal of A of finite codimension*
3) *f vanishes on an ideal of A of finite codimension*
4) *$dim\ (A \rightharpoonup f) < \infty$*
5) *$dim\ (f \leftharpoonup A) < \infty$*
6) *$dim(A \rightharpoonup f \leftharpoonup A) < \infty$*
7) *$m^* f \in A^* \otimes A^*$*

Consequently $f \in A^\circ$ if any of conditions 1) through 7) hold.

PROOF (sketch). Recall from 1.6.6 that for $a \in A$ and $f \in A^*, a \rightharpoonup f \in A^*$ is defined by $\langle a \rightharpoonup f, b\rangle = \langle f, ba\rangle$; similarly $\langle f \leftharpoonup a, b\rangle = \langle f, ab\rangle$ for all $b \in A$.

We first show 1) $\Leftrightarrow$ 4). Let R be a right ideal of A with dim $A/R < \infty$, and let $\langle f, R\rangle = 0$. Then $\langle a \rightharpoonup f, R\rangle = \langle f, Ra\rangle = 0$, and so $A \rightharpoonup f \subseteq R^\perp$, which is finite-dimensional. Thus 1) $\Rightarrow$ 4). Conversely suppose dim $(A \rightharpoonup f) < \infty$.

Then $R = (A \rightharpoonup f)^{\perp}$ is a right ideal of A and has finite codimension, since $(A \rightharpoonup f)^* \cong A/R$ as in the proof of 5.1.4, 3). Similarly 2) $\Leftrightarrow$ 5) and 3) $\Leftrightarrow$ 6).

Clearly 3) $\Rightarrow$ 1) and 2); we show 2) $\Rightarrow$ 3). Let L be a left ideal of A of finite codimension; then A/L is a finite-dimensional left A-module. Thus there is an algebra map $\phi : A \to \mathrm{End}_{\mathbf{k}}(A/L)$, a finite-dimensional algebra. Thus $I = \mathrm{Ker}\ \phi$ is an ideal of A of finite codimension. If $f(L) = 0$, then also $f(I) = 0$ since $I \subseteq L$. Thus 2) $\Rightarrow$ 3). Similarly 1) $\Rightarrow$ 3), and so 1) through 6) are equivalent.

We will show 4) $\Leftrightarrow$ 7). Assume 4) holds, and let $\{g_1, \ldots, g_n\}$ be a basis for $A \rightharpoonup f$. Thus for any $a \in A, a \rightharpoonup f = \sum_{i=1}^n h_i(a)g_i$, and the h_i are in A^*. Then for $a, b \in A$,

$$\begin{aligned} \langle m^* f, b \otimes a\rangle &= \langle f, ba\rangle = \langle a \rightharpoonup f, b\rangle \\ &= \sum_{i=1}^n \langle h_i, a\rangle\langle g_i, b\rangle = \langle \sum_{i=1}^n g_i \otimes h_i, b \otimes a\rangle. \end{aligned}$$

Thus $m^* f = \sum_i g_i \otimes h_i \in A^* \otimes A^*$, proving 7). Conversely, assume 7): $m^* f = \sum_i g_i \otimes h_i$, for $g_i, h_i \in A^*$. The above computation shows that $a \rightharpoonup f = \sum_{i=1}^n h_i(a) g_i$, for all $a \in A$; thus $A \rightharpoonup f$ is a subspace of the span of $\{g_1, \ldots, g_n\}$, and so is finite-dimensional. Thus 4) holds. □

Note that A° is a subspace of A^* since 7) is a linear condition. We can now prove 1.2.4:

9.1.2 PROPOSITION. *Let (A, m, u) be an algebra. Then $m^* A^\circ \subseteq A^\circ \otimes A^\circ$. Thus $(A^\circ, \Delta, \varepsilon)$ is a coalgebra with comultiplication $\Delta = m^*$ and counit $\varepsilon = u^*$.*

PROOF. Choose $f \in A^\circ$. As in the lemma, let $\{g_1, \cdots, g_n\}$ be a basis for $A \rightharpoonup f$. Then $m^* f = \sum_{i=1}^n g_i \otimes h_i$, for some $h_i \in A^*$. Since each $g_i \in A \rightharpoonup f$, also $A \rightharpoonup g_i \subseteq A \rightharpoonup f$, a finite-dimensional space. Thus each $g_i \in A^\circ$. Since the $\{g_i\}$ are linearly independent, we may choose $a_1, \ldots, a_n$ in A such that $g_i(a_j) = \delta_{ij}$. Now $f \leftharpoonup a_j = \sum_i g_i(a_j) h_i = h_j$, and so each $h_j \in f \leftharpoonup A$, which is finite-dimensional by the lemma. Thus also $h_j \in A^\circ$, for all j, and so $m^* f \in A^\circ \otimes A^\circ$.

Coassociativity of $\Delta = m^*$ follows by dualizing the associativity of m and restricting to A°. □

We can now prove the fact mentioned in 1.3.6.

9.1.3 THEOREM. *Let* $(B, m, u, \Delta, \varepsilon)$ *be a bialgebra. Then* $(B^\circ, \Delta^*, \varepsilon^*, m^*, u^*)$ *is also a bialgebra. If* $B = H$ *is a Hopf algebra with antipode* S, *then* H° *is a Hopf algebra with antipode* S^*.

PROOF. We know that $(B^*, \Delta^*, \varepsilon^*)$ is an algebra by 1.2.2 and that (B°, m^*, u^*) is a coalgebra by 9.1.2. We claim that B° is a subalgebra of B^*. For, choose $f, g \in B^\circ$; then $B \rightharpoonup f$ and $B \rightharpoonup g$ are finite-dimensional by 9.1.1. Now for any $a \in B$, using 4.1.10,

$$\begin{aligned} a \rightharpoonup fg &= \sum (a_1 \rightharpoonup f)(a_2 \rightharpoonup g) \\ &\subseteq \text{span of } (B \rightharpoonup f)(B \rightharpoonup g) \end{aligned}$$

which is finite-dimensional. Thus $fg \in B^\circ$ by 9.1.1. Also certainly $\varepsilon \in B^\circ$, as it vanishes on an ideal of codimension one. Thus B° is a subalgebra of B^*. It is straightforward to check B° is a bialgebra, by dualizing the diagrams for B.

Finally let $B = H$ be a Hopf algebra with antipode S; we must show $S^* H^\circ \subseteq H^\circ$. Let $a \in H, f \in H^*$. Then

$$\begin{aligned} \langle a \rightharpoonup S^* f, b \rangle &= \langle S^* f, ba \rangle = \langle f, S(ba) \rangle = \langle f, SaSb \rangle \\ &= \langle f \leftharpoonup Sa, Sb \rangle = \langle S^*(f \leftharpoonup Sa), b \rangle . \end{aligned}$$

Thus $H \rightharpoonup S^* f = S^*(f \leftharpoonup SH) \subseteq S^*(f \leftharpoonup H)$. Now if $f \in H^\circ$, then dim $(f \leftharpoonup H) < \infty$ by 9.1.1. Thus dim $(H \rightharpoonup S^* f) < \infty$, and so again by 9.1.1 we have $S^* f \in H^\circ$. The fact that S^* is the convolution inverse of id_{H° follows by dualizing the property for S and id_H. □

9.1.4 EXAMPLE. For any algebra A, recall from 1.3.5 that $\mathrm{Alg}(A, \mathbf{k})$ is the set of algebra maps from A to $\mathbf{k}$. Then $\mathrm{Alg}(A, \mathbf{k}) \subseteq A^\circ$. For if $f \in \mathrm{Alg}(A, \mathbf{k})$, $\langle m^* f, a \otimes b \rangle = f(ab) = f(a) f(b) = \langle f \otimes f, a \otimes b \rangle$. Thus $m^* f = f \otimes f \in A^* \otimes A^*$ and so $f \in A^0$. This also shows $\mathrm{Alg}(A, \mathbf{k}) = G(A^\circ)$.

9.1.5 EXAMPLE. We recall Example 1.3.6: if H is the group algebra $\mathbf{k}G$, then $H^\circ = R_{\mathbf{k}}(G)$, the Hopf algebra of representative functions on G. Note that the characterization given there follows from 9.1.1.

It can happen that H° is trivial, although $G \neq 1$; for, as is shown in [BCM, 2.7], if K is an infinite field of cardinality greater than that of k and $G = PSL_2(K)$, then $(\mathbf{k}G)^\circ = \mathbf{k}\varepsilon$. See also [A, Exercise 2.5].

9.1.6 EXAMPLE. If k has characteristic 0, then it is an old fact that $U(s\ell(n))^\circ \cong \mathcal{O}(SL_n(\mathbf{k}))$. Recently Takeuchi has given a "quantum" version of this result [Tk 92b]. Let $U_q(s\ell(n))$ and $\mathcal{O}_q(SL_n(\mathbf{k}))$ be as in the Appendix, where q is not a root of 1 and k has characteristic not 2. Then

$$U_q(s\ell(n))^\circ \cong \mathcal{O}_q(SL_n(\mathbf{k}))\#\mathbf{k}\mathbf{Z}_2^{n-1}.$$

The group $\mathbf{Z}_2^{n-1}$ is given explicitly as $\langle \gamma_1, \dots, \gamma_{n-1} \rangle$, where the γ_i are commuting Hopf algebra automorphisms of order 2 given as follows:

$$\gamma_i(E_j) = \gamma_i(F_j) = 0, \text{ for all } i, j, \text{ and } \gamma_i(K_j) = (-1)^{\delta_{ij}}.$$

9.1.7 EXAMPLE. Let $H = \mathbf{k}[x]$, polynomials in x, where we assume x is primitive. As in 5.5.4, we have the elements $f^{(n)}$ in H^*, where $\langle f^{(n)}, x^m \rangle = \delta_{n,m}$, and any $\phi \in H^*$ may be written as $\phi = \sum_{n \geq 0} a_n f^{(n)}$, where $a_n = \phi(x^n)$. If also k has characteristic 0, then by setting $f = f^{(1)}$ we have $f^n = n! f^{(n)}$ and thus $H^* \cong \mathbf{k}[[f]]$, the formal power series in f. Now choose $\phi \in H^\circ$. Then $\phi(I) = 0$ for some ideal $I \neq 0$ of $\mathbf{k}[x]$, and $I = (p(x))$ for some polynomial $p(x) = b_0 + \cdots + b_{m-1}x^{m-1} + x^m$. Since $\phi(x^n p(x)) = 0$ for all n, we see

$$(*) \qquad b_0 a_n + b_1 a_{n+1} + \cdots + b_{m-1} a_{n+m-1} + a_{n+m} = 0.$$

Thus, given $a_0, a_1, \dots, a_{m-1}$, all a_n for $n \geq m$ are determined by the recursion relation $(*)$. Conversely any such relation determines a polynomial $p(x)$. Thus H° may be identified with the linearly recursive functions on $\mathbf{k}[x]$. For more on linearly recursive functions, see [PT] and [LTa].

We can describe H° in more detail. If $\lambda \in \mathbf{k}$ is a root of $p(x)$, then $a_n = \lambda^n$, all n, gives a solution of $(*)$; that is, $\phi(x^n) = \lambda^n = \phi(x)^n$. Thus $\phi = \phi_\lambda = \sum_{n \geq 0} \lambda^n f^{(n)} \in \operatorname{Alg}(H, \mathbf{k}) = G(H^\circ)$, and all group-like elements are of this form. Note that if char $\mathbf{k} = 0$, then $\phi_\lambda = \exp(\lambda f)$. Since $\phi_\lambda \cdot \phi_\mu = \phi_{\lambda+\mu}$, it follows that $G(H^\circ) \cong (\mathbf{k}, +)$.

Now assume that $p(x) = (x - \lambda)^t$; we wish to describe $\phi \in H^*$ such that $\phi((p(x))) = 0$, that is, $\phi \in (p(x))^\perp \subseteq H^*$. Since $(p(x))$ has finite codimension in $H, (p(x))^{\perp}$ is finite dimensional. We claim it is spanned by $\{f^{(s)}\phi_\lambda \mid 0 \le s \le t-1\}$. To see this, first note that $H = \mathbf{k}[x-\lambda]$, and thus that $\{(x-\lambda)^s \mid s \ge 0\}$ is also a basis of H. Define $\mu_\lambda^{(s)} \in H^*$ by $\langle \mu_\lambda^{(s)}, (x-\lambda)^u \rangle = \delta_{s,u}$. One may then check that $\mu_\lambda^{(s)} = f^{(s)}\phi_\lambda$; the claim follows. Thus any ϕ such that $\langle \phi, (p(x)) \rangle = 0$ is of the form $\phi = \sum_{s=0}^{t-1} a_s f^{(s)} \phi_\lambda$.

When $\mathbf{k}$ is algebraically closed, we have $p(x) = \prod_{i=1}^{s}(x-\lambda_i)^{t_i}$; the ϕ corresponding to $p(x)$ will be a linear combination of terms as above. Thus

$$\begin{aligned} H^\circ &= \{\sum a_{ij} f^{(j)} \phi_{\lambda_i} \mid a_{ij}, \lambda_i \in \mathbf{k}\} \\ &\cong H_P \otimes \mathbf{k}G(H^\circ), \end{aligned}$$

where $H_P = \mathbf{k}[f^{(n)} \mid n \ge 0]$ is a divided power Hopf algebra in the $f^{(n)}$, as in 5.6.8.

In fact we have shown a special case of 5.6.4. For since H is cocommutative, H° is commutative, and thus $G(H^\circ)$ acts trivially on H_P. Thus as Hopf algebras,

$$H_P \# \mathbf{k}G(H^\circ) \cong H_P \otimes \mathbf{k}G(H^\circ) \cong H^\circ.$$

When $\mathbf{k}$ has characteristic 0, $P(H^\circ) = \mathbf{k}f$ and so $H_P = \mathbf{k}[f] \cong U(P(H^\circ))$, a special case of 5.6.5.

§9.2 SubHopfalgebras of H° and density

In this section we consider some subHopfalgebras of H° and how large they are relative to H^*.

9.2.1 LEMMA. *Let I be a Hopf ideal of H and let*

$$H_I^\circ = \{f \in H^\circ \mid f(I^n) = 0, \text{ for some } n > 0\}.$$

Then H_I° is a subHopfalgebra of H°.

PROOF. We first check that H_I° is a subalgebra of H^*. Choose $f, g \in H_I^\circ$ and $n, m > 0$ such that $f(I^n) = 0$ and $g(I^m) = 0$. Since I is a coideal of H,

$$\Delta I^{n+m} \subseteq (I \otimes H + H \otimes I)^{n+m} \subseteq \sum_{j=0}^{n+m} I^j \otimes I^{n+m-j}.$$

Thus $\langle f \otimes g, \Delta I^{n+m} \rangle = 0 = (f * g)(I^{n+m})$. Thus $f * g \in H_I^\circ$. Also $\varepsilon(I) = 0$ and so $\varepsilon \in H_I^\circ$. Since H_I° is clearly a subspace, it is a subalgebra.

We next show that it is a subcoalgebra. Choose $f \in H_I^\circ$ and write $\Delta f = \sum_{i=1}^n g_i \otimes g_i'$, for $g_i, g_i' \in H^\circ$. We may assume that $g_1, \ldots, g_m \in H_I^\circ$ and that $g_{m+1}, \ldots, g_n$ are linearly independent mod H_I°. Choose $t > 0$ such that $f(I^t) = 0$ and $g_i(I^t) = 0$, all $i = 1, \ldots, m$. Now choose $h \in I^t, k \in H$. Since I is an ideal,

$$0 = f(hk) = \langle \Delta f, h \otimes k \rangle = \sum_{i=1}^{n} g_i(h) g_i'(k) = \sum_{i=m+1}^{n} g_i(h) g_i'(k).$$

This says that the function $g = \sum_{m+1}^n g_i'(k) g_i$ vanishes on I^t, and so $g \in H_I^\circ$, for each $k \in H$. But $g_{m+1}, \ldots, g_n$ are independent mod H_I°; thus $g_i'(k) = 0$, all $i = m+1, \ldots, n$, for all k. Thus $g_i' = 0$ for all $i \geq m+1$ and so $\Delta f \in H_I^\circ \otimes H^\circ$. Similarly $\Delta f \in H^\circ \otimes H_I^\circ$, and so $\Delta f \in H_I^\circ \otimes H_I^\circ$. Thus H_I° is a subcoalgebra.

Finally, $SI \subseteq I$ since I is a Hopf ideal. Thus if $f \in H_I^\circ$, and so $f(I^n) = 0$ for some n, then $\langle Sf, I^n \rangle = \langle f, SI^n \rangle \subseteq \langle f, (SI)^n \rangle \subseteq \langle f, I^n \rangle = 0$. Thus $Sf \in H_I^\circ$. □

The largest such H_I° is H° itself, which occurs when $I = 0$. At the other extreme, the smallest H_I° occurs when we use $I = \omega = \operatorname{Ker} \varepsilon$, the *augmentation ideal* of H.

9.2.2 DEFINITION. $H' = H_\omega^\circ = \{f \in H^\circ \mid f(\omega^n) = 0, \text{ some } n > 0\}$.

H' is a subHopfalgebra of H° by the lemma. One might ask when H' consists of all $f \in H^*$ such that $f(\omega^n) = 0$ for some n; that is, we do not need the additional hypothesis that f is in H°. This will certainly be true if all ω^n have finite codimension. A sufficient condition for this to happen is that H is a finitely-generated k-algebra:

9.2.3 LEMMA. *If H is k-affine, then each ω^n has finite codimension.*

PROOF. Write $H = \mathrm{k}\langle a_1, \ldots, a_m \rangle$. Now $h - \varepsilon(h)1 \in \omega$, for all $h \in H$, so we may assume that $a_1 = 1$ and that $a_2, \ldots, a_m \in \omega$. Then $\omega = \sum_{i=2}^m a_i H$, so is finitely-generated as a right (or left) ideal. It follows that ω^n is a finitely-generated right ideal, for all $n \geq 1$: for example, if $n = 2$, $\omega^2 = (\sum_{i=2}^m a_i H)(\sum_{j=2}^m a_j H) = \sum_{i,j=2}^m a_i a_j H$, since any $h \in H$ can be written as a polynomial in the $a_i, i > 2$, plus a scalar $\alpha 1$ which commutes with all a_j.

Thus for all $n > 0$, ω^n / ω^{n+1} is a finitely-generated right $H/\omega \cong \mathrm{k}$-module, and so is finite-dimensional. Thus each ω^n has finite codimension. □

It is certainly false in general, however, that all ω^n have finite codimension: if $H = \mathrm{k}[x_1, x_2, \ldots, x_m, \ldots]$, polynomials with each x_i primitive, then $\dim H/\omega^n = \infty$ for all $n > 1$.

Before giving some examples, we compare the group-like and primitive elements of H' and H°. In particular if $G(H^\circ)$ is non-trivial, then H' will be strictly smaller than H°.

9.2.4 LEMMA. *For any $H, G(H') = \{\varepsilon\}$ and $P(H') = P(H^\circ)$.*

PROOF. First choose $f \in G(H')$; then $f(\omega^t) = 0$ for some $t > 0$. Thus $f(\omega)^t = f(\omega^t) = 0$ since $f \in \mathrm{Alg}(H, k)$ and so $f(\omega) = 0$. It follows that $f \in \mathrm{k}\varepsilon$. Since $f \in G(H')$, we see $f = \varepsilon$.

Now if $f \in P(H^\circ)$, then $\Delta f = m^* f = f \otimes \varepsilon + \varepsilon \otimes f$. Using 1.6.6, $\langle f, \omega^2 \rangle = \langle \omega \rightharpoonup f, \omega \rangle = \sum \langle f_2, \omega \rangle \langle f_1, \omega \rangle = 0$. Thus $f \in H'$. □

In fact a much stronger relationship holds between H' and ε than the statement in 9.2.4. We recall the definition of irreducible component from 5.6.1.

9.2.5 PROPOSITION. *For any $H, H' = (H^\circ)_\varepsilon$, the irreducible component of ε in H°. In particular H' is connected. Thus if also H is commutative and* k *has characteristic* $0, H' \cong U(P(H^\circ))$.

PROOF. We generalize the proof of [A, 4.3.13], without the hypotheses that H is affine and commutative. First we show that H' is irreducible. To see this it suffices to show that $\varepsilon \in C$ for any simple subcoalgebra C of H'. Since C is finite-dimensional, we may choose n such that $C \subseteq (\omega^n)^\perp$. Letting $\perp_H$ denote the annihilator in H, we then have $C^{\perp_H} \supseteq (\omega^n)^{\perp\perp_H} \supseteq \omega^n$. Since C is simple, $C^{\perp_H}$ is a maximal ideal of H as in 5.1.4, and thus is prime. It follows that $\omega \subseteq C^{\perp_H}$, and so $\omega = C^{\perp_H}$. Since C was finite-dimensional, $C = C^{\perp_H\perp} = \omega^\perp = \mathrm{k}\varepsilon$. Thus $\varepsilon \in C$. Consequently $H' \subseteq (H^\circ)_\varepsilon$.

To see the other containment, let C be any finite-dimensional subcoalgebra of $(H^\circ)_\varepsilon$. Then $\varepsilon \in C$, and so $\omega = \varepsilon^{\perp_H} \supset C^{\perp_H} = I$; here I is an ideal of H of finite codimension as in 5.1.4. If J is an ideal of H with $I \subseteq J \neq H$, then $J^\perp \subseteq I^\perp = C \subseteq (H^\circ)_\varepsilon$. $J^\perp$ is a subcoalgebra of $(H^\circ)_\varepsilon$ by 5.1.4, and so $\varepsilon \in J^\perp$. Then $\omega = \varepsilon^{\perp_H} \supseteq J^{\perp\perp_H} \supseteq J$. Thus H/I is a finite-dimensional algebra with unique maximal ideal $\bar{\omega}$, the image of ω. Thus $\bar{\omega} = \mathrm{Jac}(H/I)$, so is nilpotent. But then $\omega^n \subseteq I$ for some n, and so $C = I^{\perp} \subseteq (\omega^n)^{\perp}$. Thus $C \subseteq H'$. Since $(H^\circ)_\varepsilon$ is "locally finite" by 5.1.1, $(H^\circ)_\varepsilon \subseteq H'$.

The last statement follows by Kostant's theorem 5.6.5 and the fact noted above that $P(H^\circ) = P(H')$. □

When H is not commutative, H' may be too small to be very useful. See [ChM] for a suggested replacement for H'.

9.2.6 EXAMPLE. We look again at Example 9.1.7. Using 9.2.5 we see that $\mathbf{k}[x]' \cong \mathbf{k}[f]$. More generally, let $\mathbf{g}$ be any finite-dimensional nilpotent Lie algebra over $\mathbf{k}$ of characteristic 0 and let $H = U(\mathbf{g})$. Then $H' = \mathcal{O}(N)$, where N is the unipotent algebraic group with $\mathbf{g} = \text{Lie } N$ and $\mathcal{O}(N)$ is the Hopf algebra of regular functions on N [H 81, Chap. XVI, 4.2].

9.2.7 EXAMPLE. Let $H = \mathbf{kZ}$, the group algebra of $\mathbf{Z}$ over an algebraically closed field k of characteristic 0. If $f \in P(H^\circ)$, then one may check that $f(n) = nf(1)$, for all $n \in \mathbf{Z}$, and so f is determined by $f(1)$. Thus $f = f(1)f_0$, where $f_0(1) = 1$, and so $P(H^\circ) = \mathbf{k}f_0$. It follows that $H' = \mathbf{k}[f_0]$.

Now for each $0 \neq \lambda \in \mathbf{k}, f_\lambda \in H^\circ$ given by $f_\lambda(n) = \lambda^n$ is in $G(H^\circ)$; moreover every $f \in G(H^\circ)$ is of this form. Since $f_\lambda \cdot f_\mu = f_{\lambda\mu}$ it follows that $G(H^\circ) \cong \mathbf{k}^\bullet$, the multiplicative group of $\mathbf{k}$. Compare this with the fact that in 9.1.7, $G(H^\circ) \cong (\mathbf{k}, +)$.

In this case we can also show 5.6.4 directly. That is,

$$H^\circ \cong \mathbf{k}[f_0] \otimes \mathbf{k}G(H^\circ) = U(P(H^\circ)) \otimes \mathbf{k}G(H^\circ).$$

9.2.8 EXAMPLE. Let G be an affine algebraic $\mathbf{k}$-group, and let $H = \mathcal{O}(G)$. If $\mathbf{k}$ is algebraically closed, then $\mathcal{O}(G)^\circ \cong \mathcal{O}(G)' \# kG(H^\circ)$, using 5.6.4 and 9.2.5. In fact, in this case $G(H^\circ) \cong G$; see §9.3. $\mathcal{O}(G)'$ is called the *hyper-algebra* of G; when $\mathbf{k}$ has characteristic $0, \mathcal{O}(G)' \cong U(P(H^\circ)) = U(\text{Lie } G)$. See also [A, Sec. 4.3].

We now consider how large these duals H° and H' are compared to H^*. Recall a subspace U of H^* is *dense* (in the finite-discrete topology on $H^* = \text{Hom}(H, \mathbf{k})$, where $\mathbf{k}$ has the discrete topology) if it separates the points of H. Note that if U is dense in H^*, then H imbeds in U^*.

9.2.9 DEFINITION. An algebra A is *residually finite-dimensional* over $\mathbf{k}$ if there exists a family $\{\pi_\alpha\}$ of finite-dimensional k-representations of A such that $\cap_\alpha \text{Ker } \pi_\alpha = (0)$.

We note that in the Hopf algebra literature, this condition is usually called *proper*. We propose the present terminology (which comes from group theory) for two reasons: first, it is more descriptive, and second, because proper is one of the more overused terms in mathematics.

However, whatever its name, 9.2.9 is what is needed to guarantee density of H°.

9.2.10 PROPOSITION. *Let H be any Hopf algebra.*
1) *H° is dense in $H^* \Leftrightarrow H$ is residually finite-dimensional.*
2) *If H is an affine* k*-algebra, then H' is dense in $H^* \Leftrightarrow \cap_{n\geq 0}\omega^n = 0$.*

PROOF. 1) Assume H° is dense in H^* and consider the set $\{I_\alpha\}$ of ideals of H of finite codimension; it suffices to show $\cap_\alpha I_\alpha = 0$. If not, choose $0 \neq h \in \cap_\alpha I_\alpha$. Since H° separates the points of H, there exists $f \in H^\circ$ such that $f(h) \neq 0$. Since $f \in H^\circ$, there exists some I_β with $f(I_\beta) = 0$, a contradiction.

Conversely, assume $\cap_\alpha I_\alpha = 0$, for the set $\{I_\alpha\}$ as above. If $h_1 \neq h_2$ in H, then $h = h_1 - h_2 \neq 0$, and so there exists β with $h \notin I_\beta$. Then $\bar{0} \neq \bar{h}$ in H/I_β, a finite-dimensional space. Thus there exists $g \in (H/I_\beta)^*$ such that $g(\bar{h}) \neq 0$. If $\pi_\beta : H \to H/I_\beta$ denotes the quotient map, set $f = g \circ \pi_\beta$; then $f(I_\beta) = 0$ and so $f \in H^\circ$. But $f(h) = g(\bar{h}) \neq 0$. Thus H° separates the points of H, and so is dense in H^*.
2) By 9.2.3 we know that each ω^n has finite codimension in H. The proof now follows as in 1), replacing the set $\{I_\alpha\}$ by the set of $\{\omega^n, \text{ all } n\}$. □

It is known that H is residually finite-dimensional in a number of cases. It is true whenever H is commutative and affine, by using the Krull Intersection Theorem; see [S, 6.1.3, or A, 2.3.19]. For group algebras, a theorem of R. Steinberg says that kG is residually finite-dimensional $\Leftrightarrow$ G is residually linear. For enveloping algebras in characteristic 0, a result of Harish-Chandra says that $U(\mathfrak{g})$ has a faithful finite-dimensional representation whenever $\mathfrak{g}$ is finite-dimensional. Michaelis has extended this fact to any characteristic [Mh]; see [PQ] for some related results. A new result of A. Joseph and G. Letzter shows that $U_q(\mathfrak{g})$ is residually finite-dimensional if $\mathfrak{g}$ is finite-dimensional and semisimple [JL].

The cases when $\cap_{n \geq 0} \omega^n = 0$ are more restrictive. If char$\mathbf{k} = 0$, it is true if either $H = U(\mathfrak{g})$ for $\mathfrak{g}$ a finite-dimensional nilpotent Lie algebra [H 81, p. 230] or $H = \mathbf{k}G$ for G a residually torsion-free nilpotent group [Ps]. When H is commutative and affine, it is true provided H has no elements $a \notin \mathbf{k}$ which are algebraic and separable over $\mathbf{k}$ [S 71, A 4.6.4]. Recall such Hopf algebras were considered in 5.7.4.

We apply these ideas, as well as some results from Chapter 5, to prove an old result of Cartier; see [A, 2.5.4], [S, 13.1.2], and [W, 11.4]. Recall an algebra is *reduced* if it has no non-zero nilpotent elements.

9.2.11 COROLLARY. *If* $\mathbf{k}$ *has characteristic* 0 *and* H *is commutative, then* H *is reduced.*

PROOF. We follow the proof in [A] and [S]. Let $\bar{\mathbf{k}}$ be the algebraic closure of $\mathbf{k}$; since $H \hookrightarrow H \otimes_{\mathbf{k}} \bar{\mathbf{k}}$, it suffices to prove the result when $\mathbf{k}$ is algebraically closed. We may also assume that H is affine, for if $0 \neq h \in H, h$ lies in a finite-dimensional subcoalgebra C by 5.1.1. Since $S^2 = id$ by 1.5.12, the subalgebra K of H generated by $C + S(C)$ is an affine Hopf algebra containing h. Thus we may assume that H° is dense in H^*, since H is residually finite-dimensional as noted above. Then H imbeds in $(H^\circ)^*$, and it suffices to show $(H^\circ)^*$ is reduced.

Now H° is a cocommutative pointed Hopf algebra, since $\mathbf{k}$ is algebraically closed, and thus 5.6.4 and 5.6.5 apply to give $H^\circ \cong \oplus_{g \in G(H^\circ)} U(P(H^\circ))$ as coalgebras. Hence $(H^\circ)^* \cong \prod_g U(P(H^\circ))^*$. But now by 5.5.4, $U(P(H^\circ))^*$ is a power series ring, so an integral domain. Thus $(H^\circ)^*$ is reduced. □

9.2.12 REMARK. Although 9.2.11 is certainly false if $\mathbf{k}$ has characteristic $p \neq 0$, something can still be said. For, assume that H is affine (and commutative) and that $\mathbf{k}$ is algebraically closed. Then there is a one-to-one correspondence between maximal ideals M of H and group-like elements $g \in G(H^\circ) = \text{Alg}(H, \mathbf{k})$, via $g \mapsto M = \text{Ker}\, g$ and $M \mapsto \{g : H \to H/M \cong \mathbf{k}\}$. Thus $J = \text{Jac}\,(H) = \cap M = (\mathbf{k}G(H^\circ))^{\perp_H}$. Since $\mathbf{k}G(H^\circ)$ is a subalgebra of H°, J is a coideal of H, by arguments analogous to those in 5.1.4. By the Nullstellensatz $J = N$, the nilradical of H, and thus N is a Hopf ideal of H. Thus $\bar{H} = H/N$ is a Hopf algebra and is reduced.

If H is not commutative, then J and N are not always coideals, as is noted in the discussion after 5.2.8.

§9.3 Classical duality

We discuss here the duality between groups and commutative Hopf algebras; a basic reference here is [A, Sec. 3.4 and 4.1]. For a given field $\mathbf{k}$, we have already seen that for any group G, we can construct a commutative Hopf algebra from G, namely the Hopf algebra $R_{\mathbf{k}}(G)$ of representative functions on G. This gives a correspondence

$$\Phi : \mathcal{G} \to \mathcal{H}_{\mathbf{k}} \text{ given by } G \mapsto R_{\mathbf{k}}(G),$$

where $\mathcal{G}$ is the category of groups and $\mathcal{H}_{\mathbf{k}}$ is the category of commutative Hopf algebras over $\mathbf{k}$. Conversely, there is a correspondence

$$\Psi : \mathcal{H}_{\mathbf{k}} \to \mathcal{G} \text{ given by } H \mapsto G(H^\circ),$$

the group of group-like elements in H°. In fact Φ and Ψ are contravariant functors which are adjoint to each other.

For an appropriate field $\mathbf{k}$, it is of interest to characterize the subcategory of $\mathcal{H}_{\mathbf{k}}$ which corresponds to a particular subcategory of $\mathcal{G}$. We consider a few important cases:

9.3.1 FINITE GROUPS. Assume that $\mathbf{k}$ is algebraically closed. If G is finite, we know that $R_{\mathbf{k}}(G) = (\mathbf{k}G)^* \cong \mathbf{k}^{(n)}$, for $n = |G|$, and thus $R_{\mathbf{k}}(G)$ is finite-dimensional and semisimple. Conversely if H is finite-dimensional and semisimple, then $H \cong (\mathbf{k}G)^*$ by 2.3.1. Thus the appropriate subcategory here of $\mathcal{H}_{\mathbf{k}}$ consists of (commutative) finite-dimensional semisimple Hopf algebras.

9.3.2 AFFINE ALGEBRAIC GROUPS. Again let $\mathbf{k}$ be algebraically closed. If A is a commutative affine $\mathbf{k}$-algebra, we say $X_A = \operatorname{Alg}(A, \mathbf{k})$ is an affine algebraic $\mathbf{k}$-variety. A group G is an *affine algebraic* $\mathbf{k}$*-group* if $G = X_H$, for some affine (commutative) algebra H, and the multiplication m and inverse map s of G are morphisms of algebraic varieties. The maps m, s, and the imbedding $1_G \hookrightarrow G$ correspond to $\mathbf{k}$-algebra morphisms

$$\Delta : H \to H \otimes H, \quad S : H \to H, \quad \varepsilon : H \to \mathbf{k}$$

and it follows that H is a Hopf algebra in this situation. By construction $G = G(H^\circ)$.

Now $\Phi(G)$ is the subHopfalgebra of $R_{\mathbf{k}}(G)$ consisting of those functions $f : G \to \mathbf{k}$ which are morphisms of algebraic varieties; write $\Phi(G) = H(G)$. Note that $H(G)$ is reduced, as it is a subalgebra of $R_{\mathbf{k}}(G)$; in fact $H(G) = H/N$, where N is the nilradical of H, and thus $H(G) \cong H$ if H is reduced. The two correspondences

$$\Phi : G \mapsto H(G) \qquad \text{and } \Psi : H \to X_H$$

give an anti-equivalence between the category of affine algebraic $\mathbf{k}$-groups and the category of affine commutative reduced $\mathbf{k}$-Hopf algebras.

Note that by 9.2.11, all commutative Hopf algebras are reduced when $\mathbf{k}$ has characteristic 0.

9.3.3 COMPACT TOPOLOGICAL GROUPS. For a compact topological group G, let $H = \mathcal{R}_{\mathbf{C}}(G)$ be the Hopf algebra of continuous complex-valued representative functions, as in 2.4.5. Then H has the following properties:

1) H has an involution $*$ which commutes with Δ, ε, and s (namely $f^*(x) = \overline{f(x)}$, complex conjugation, for all $f \in H, x \in G$)
2) $G(H^\circ)$ separates the points of H
3) there exists an integral ϕ on H such that $\phi(ff^*) > 0$ if $f \neq 0$ in H (namely, $\phi(f)$ is the Haar integral as in 2.4.5).

Conversely, given any $\mathbf{C}$-Hopf algebra H satisfying these three properties, then $G = G(H^\circ)$ is a compact topological group. Moreover Φ and Ψ give a correspondence between the category of compact topological groups and the category of commutative $\mathbf{C}$-Hopf algebras satisfying 1), 2), and 3), such that

$$G(\mathcal{R}(G)^\circ) \cong G \qquad \text{and } \mathcal{R}(G(H^\circ)) \cong H.$$

These facts are known as the Tannaka Duality Theorem; see [Che, p. 211]. A version of the theorem over $\mathbf{R}$ is proved in [H 65], and stated in [A, 3.4.3].

In fact Tannaka's original theorem [Tn] was somewhat different; he proved that the group could be recovered from properties of the category of its finite-dimensional unitary representations. Among other things the category had to be closed under tensor products. This approach of recovering a group or a Hopf algebra from a suitable category of its representations was studied algebraically in [S-R] and [DM], where such categories are called "Tannakian". This is a subject of continuing interest in quantum groups, and we shall return to it in §10.4.

§9.4 Duality for actions

The duality theorems we consider next are of a different nature than those in §9.3. They have the following general form: given an H-module algebra A, one can form the smash product $A\#H$. This new algebra is an H°-module algebra, by letting H° act trivially on A and by the usual $\rightharpoonup$ action on H; thus we may form $(A\#H)\#H^\circ$. The idea is to show that

$$(A\#H)\#H^\circ \cong A \otimes \mathcal{L},$$

where $\mathcal{L}$ is some large (dense) subring of $\mathrm{End}_{\mathbf{k}}(H)$, for suitable assumptions about H, H° and their actions. In particular when H is finite-dimensional, $(A\#H)\#H^*$ is Morita-equivalent to A; even when H is infinite-dimensional, such duality results are useful in comparing the ideals of A with those of $A\#H$.

Results of this type have their origin in operator algebras, in work of Takesaki, Landstad, Nakagami, and Stratila-Voiculesu-Zsido, for actions and coactions of locally compact groups on von Neumann algebras [NaT]. There are more recent versions for C^*-algebras with applications to the K-theory of these algebras. An algebraic version of duality for actions and coactions of finite groups was given by [CM]; this was extended to infinite groups in [Q 85]. We will give some applications of the "Duality Theorem" to graded algebras in §9.5.

The main result of this section, due to Blattner and Montgomery [BM 85], is a duality theorem for arbitrary Hopf algebras. The proof we present here uses a simplification due to R.J. Blattner, presented at the AMS Special Session on Hopf algebras in San Francisco in 1991.

We need some preliminary steps. First, for any H, let U be a subHopf-algebra of H°. By 1.6.5, 1.6.6 and 4.1.10, we know that H is a left U-module algebra and that U is a left H-module algebra via $\rightharpoonup$; thus we may form $U\#H$ and $H\#U$. Moreover U acts on the right of H via $\leftharpoonup$.

9.4.1 DEFINITION. Let $U \subset H^\circ$ be as above. Define the maps λ, ρ by:
1) $\lambda : H\#U \to \mathrm{End}_{\mathbf{k}}H$ via $\lambda(h\#f)(k) = h(f \rightharpoonup k)$,
2) $\rho : U\#H \to \mathrm{End}_{\mathbf{k}}H$ via $\rho(f\#h)(k) = (k \leftharpoonup f)h$,
for all $h, k, \in H, f \in U$.

9.4.2 LEMMA. *λ is an algebra morphism and ρ is an anti-algebra morphism. If also H has bijective antipode, then λ and ρ are injective.*

PROOF. We consider λ, as the argument for ρ is similar. It is easy to check that λ is an algebra map; the difficulty is in seeing that it is injective. To see this we follow the argument in [M 85].

We construct maps $\lambda' : H\#U \to \mathrm{End}_{\mathbf{k}}(H)$ and $\psi : \mathrm{End}_{\mathbf{k}}(H) \to \mathrm{End}_{\mathbf{k}}(H)$ such that $\lambda' = \psi \circ \lambda$ and λ' is injective; it will then follow that λ is injective. These maps are defined as follows:

$$\lambda'(h\#f)(k) = \langle f, k\rangle h$$

$$\psi(\sigma)(k) = \sum \sigma(k_2)\bar{S}k_1,$$

for $h, k \in H, f \in U, \sigma \in \mathrm{End}_{\mathbf{k}}(H)$, and where $\bar{S}$ is the inverse of S.

To see λ' is injective, choose $w = \sum_{j=1}^{n} h_j\#f_j \in \mathrm{Ker}\ \lambda'$, where the $\{f_j\}$ are linearly independent. Then we may choose $k_1, \ldots, k_n \in H$ so that $\langle f_i, k_j\rangle = \delta_{ij}$, all $1 \le i, j \le n$. Then $0 = \lambda'(w)(k_j) = h_j$, all j, and so $w = 0$ and λ' is injective.

Now we check that $\psi \circ \lambda = \lambda'$. For $h, k \in H$ and $f \in U$,

$$\begin{aligned} \psi \circ \lambda(h\#f)(k) &= \sum \lambda(h\#f)(k_2)\bar{S}k_1 \\ &= \sum h\langle f, k_3\rangle k_2\bar{S}k_1 \\ &= h\langle f, k\rangle = \lambda'(h\#f)(k). \end{aligned}$$

This proves the lemma. □

9.4.3 COROLLARY. *If H is a finite-dimensional Hopf algebra and H^* acts on H via $\rightharpoonup$, then $H\#H^* \cong \mathrm{End}_{\mathbf{k}} H$ as algebras.*

Recall from 4.1.10 that $H\#H^*$ is sometimes called the Heisenberg double. When H is infinite-dimensional, $H\#H^\circ$ is not always a large subring of $\mathrm{End}(H)$: recall from 9.1.5 that it is possible for H° to be trivial. The next result gives a sufficient condition for $H\#H^\circ$ to be dense in $\mathrm{End}(H)$.

9.4.4 THEOREM. *Let H be a residually finite-dimensional Hopf algebra with bijective antipode S, and let U be a dense subHopfalgebra of H°. Then $\lambda(H\#U)$ is a dense subring of $\mathrm{End}_{\mathbf{k}} H$.*

If in addition U is pointed cocommutative, then $H\#U$ is a simple algebra.

PROOF. For simplicity of notation, we identify $H\#U$ with its image $\lambda(H\#U)$ in $\operatorname{End} H$; note λ is injective by the lemma. The idea of the proof is to show that H is a simple left $H\#U$-module, and that the centralizer of $H\#U$ in $\operatorname{End}(H)$ is just $\mathbf{k}$ itself. Thus the Chevalley-Jacobson Density Theorem will apply to prove the theorem.

First, choose any $0 \neq h \in H$; we claim that $(H\#U)h = H$. Write $\Delta h = \sum_{i=1}^r a_i \otimes b_i$, where the b_i are linearly independent. Since H is residually finite-dimensional and U is dense, we may choose $f_1, \ldots, f_r \in U$ so that $\langle f_j, b_i \rangle = \delta_{ij}$. Then for any $k_1, \ldots, k_r \in H$,

$$(*) \qquad (\sum_{j=1}^r k_j \# f_j)(h) = \sum_{j=1}^r k_j (f_j \rightharpoonup h) = \sum_{i,j=1}^r \langle f_j, b_i \rangle k_j a_i = \sum_{i=1}^r k_i a_i.$$

Thus by suitable choice of k_j, we see that all $a_i \in (H\#U)h$. By the counit property, $\varepsilon(a_i) \neq 0$ for some i since $h \neq 0$. Thus $\varepsilon(a) = 1$ for some $a \in (H\#U)h$; replacing h by a, we may assume $\varepsilon(h) = 1$.

Now use $k_j = \bar{S} b_j$, all $1 \leq j \leq r$, in $(*)$ to see that $(\sum_j \bar{S} b_j \# f_j)(h) = \sum (\bar{S} b_j) a_j = \varepsilon(h) 1 = 1$. Thus $1 \in (H\#U)h$ and so $(H\#U)h = H$. It follows that H is a simple $H\#U$-module.

Now choose $\sigma \in \operatorname{End}(H)$ which centralizes $H\#U$. Since σ commutes with the left regular representation of H on itself, $\sigma(h) = h\sigma(1)$. Thus it will suffice to show $\sigma(1) \in \mathbf{k}1$.

We may assume $\sigma(1) \neq 0$. Write $\Delta\sigma(1) = \sum_{i=1}^m c_i \otimes d_i$, where the d_i are linearly independent and $d_1 = 1$. Then for any $f \in U$,

$$(**) \qquad \sum_{i=1}^m \langle f, d_i \rangle c_i = f \rightharpoonup \sigma(1) = \sigma(f \rightharpoonup 1) = \sigma(f(1)1) = f(1)\sigma(1)$$

since σ commutes with $\lambda(1\#f)$. Choose $f_1, \ldots, f_m \in U$ such that $\langle f_j, d_i \rangle = \delta_{ij}$. Then $(**)$ gives $c_j = \delta_{j1}\sigma(1)$ for $1 \leq j \leq m$, which implies that $\Delta\sigma(1) = \sigma(1) \otimes 1$. Now the counit property gives $\sigma(1) = \varepsilon(\sigma(1))1 \in \mathbf{k}1$. Thus $H\#U$ is dense in $\operatorname{End}(H)$.

Finally if U is also pointed cocommutative, then $H\#U$ is simple by a theorem of McConnell and Sweedler [McS]. □

The next definition connects the right action ρ and the left action λ from 9.4.1.

9.4.5 DEFINITION. Let H be a Hopf algebra and let U be a subHopfalgebra of H°. Then U *satisfies the RL-condition with respect to* H if $\rho(U\#1) \subseteq \lambda(H\#U)$.

That is, the right action of any $f \in U$ on H can be expressed as a left action of some $w \in H\#U$.

9.4.6 EXAMPLE. If H is finite-dimensional, then the RL-condition is satisfied for $U = H^*$ since $\lambda(H\#H^*) = \operatorname{End}_{\mathbf{k}}(H)$ by 9.4.3.

9.4.7 EXAMPLE. If H is cocommutative, then the RL-condition is always satisfied, for any U, since in that case $\rho(f\#1) = \lambda(1\#f)$, all $f \in U$.

As we shall see in 9.4.15, the RL-condition is closely related to whether or not the adjoint action of U on itself is locally finite. This local-finiteness property certainly holds in Examples 9.4.6 and 9.4.7; when H is cocommutative, U is commutative and so the adjoint action is trivial.

We can now state our main theorem. First note that if U is a subHopfalgebra of H° and A is a U-comodule algebra, then A becomes a left H-module algebra as in 1.6.4 by

$$h \cdot a = \sum \langle a_1, h\rangle a_0, \tag{9.4.8}$$

where the structure map $A \to A \otimes U$ is given by $a \mapsto \sum a_0 \otimes a_1$.

9.4.9 THEOREM [BM 85]. *Let H be a Hopf algebra and U a subHopfalgebra of H° such that both H and U have bijective antipodes, and assume that U satisfies the RL-condition with respect to H.*

Let A be a U-comodule algebra, so that A is an H-module algebra as above. Let U act on $A\#H$ by acting trivially on A and via $\rightharpoonup$ on H. Then

$$(A\#H)\#U \cong A \otimes (H\#U).$$

Before proving the theorem, we need one more lemma, which gives "commutation relations" for the images of λ and ρ in $\operatorname{End}(H)$. It is the essential observation for Blattner's proof.

9.4.10 LEMMA. *Let H, U, λ and ρ be as in 9.4.1. Then:*

1) $\lambda(h\#f)\rho(g\#k) = \sum \rho(g_2\#(f_2 \rightharpoonup k))\, \lambda((h \leftharpoonup Sg_1)\#f_1)$

2) $\rho(g\#k)\lambda(h\#f) = \sum \lambda((h \leftharpoonup g_1)\#f_1)\, \rho(g_2\#(Sf_2 \rightharpoonup k))$

for all $h, k \in H$, $f, g \in U$.

PROOF. We prove 1), as 2) is similar. First, obviously $\lambda(h\#1)$ and $\rho(1\#k)$ commute, as do $\lambda(1\#f)$ and $\rho(g\#1)$. Now

$$\begin{aligned}\rho(f\#1)\lambda(k\#1)m &= (km) \leftharpoonup f = \sum (k \leftharpoonup f_1)(m \leftharpoonup f_2)\\ &= \sum \lambda((k \leftharpoonup f_1)\#1)\,\rho(f_2\#1)m,\end{aligned}$$

and so

$$\begin{aligned}\sum \rho(g_2\#1)\lambda(h \leftharpoonup Sg_1\#1) &= \sum \lambda(((h \leftharpoonup Sg_1) \leftharpoonup g_2)\#1)\,\rho(g_3\#1)\\ &= \lambda(h\#1)\rho(g\#1).\end{aligned}$$

Similarly $\lambda(1\#f)\rho(1\#k) = \sum \rho(1\#(f_2 \rightharpoonup k))\lambda(1\#f_1)$. Then

$$\begin{aligned}\lambda(h\#f)\rho(g\#k) &= \lambda(h\#1)\lambda(1\#f)\rho(1\#k)\rho(g\#1)\\ &= \sum \lambda(h\#1)\rho(1\#(f_2 \rightharpoonup k))\lambda(1\#f_1)\rho(g\#1)\\ &= \sum \rho(1\#(f_2 \rightharpoonup k))\lambda(h\#1)\rho(g\#1)\lambda(1\#f_1)\\ &= \sum \rho(1\#(f_2 \rightharpoonup k)\,\rho(g_2\#1)\lambda((h \leftharpoonup Sg_1)\#1)\lambda(1\#f_1)\\ &= \sum \rho(g_2\#(f_2 \rightharpoonup k))\lambda((h \leftharpoonup Sg_1)\#f_1)\end{aligned}$$

as desired. □

PROOF OF 9.4.9. For each $f \in U$, define $w(f) = \lambda^{-1} \circ \rho(\bar{S}f\#1) \in H\#U$; this can be done since the RL-condition holds and since S is bijective. Since ρ and $\bar{S}$ are anti-algebra maps, w is an algebra map. We now define:

$$\begin{aligned}\Phi:\ & (A\#H)\#U \to A \otimes (H\#U)\\ & (a\#h)\#f \mapsto \sum a_0 \otimes w(a_1)(h\#f)\end{aligned}$$

and

$$\begin{aligned}\Psi:\ & A \otimes (H\#U) \to (A\#H)\#U\\ & a \otimes (h\#f) \mapsto \sum ((a_0\#1)\#1)\overline{w(Sa_1)(h\#f)}\end{aligned}$$

where $\overline{k\#g} = (1\#k)\#g$.

We claim that Φ is an isomorphism, with inverse Ψ.

To see that Φ is an algebra morphism, first identify $H\#U$ with $\lambda(H\#U) \subseteq \mathrm{End}H$. Then

$$\Phi : (a\#h)\#f \mapsto \sum a_0 \otimes \rho(\bar{S}a_1\#1)\lambda(h\#f).$$

Thus we may write $\Phi((a\#h)\#f) = \phi(a)\psi(h\#f)$, where $\phi : A \to A \otimes \mathrm{End}H$ by $a \mapsto \sum a_0 \otimes \rho(\bar{S}a_1\#1)$ and $\psi : H\#U \to A \otimes \mathrm{End}H$ by $h\#f \mapsto 1 \otimes \lambda(h\#f)$. Note that both ϕ and ψ are algebra morphisms. We claim that

$$(*) \qquad \psi(h\#f)\phi(a) = \sum \phi(h_1 \cdot a)\psi(h_2\#f).$$

For,

$$\begin{aligned}
\psi(h\#f)\phi(a) &= \sum a_0 \otimes \lambda(h\#f)\rho(\bar{S}a_1\#1) \\
&= \sum \langle S(\bar{S}a_1)_1, h_1\rangle a_0 \otimes \rho((\bar{S}a_1)_2\#1)\lambda(h_2\#f) \qquad \text{by 9.4.10, 1)} \\
&= \sum \langle S\bar{S}a_{12}, h_1\rangle a_0 \otimes \rho(\bar{S}a_{11}\#1)\lambda(h_2\#f) \\
&= \sum \langle a_1, h_1\rangle a_{00} \otimes \rho(\bar{S}a_{01}\#1)\lambda(h_2\#f) \\
&= \sum \langle a_1, h_1\rangle \phi(a_0)\psi(h_2\#f) \\
&= \sum \phi(h_1 \cdot a)\psi(h_2\#f) \qquad \text{by 9.4.8.}
\end{aligned}$$

This shows the claim. It now follows that Φ is an algebra map. For, if $a, b \in A$, $h, k \in H$, and $f, g \in U$, we have

$$\begin{aligned}
\Phi((a\#h)\#f)\Phi((b\#k)\#g) &= \phi(a)\psi(h\#f)\phi(b)\psi(k\#g) \\
&= \sum \phi(a)\phi(h_1 \cdot b)\psi(h_2\#f)\psi(k\#g) \qquad \text{by } (*) \\
&= \sum \phi(a(h_1 \cdot b))\psi(h_2(f_1 \rightharpoonup k)\#f_2 g) \\
&= \Phi(\sum (a(h_1 \cdot b)\#h_2(f_1 \rightharpoonup k))\#f_2 g) \\
&= \Phi(\sum ((a\#h)(b\#f_1 \rightharpoonup k))\#f_2 g) \\
&= \Phi((a\#h)\#f)((b\#k)\#g)).
\end{aligned}$$

Finally we check that Φ and Ψ are inverses. Now

$$\begin{aligned} \Psi \circ \Phi((a\#h)\#f) &= \Psi(\sum a_0 \otimes w(a_1)(h\#f)) \\ &= \sum((a_{00}\#1)\#1)\overline{w(Sa_{01})}\,\overline{w(a_1)(h\#f)} \\ &= \sum((a_0\#1)\#1)\overline{w((Sa_1)a_2)(h\#f)} \\ &= ((a\#1)\#1)((1\#h)\#f) = (a\#h)\#f. \end{aligned}$$

Thus $\Psi \circ \Phi = id$. A similar computation shows $\Phi \circ \Psi = id$. The theorem is proved. □

We combine 9.4.9 with 9.4.4:

9.4.11 COROLLARY. *Let H be a residually finite-dimensional Hopf algebra and U a dense subHopfalgebra of H° such that H and U have bijective antipodes and such that U satisfies the RL-condition with respect to H. Then for any U-comodule algebra A,*

$$(A\#H)\#U \cong A \otimes \mathcal{L},$$

where $\mathcal{L}$ is a dense subring of $End_{\mathbf{k}}H$.

PROOF. Simply set $\mathcal{L} = \lambda(H\#U)$; it is dense by 9.4.4. □

In particular the corollary applies to any H-module algebra A if H is finite-dimensional and $U = H^*$. Thus $(A\#H)\#H^* \cong A \otimes \mathrm{End}H$, for any finite-dimensional H. This fact was shown independently in [vdB].

Recall from 1.6.4 that the hypothesis that A be a U-comodule algebra is a local finiteness condition on the action of H; we need a slight extension of this fact.

9.4.12 LEMMA. *A left H-module algebra A is a U-comodule algebra, for U a subcoalgebra of H°, if and only if for each $a \in A$ there exist $f_1, \ldots, f_r \in U$ such that $(\bigcap_{j=1}^r Ker\ f_j) \cdot a = 0$.*

PROOF. If A is a U-comodule algebra, via $\rho : A \to A \otimes U$, write $\rho(a) = \sum_{i=1}^r a_i \otimes f_i$. Then as in 1.6.4, $h \cdot a = \sum \langle f_i, h\rangle a_i$, for all $h \in H$. Thus clearly $(\bigcap_{j=1}^r \mathrm{Ker}\ f_j) \cdot a = 0$.

Conversely, assume that for $a \in A$ we are given $f_1, \dots, f_r \in U$ such that $(\cap_j \text{Ker } f_j) \cdot a = 0$; we may assume that $\{f_j\}$ are linearly independent. Choose $h_1, \dots, h_r \in H$ such that $\langle f_i, h_j \rangle = \delta_{ij}$ for $1 \le i, j \le r$. Then for all $k \in H$,

$$k - \sum_j \langle f_j, k \rangle h_j \in \bigcap_j \text{Ker } f_j$$

and thus $k \cdot a = \sum_j \langle f_j, k \rangle h_j \cdot a$. We may now define $\rho : A \to A \otimes U$ by $a \mapsto \sum h_j \cdot a \otimes f_j$, as in 1.6.4. □

We can now give an application to actions of Lie algebras.

9.4.13 COROLLARY. *Let* $\mathbf{g}$ *be a nilpotent Lie algebra of dimension* n *over* $\mathbf{k}$ *of characteristic* 0. *Let* $\mathbf{g}$ *act on the algebra* B *as locally nilpotent derivations (that is, each* $b \in B$ *is annihilated by some power of the augmentation ideal* $\omega_{U(\mathbf{g})}$*). Then*

$$(B \# U(\mathbf{g})) \# U(\mathbf{g})' \cong B \otimes \mathbf{A}_n,$$

where $\mathbf{A}_n$ *is the* n^{th} *Weyl algebra.*

PROOF. Let $H = U(\mathbf{g})$. Now as noted after 9.2.10, $\cap_n \omega^n = 0$ since $\mathbf{g}$ is nilpotent and char $\mathbf{k} = 0$. Thus $H' = U(\mathbf{g})'$ is dense in H^* (and so in H°) by 9.2.10; in particular H is residually finite-dimensional. The RL-condition is trivially satisfied since H is cocommutative. The fact that $\mathbf{g}$ acts as locally nilpotent derivations implies that B is an H'-comodule algebra. For, given $b \in B$, we know that $\omega^m \cdot b = 0$ for some m. Now ω^m has finite codimension r by 9.2.3. Thus we may find $f_1, \dots, f_r \in H^\circ$ such that $\omega^m = \cap_{i=1}^r \text{Ker } f_i$. Since $f_i(\omega^m) = 0$, necessarily $f_i \in H' = U$. Now use 9.4.12.

By 9.4.4 we know that $H \# H'$ is a dense subring of $\text{End}(H)$. However, in fact $H \# H' \cong \mathbf{A}_n$, the nth Weyl algebra. This fact seems to be well-known, and depends on the fact that $H' = U(\mathbf{g})' \cong \mathcal{O}(N)$, where N is the unipotent algebraic group with $\mathbf{g} = \text{Lie } N$, mentioned in 9.2.6. For a proof that $H \# H' \cong \mathbf{A}_n$, see [BM 85, 4.3]. □

The next example shows that 9.4.13, and so 9.4.9, can fail if the action of H on A is not locally finite.

9.4.14 EXAMPLE [BM 85, 4.1]. This example is due to S.P. Smith. Let char $\mathbf{k} = 0$ and let $H = \mathbf{k}[x]$. From 9.2.6 we know that $H' = \mathbf{k}[y]$, and it follows that $H \# H' \cong \mathbf{k}\langle x, y \rangle / (yx - xy - 1) \cong \mathbf{k}[x, \frac{d}{dx}]$, the Weyl algebra $\mathbf{A}_1$. Let A

be the algebra $\mathbf{k}[z]$; then A becomes an H-module algebra by letting x act as $z^2(d/dz)$; this action is not locally finite. Now $(A\#H)\#H' \cong \mathbf{k}\langle x,y,z\rangle$ with the relations $[y,x]=1, [y,z]=0$, and $[x,z]=z^2$. If this algebra were isomorphic to $A\otimes(H\#H')\cong A\otimes \mathbf{A}_1$, then its center would be isomorphic to $A=\mathbf{k}[z]$. However, as is shown in [BM 85], its center consists only of scalars.

In fact the RL-condition 9.4.5 is itself a kind of local finiteness property. For any Hopf algebra H, we say that H is (*left*) *ad-locally finite* if for every $h\in H$, $\{(ad_\ell H)(h)\}$ is finite-dimensional.

9.4.15 PROPOSITION. *Let H be any Hopf algebra and let U be a subHopf-algebra of H°.*
1) *If U satisfies the RL-condition with respect to H, then U is ad-locally finite.*
2) *If U is ad-locally finite and S_U is bijective, then for any $u\in U$, there exists $w\in U^\circ\#U$ such that the right action of u on U° equals the left action of w on U°.*
Consequently if U is residually finite-dimensional, S_U is bijective, and $H=U^\circ$, then U° satisfies the RL-condition with respect to $H \Leftrightarrow U$ is ad-locally finite.

PROOF. First note that for any $u,v\in U$ and $h,k\in H$,

$$(*)\qquad \langle v, k(u\rightharpoonup h)\rangle = \langle (v\leftharpoonup k)u, h\rangle$$

For, recall from 1.6.5 and 1.6.6 that $\rightharpoonup$ and $\leftharpoonup$ are the transposes of, respectively, right and left multiplication. Thus

$$\begin{aligned}\langle v, k(u\rightharpoonup h)\rangle &= \sum\langle v_1,k\rangle\langle v_2, u\rightharpoonup h\rangle\\ &= \sum\langle v_1,k\rangle\langle v_2u,h\rangle\\ &= \langle (v\leftharpoonup k)u,h\rangle.\end{aligned}$$

We now prove 1). Assume the RL-condition 9.4.5 for U and H. Thus given $u\in U$, there exists $w=\sum_{i=1}^{\ell} k_i\#f_i\in H\#U$ such that for all $h\in H$,

$$h\leftharpoonup u = w\rightharpoonup h = \sum k_i(f_i\rightharpoonup h).$$

Thus for all $v \in U$,

$$\langle v, h \leftharpoonup u\rangle = \langle v, \sum k_i(f_i \rightharpoonup h)\rangle = \langle \sum_i (v \leftharpoonup k_i) f_i, h\rangle$$

by $(*)$. Also $\langle v, h \leftharpoonup u\rangle = \langle uv, h\rangle$, for all $h \in H$. It follows that

$$(**) \qquad uv = \sum_i (v \leftharpoonup k_i) f_i$$

for all $v \in U$. Thus

$$(***) \qquad \begin{aligned} (ad\, v)(u) &= \sum v_1 u(Sv_2) = \sum_i v_1(Sv_2 \leftharpoonup k_i) f_i \\ &= \sum_i v_1 \langle Sv_3, k_i\rangle (Sv_2) f_i = \sum_i \langle Sv, k_i\rangle f_i \end{aligned}$$

and so $(ad\, U)(u) \subseteq$ span $\{f_1, \dots, f_\ell\}$. Thus U is ad-locally-finite.
2) Conversely, assume that U is ad-locally-finite. Thus given $u \in U$, we may choose independent elements $f_1, \dots, f_\ell \in U$ such that $(ad\, U)(u) =$ span $\{f_1, \dots, f_\ell\} = V$; note that also $(ad\, U)(f_i) \subseteq V$ since $(ad\, a)(ad\, b) = ad(ab)$. Thus for any $v \in U$,

$$(ad\, v)(u) = \sum_{i=1}^{\ell} \langle Sv, z_i\rangle f_i$$

for some $z_i, \dots, z_\ell \in U^*$. Note we have used here that S is bijective to be able to write Sv in $\langle\ ,\ \rangle$, since for any $z \in U^*$, $z(v) = z(\bar{S}Sv) = z \circ \bar{S}(Sv)$.

We claim that each $z_i \in U^\circ$; this follows since each $z_i(I) = 0$, where I is the kernel of the map $U \to \operatorname{End} V$ given by $v \mapsto ad\, v$, and I has finite codimension. Thus as in $(***)$ we see that

$$(ad\, v)(u) = \sum_{i=1}^{\ell} \langle Sv, z_i\rangle f_i = \sum_{i=1}^{\ell} v_1(Sv_2 \leftharpoonup z_i) f_i.$$

Then

$$u(Sv) = \sum (Sv_1) v_2 u Sv_3 = \sum_i (Sv_1) v_2 (Sv_3 \leftharpoonup z_i) f_i = \sum_i (Sv \leftharpoonup z_i) f_i.$$

Since S_U is bijective, this is equivalent to

$$uv = \sum_i (v \leftharpoonup z_i) f_i$$

and thus, reversing the derivation of $(**)$ in 1) we have

$$z \leftharpoonup u = \sum z_i(f_i \rightharpoonup z) = w \rightharpoonup z$$

for all $z \in U^\circ$, where $w = \sum_i z_i \# f_i \in U^\circ \# U$. This proves 2).

The last statement is now clear since if U is residually-finite-dimensional then $U \hookrightarrow U^{\circ\circ}$ by 9.3.10. Thus if $H = U^\circ, U \hookrightarrow H^\circ = U^{\circ\circ}$, and the result follows from 1) and 2). □

We give an application of the proposition to duals of enveloping algebras.

9.4.16 COROLLARY. *Let $U = U(\mathfrak{g})$, for $\mathfrak{g}$ a finite-dimensional Lie algebra, and let A be a U-comodule algebra. Then*

$$(A\#U(\mathfrak{g})^\circ)\#U(\mathfrak{g}) \cong A \otimes \mathcal{L},$$

for $\mathcal{L}$ a simple dense subring of End(U°).

PROOF. As noted after 9.2.10, $U(\mathfrak{g})$ is residually finite-dimensional; we also know that $U(\mathfrak{g})$ is ad-locally finite. Thus by 9.4.15, U satisfies the RL-condition with respect to $H = U^\circ$. Now apply 9.4.4 and 9.4.9. □

It would be nice to be able to apply the duality theorem to $U_q(\mathfrak{g})$; however this Hopf algebra is not ad-locally-finite in general. Thus one would need to prove a duality theorem with a less restrictive hypothesis than the RL-condition.

We close this section by considering what happens if $A\#H$ is replaced by a crossed product $A\#_\sigma H$. When H is finite-dimensional, the duality theorem is still true [BM 89]; it follows from facts about Galois extensions, extending the proof of [vdB] for smash products. A version of 9.4.9 when H is infinite-dimensional has been proved by Koppinen [Ko 92]. Here we prove the finite-dimensional case, as it is an application of facts from Chapters 7 and 8.

9.4.17 COROLLARY [BM 89]. *Let H be finite-dimensional and let $A\#_\sigma H$ be a crossed product. Then $(A\#_\sigma H)\#H^* \cong M_n(A) \cong A \otimes M_n(\mathrm{k})$, where $n = \dim H$.*

PROOF. By 8.2.5, $A \hookrightarrow A\#_\sigma H$ is right H-Galois. Thus $(A\#_\sigma H)\#H^* \cong$

$\mathrm{End}((A\#_\sigma H)_A)$ by 8.3.3, 2a), replacing H by H^*. By 7.2.11 $A\#_\sigma H$ is a free right A-module of rank $n = \dim H$; thus $\mathrm{End}((A\#_\sigma H)_A) \cong M_n(A)$. □

Some other extensions of the duality theorem have been given by Beattie [Ba 92a, Ba 92b].

§9.5 Duality for graded algebras

In this last section we illustrate the duality theorem for the case of graded algebras, and give an application to describing the prime ideals of these algebras. We first consider the case of an algebra A graded by a finite group G. Thus $A = \oplus_{x\in G} A_x$, and A is a $(\mathbf{k}G)^*$-module algebra by 4.1.7. In this case 9.4.4 and 9.4.9, or 9.4.17, give the following:

9.5.1 THEOREM [CM]. *Let A be graded by the finite group G, of order n. Then*

$$(A\#(\mathbf{k}G)^*)\#\mathbf{k}G \cong A \otimes M_n(\mathbf{k}) \cong M_n(A).$$

It is of interest to describe explicitly the image of $(A\#1)\#1$ in $M_n(A)$ given by the imbedding Φ of 9.4.9 (it is *not* $A \cdot 1$!). The proof in [CM] did not give this explicitly, and the proof in [BM], though more explicit, was more complicated than in 9.4.9. In fact for the case of graded rings, Φ reduces to the map given by Quinn [Q 85], as we shall see.

First, let $\{p_x \mid x \in G\}$ be the basis of $(\mathbf{k}G)^*$ as in 1.3.7. It follows that the standard left and right actions of G on $(\mathbf{k}G)^*$ are given by

$$y \rightharpoonup p_x = p_{xy^{-1}} \quad \text{and} \quad p_x \leftharpoonup y = p_{y^{-1}x},$$

for all $x, y \in G$. Thus the left action of G on $A\#(\mathbf{k}G)^*$ is given by

$$y \cdot (a\#p_x) = a\#(y \rightharpoonup p_x) = a\#p_{xy^{-1}}.$$

We describe the maps λ and ρ of 9.4.1. To do this, we identify $\mathrm{End}_{\mathbf{k}} H = \mathrm{End}_{\mathbf{k}}(\mathbf{k}G)^*$ with $M_n(\mathbf{k})$ by choosing a basis of "matrix units" $\{e_{x,y} \mid x, y \in G\}$. These act as follows on our basis $\{p_x\}$ of H:

$$e_{x,y} \cdot p_z = \delta_{y,z} p_x$$

for all $x, y, z \in G$.

9.5.2 LEMMA. *For $H = (\mathrm{k}G)^*, U = H^* = \mathrm{k}G$, and $\{p_x\}, \{e_{x,y}\}$ as above the maps $\lambda : H\#H^* \to M_n(\mathrm{k})$ and $\rho : H^*\#1 \to M_n(\mathrm{k})$ are given as follows:*
1) $\rho(y\#1)\cdot p_z = p_{y^{-1}z}$, and thus $\rho(y\#1) = \sum_z e_{y^{-1}z,z}$
2) $\lambda(p_x\#y)\cdot p_z = \delta_{x,zy^{-1}}p_x$, and thus $\lambda(\sum_x p_{y^{-1}x}\#x^{-1}yx) = \sum_z e_{y^{-1}z,z}$.
Thus the RL-condition is given explicitly by

$$\rho(y\#1) = \lambda(\sum_x p_{y^{-1}x}\#x^{-1}yx)$$

for all $x, y \in G$, and the map $\Phi' = (id \otimes \lambda)\circ\Phi : (A\#1)\#1 \to A \otimes M_n(\mathrm{k})$ is given by

$$\Phi' : (a_x\#1)\#1 \mapsto a_x \otimes (\sum_z e_{xz,z})$$

for each $a_x \in A_x$, where Φ is as in 9.4.9.

PROOF. 1) This follows from the definition of ρ, since $\rho(y\#1)\cdot p_z = p_z \leftharpoonup y = p_{y^{-1}z}$ from above. It is then clear that $\rho(y\#1) = \sum e_{y^{-1}x,x}$.

2) By the definition of λ, $\lambda(p_x\#y)p_z = p_x(y \rightharpoonup p_z) = p_xp_{zy^{-1}} = \delta_{x,zy^{-1}}p_x$. Then

$$\lambda(\sum_x p_{y^{-1}x}\#x^{-1}yx)p_z = \sum_x \delta_{y^{-1}x,zx^{-1}y^{-1}x}p_{y^{-1}x} = p_{y^{-1}z} = \rho(y\#1)p_z.$$

This verifies the RL-condition as well as 2).

Recall from 9.4.9 that $\Phi : (a_x\#1)\#1 \mapsto \sum a_0 \otimes w(a_1)(1\#1)$. In our case the comodule structure map $A \to A \otimes \mathrm{k}G$ is given by $a_x \mapsto a_x \otimes x$, so we must only find $w(x)$. Now

$$w(x) = \lambda^{-1}\circ\rho(\bar{S}x\#1) = \lambda^{-1}\circ\rho(x^{-1}\#1)$$

and thus

$$\Phi' = (id\otimes\lambda)\circ\Phi : (a_x\#1)\#1 \to a_x \otimes \rho(x^{-1}\#1) = a_x \otimes \sum_z e_{xz,z}.$$

□

We now turn to prime ideals of A; these results come from [CM]. Recall that I is a *graded ideal* of A if $I = \oplus_x I_x$, where $I_x = I \cap A_x$; more generally for any ideal I of A one has an *associated graded ideal* $I_G = \oplus_x I_x \subseteq I$. Equivalently, an ideal is graded if it is a $(\mathrm{k}G)^*$-submodule of A. A graded ideal I of A is *graded prime* if whenever $JK \subseteq I$, for J, K graded ideals of A, then either $J \subseteq I$ or $K \subseteq I$. In fact I is a graded prime ideal $\Leftrightarrow I = P_G$, the associated graded ideal of an ordinary prime ideal P of A (to see this, note that ($\Leftarrow$) is trivial, and the other direction follows by applying Zorn's lemma to the set $\mathcal{S}$ of ideals J of A with $J_G = I$).

We say that A is a *graded prime ring* if (0) is a graded prime ideal, and A is *graded semiprime* if A has no non-zero nilpotent graded ideals. A result of Cohen and Rowen [CR, 1.2] implies that if A is graded semiprime and I is an ideal of A with $I \cap A_1 = 0$, then $I_G = 0$. For, they show by a counting argument that if $I \cap A_1 = 0$, then $(I_G)^{|G|} = 0$; thus I_G would be a graded nilpotent ideal of A.

We first compare prime ideals of A and $A\#(\mathrm{k}G)^*$.

9.5.3 PROPOSITION [CM]. *Let A be graded by the finite group G, and consider the G-action on $A\#(\mathrm{k}G)^*$ as described before 9.5.2.*

1) *For any ideal $\mathcal{I}$ of $A\#(\mathrm{k}G)^*, I = \mathcal{I} \cap A$ is a graded ideal of A.*

2) *If also $\mathcal{I}$ is G-stable, then $\mathcal{I} = (\mathcal{I} \cap A)\#(\mathrm{k}G)^*$.*

3) *If A is graded semiprime, then $A\#(\mathrm{k}G)^*$ is semiprime.*

4) *If P is a prime ideal of A, then there exists a prime $\mathcal{P}$ of $A\#(\mathrm{k}G)^*$ such that $\mathcal{P} \cap A = P_G$. $\mathcal{P}$ is unique up to its G-orbit $\{\mathcal{P}^x \mid x \in G\}$, and $P_G\#(\mathrm{k}G)^* = \cap_x \mathcal{P}^x$.*

5) *If $\mathcal{P}$ is any prime ideal of $A\#(\mathrm{k}G)^*$, then $\mathcal{P} \cap A = P_G$ for some prime P of A.*

PROOF. 1) In any smash product $A\#H$, it is true that the intersection of an ideal of $A\#H$ with A is H-stable, since for all $a \in A, h \in H$,

$$h \cdot a = \sum h_1 a S h_2.$$

2) This is a special case of 8.3.11; however we give a simple direct proof. Choose $w = \sum_i a_i p_{x_i} \in \mathcal{I}$; it suffices to show $a_i \in \mathcal{I} \cap A$, for all i. Now $w p_{x_j} = a_j p_{x_j} \in \mathcal{I}$ for each j, and so $\sum_{y \in G} y \cdot (a_j p_{x_j}) = a_j (\sum_y p_{x_j y^{-1}}) = a_j \in \mathcal{I}$, since $\mathcal{I}$ is G-stable.

3) Let $\mathcal{I}$ be a nilpotent ideal of $A\#(\mathrm{k}G)^*$; then $\mathcal{J} = \sum_x \mathcal{I}^x$ is a G-stable nilpotent ideal. Thus $\mathcal{J} \cap A$ is a graded nilpotent ideal of A, and so $\mathcal{J} \cap A = 0$. But by 2), $\mathcal{J} = (\mathcal{J} \cap A)\#(\mathrm{k}G)^* = 0$, and so $\mathcal{I} = 0$. Thus $A\#(\mathrm{k}G)^*$ is semiprime.

4) Since P is prime, P_G is a graded prime; it follows that $P_G\#(\mathrm{k}G)^*$ is a G-prime ideal of $A\#(\mathrm{k}G)^*$. For if $\mathcal{A}, \mathcal{B}$ are G-stable ideals of $A\#(\mathrm{k}G)^*$ with $\mathcal{AB} \subseteq P_G\#(\mathrm{k}G)^*$, then $(\mathcal{A} \cap A)(\mathcal{B} \cap A) \subseteq P_G$, and thus either $\mathcal{A} \cap A \subseteq P_G$ or $\mathcal{B} \cap A \subseteq P_G$. Assume $\mathcal{A} \cap A \subseteq P_G$; by 2) we see

$$\mathcal{A} = (\mathcal{A} \cap A)\#(\mathrm{k}G)^* \subseteq P_G\#(\mathrm{k}G)^*.$$

Thus $P_G\#(\mathrm{k}G)^*$ is G-prime.

Thus there exists a prime ideal $\mathcal{P}$ of $A\#(\mathrm{k}G)^*$ such that $P_G\#(\mathrm{k}G)^* = \cap_x \mathcal{P}^x$ (this is a standard fact about finite group actions, using Zorn's lemma). Moreover it is easy to see that $\mathcal{P}$ is unique up to its G-orbit.

5) $\mathcal{P} \cap A$ is a graded ideal of A, and it is also a graded prime : if $IJ \subseteq \mathcal{P} \cap A$, for I, J graded ideals of A, then $(I\#(\mathrm{k}G)^*)(J\#(\mathrm{k}G)^*) \subseteq \mathcal{P}$, and so, say, $I\#(\mathrm{k}G)^* \subseteq \mathcal{P}$. Then $I \subseteq \mathcal{P}$. But now since $\mathcal{P} \cap A$ is graded prime, it equals P_G for some prime P of A. □

We can now give an application of the duality theorem. The idea is to use known results about group actions to obtain results about graded rings, since it is easier to "go up" through a group action than "go down" through a grading.

9.5.4 THEOREM[CM]. *Let A be graded by the finite group G, and let Q be a graded prime ideal of A. Then*

1) *a prime ideal P of A is minimal over $Q \Leftrightarrow P_G = Q$,*

2) *there are finitely many such minimal primes, say $P_1, \ldots, P_m$, and $m \leq |G|$,*

3) *if $I = P_1 \cap \cdots \cap P_m$, then $I^{|G|} \subseteq Q$, and $I = Q$ if A/Q has no $|G|$-torsion.*

PROOF. By passing to A/Q we may assume that A is a graded prime ring and that $Q = 0$. Let $R = A\#(\mathrm{k}G)^*$; by 9.5.3, 4) we have $0 = \cap_x \mathcal{P}^x$ for some prime $\mathcal{P}$ of R and thus R is a G-prime ring. We now apply a theorem of Lorenz and Passman [P, 16.2] to see the following:

1) a prime ideal P' of $R\#\mathrm{k}G$ is minimal $\Leftrightarrow P' \cap R = 0$,

2) there are $m \leq |G|$ such minimal primes, call them $P'_1, \ldots, P'_m$, and

3) $I' = P'_1 \cap \cdots \cap P'_m$ is the unique largest nilpotent ideal of $R\#\mathrm{k}G$ and $I'^{|G|} = 0$.

The duality theorem 9.5.1 gives $R\#\mathrm{k}G \cong M_n(A)$; since for any ideal J' of $M_n(A)$, we know $J' = M_n(J)$ for $J = J' \cap A \cdot 1$, there exists a one-to-one correspondence $f : J' \to J$, preserving inclusions and intersections, between ideals of $(A\#(kG)^*)\#kG$ and ideals of A. Thus there are $m \le |G|$ minimal primes of A, say $P_1, \dots, P_m$, and if $I = P_1 \cap \dots \cap P_m$ then $I^{|G|} = 0$. If A has no $|G|$-torsion, then R has no $|G|$-torsion, and is semiprime by 9.5.3, 3). Thus $M_n(A) \cong R\#kG$ is semiprime by [FM]. Thus A is semiprime, and so $I = 0$. Thus 2) and 3) are proved.

To show 1), we use the imbedding of A in $A \otimes M_n(k)$ as given in 9.5.2. By 9.5.2, 2), we have $\lambda(p_x\#1) \cdot p_z = \delta_{x,z}p_z$ and thus $\lambda(p_x\#1) = e_{x,x}$. Thus $\Phi'((a_x\#p_y)\#1) = a_x \otimes (\sum_z e_{xz,z})e_{y,y} = a_x e_{xy,y}$. It follows that $\Phi'(R\#1)$ will be a sum of such terms, and thus that $f(\Phi'(R\#1)) = [\Phi'(R\#1)] \cap A \cdot 1 = A_1 \cdot 1$, since $1 = \sum e_{y,y}$ in $M_n(A)$.

For simplicity, we now identify $R\#kG$ with $M_n(A)$. Then:

P is a minimal prime in A

$\Leftrightarrow$ $M_n(P) = P'$ is a minimal prime in $M_n(A) = R\#kG$

$\Leftrightarrow$ $P' \cap (R\#1) = 0$

$\Leftrightarrow$ $f(P') \cap f(R\#1) = 0$

$\Leftrightarrow$ $P \cap A_1 = 0$.

Since A is graded prime, $P \cap A_1 = 0$ implies that $P_G = 0$ by the result of [CR] mentioned before 9.5.3. □

In [CM] and in [MS 86], 9.5.4 is applied to study the relationship between primes of A and primes of A_1. We mention one such result, from [CM]:

9.5.5 COROLLARY (*Incomparability*). *Let A be graded by the finite group G. If $P \subseteq Q$ are prime ideals of A with $P \neq Q$, then $P \cap A_1 \neq Q \cap A_1$.*

PROOF. By passing to A/P_G, we may assume A is graded prime and $P_G = 0$. But Q is not a minimal prime, and so $Q \cap A_1 \neq 0$ by the argument at the end of 9.5.4. □

We finish this Chapter by briefly considering the case of an algebra A graded by an infinite group G. In this case A is a kG-comodule algebra, and thus is a $(kG)^\circ$-module algebra (though we remind the reader that not all $(kG)^\circ$-module algebras are G-graded). As a speical case of 9.4.15, we have a necessary and sufficient condition for the RL-condition to be satisfied.

9.5.6 COROLLARY[BM 85]. *Let G be a residually* k*-linear group. Then*

$U = \mathrm{k}G$ satisfies the RL-condition with respect to $H = (\mathrm{k}G)^\circ \Leftrightarrow G$ is an FC-group; that is, every conjugacy class in G is finite.

This is clear, since $\mathrm{k}G$ is ad-locally-finite $\Leftrightarrow G$ is ad-locally-finite. The corollary is not surprising in view of the relationship between λ and ρ in 9.5.2; however the $\{p_x\}$ are not in $(\mathrm{k}G)^\circ$, in general.

9.5.7 COROLLARY. *Let G be a residually* k*-linear FC-group, and let A be a G-graded algebra. Then*

$$(A\#(\mathrm{k}G)^\circ)\#\mathrm{k}G \cong A \otimes ((\mathrm{k}G)^\circ \# \mathrm{k}G),$$

and moreover $(\mathrm{k}G)^\circ\#\mathrm{k}G$ is a simple dense ring of k*-linear transformations on $(\mathrm{k}G)^\circ$.*

PROOF. 9.4.4, 9.4.9, and 9.5.6. □

For example, any finitely-generated abelian group G satisfies the hypotheses of the corollary.

If the RL-condition does not hold, then we do not know when a decomposition such as that in 9.5.7 holds. However, for many applications 9.5.7 is stronger than is needed: it is not actually necessary to be able to separate off A as a tensor factor. The imbedding due to Quinn mentioned earlier in fact generalizes to an arbitrary group G; he imbeds A into $M_G(A)$, the row-and column-finite matrices over A with rows and columns indexed by G. As in 9.5.2, p_x can be identified with $e_{x,x} \in M_G(A)$ and $y \in G$ can be identified with $\sum_x e_{x,xy}$; the "infinite sum" is a well-defined element of $M_G(A)$. Finally for any $a \in A$, the map $a \mapsto \sum_{x,z} a_x e_{xz,z}$ gives an isomorphic copy of A in $M_G(A)$. Using this imbedding he proves a theorem giving necessary and sufficient conditions for A to be a prime ring [Q 85].

The imbedding of A into $M_G(A)$ is used to study chains of prime ideals when G is polycyclic-by-finite in [ChQ]. A slightly different construction, although with a similar flavor, has been used by Beattie to study the graded Jacobson radical of A, for an arbitrary group G [Ba 88]. Recently a coalgebra version of duality has been given in [DNSvO].

Chapter 10

New Constructions from Quantum Groups

Quantum groups were introduced independently in work of Jimbo and Drinfeld, who defined the quantum enveloping algebras $U_q(\mathbf{g})$, and by Woronowicz, who defined the quantum group corresponding to the functions on $SU(2)$ as well as an analog of $U_q(s\ell(2))$. Although Drinfeld in [Dr 86] defines a quantum group as any non-commutative, non-cocommutative Hopf algebra, in fact most of the work in the area has focused on the two basic examples above, that is, $U_q(\mathbf{g})$ for $\mathbf{g}$ a semisimple Lie algebra, and the quantum coordinate rings $\mathcal{O}_q(G)$ for G an affine algebraic group (we remind the reader that definitions of these algebras are in the Appendix). Nevertheless, a number of new definitions and constructions can be described in purely Hopf-algebraic terms, and it is these ideas which we shall discuss here.

§10.1 Quasitriangular and almost cocommutative Hopf algebras

We wish to generalize the notion of cocommutative Hopf algebras, while still keeping many of their nice properties.

10.1.1 DEFINITION [Dr 89a]. A Hopf algebra H is *almost cocommutative* if the antipode S of H is bijective and if there exists an invertible element $R \in H \otimes H$ such that for all $h \in H$,

$$\tau(\Delta(h)) = R\Delta(h)R^{-1},$$

where τ is the usual twist map.

In summation notation, this says that $\sum h_2 \otimes h_1 = R(\sum h_1 \otimes h_2)R^{-1}$.

Our first lemma generalizes a basic fact about cocommutative Hopf algebras.

10.1.2 LEMMA. *Let H be an almost cocommutative Hopf algebra, and let V, W be left H-modules. Then $V \otimes W \cong W \otimes V$ as left H-modules.*

PROOF. Define $\phi : V \otimes W \to W \otimes V$ via $v \otimes w \mapsto R^{-1}(w \otimes v)$. Then for all $h \in H$,

$$\begin{aligned} \phi(h \cdot (v \otimes w)) &= R^{-1}(\sum h_2 w \otimes h_1 v) &= R^{-1}\tau(\Delta h)(w \otimes v) \\ &= \Delta(h) R^{-1}(w \otimes v) &= h \cdot \phi(v \otimes w). \end{aligned}$$

□

10.1.3 EXAMPLE. Any cocommutative Hopf algebra H is almost cocommutative: simply let $R = 1 \otimes 1$. In fact, however, such an H can also be almost cocommutative for a non-trivial R, as we will see in 10.1.16. On the other hand, we have seen after 1.8.1 that if $H = (\mathbf{k}G)^*$ for G a finite non-abelian group, then there exist V, W such that $V \otimes W \not\cong W \otimes V$. Thus such an H cannot be almost cocommutative, for any R.

Recall from 1.5.12 that if H is cocommutative, then $S^2 = id$. This fact can also be generalized:

10.1.4 PROPOSITION [Dr 89a]. *Let H be almost cocommutative. Then S^2 is an inner automorphism of H.*

More precisely, write $R = \sum_i a_i \otimes b_i$ and set $u = \sum_i (Sb_i)a_i$. Then u is invertible in H, and for all $h \in H$,

$$S^2 h = uhu^{-1} = (Su)^{-1} h (Su).$$

Consequently $u(Su)$ is a central element of H.

PROOF. We first show that for any $h \in H$,

$$(*) \qquad uh = (S^2 h)u.$$

By 10.1.1, $(R \otimes 1)(\sum h_1 \otimes h_2 \otimes h_3) = (\sum h_2 \otimes h_1 \otimes h_3)(R \otimes 1)$; that is

$$\sum_i a_i h_1 \otimes b_i h_2 \otimes h_3 = \sum_i h_2 a_i \otimes h_1 b_i \otimes h_3.$$

Thus

$$\begin{aligned}
\sum_i (S^2h_3)S(b_ih_2)a_ih_1 &= \sum_i (S^2h_3)S(h_1b_i)h_2a_i \\
\sum_i S(h_2Sh_3)(Sb_i)a_ih_1 &= \sum_i (S^2h_3)(Sb_i)(Sh_1)h_2a_i \\
\sum_i (Sb_i)a_ih &= \sum_i (S^2h)(Sb_i)a_i \\
uh &= (S^2h)u.
\end{aligned}$$

We next show that u is invertible. Write $R^{-1} = \sum_j c_j \otimes d_j$ and set $v = \sum_j (S^{-1}d_j)c_j$. Then, using $(*)$,

$$\begin{aligned}
uv &= \sum_j u(S^{-1}d_j)c_j = \sum_j (Sd_j)uc_j \\
&= \sum_j (Sd_j)(\sum_i (Sb_i)a_i)c_j \\
&= \sum_{j,i} S(b_id_j)a_ic_j.
\end{aligned}$$

Since $\sum_{i,j} a_ic_j \otimes b_id_j = RR^{-1} = 1 \otimes 1$, it follows that $uv = 1$. Now apply $(*)$ with $h = v$ to see that $(S^2v)u = 1$. Thus u is invertible, and so $S^2h = uhu^{-1}$. Applying S to this expression and replacing Sh by h gives the second formula for S^2. Now clearly $u(Su)$ is central in H. □

We come to a very important class of quantum groups.

10.1.5 DEFINITION [Dr 86]. A Hopf algebra H is *quasitriangular* (QT) if H is almost cocommutative, via $R = \sum_i a_i \otimes b_i \in H \otimes H$ as in 10.1.1, and

(10.1.6) $$(\Delta \otimes id)R = R^{13}R^{23}$$

(10.1.7) $$(id \otimes \Delta)R = R^{13}R^{12},$$

where $R^{12} = \sum_i a_i \otimes b_i \otimes 1$, $R^{23} = \sum_i 1 \otimes a_i \otimes b_i$, and $R^{13} = \sum_i a_i \otimes 1 \otimes b_i$. H is *triangular* if also $R^{-1} = \tau(R)$.

10.1.8 PROPOSITION [Dr 89a]. *If (H, R) is quasitriangular, then the following additional properties hold:*

(10.1.9) $$R^{12}R^{13}R^{23} = R^{23}R^{13}R^{12}$$

(10.1.10) $$(S\otimes id)R = R^{-1} = (id\otimes S^{-1})R, \text{ and } (S\otimes S)R = R$$

(10.1.11) $$(\varepsilon\otimes id)R = 1 = (id\otimes\varepsilon)R.$$

PROOF. First, $R^{12}R^{13}R^{23} = R^{12}(\Delta\otimes id)(R) = (\tau\circ\Delta\otimes id)(R)R^{12} = R^{23}R^{13}R^{12}$. Now apply $\varepsilon\otimes id^2$ to 10.1.6 to see $(\varepsilon\otimes id)(R) = 1$, and $id^2\otimes\varepsilon$ to 10.1.7 to see $(id\otimes\varepsilon)(R) = 1$. Thus we have shown 10.1.9 and 10.1.11. For 10.1.10,

$$\begin{aligned} R(S\otimes id)(R) &= (m\otimes id)(id\otimes S\otimes id)(R^{13}R^{23}) \\ &= (m\otimes id)(id\otimes S\otimes id)(\Delta\otimes id)(R) \\ &= (\varepsilon\otimes id)(R) = 1 \end{aligned}$$

and thus $(S\otimes id)(R) = R^{-1}$. Now (H^{cop}, R^{21}) is a quasitriangular Hopf algebra with antipode S^{-1}. Thus $(S^{-1}\otimes id)(R^{21}) = (R^{21})^{-1}$, or equivalently $(id\otimes S^{-1})(R) = R^{-1}$. Finally $(S\otimes S)R = (id\otimes S)(S\otimes id)(R) = (id\otimes S)(R^{-1}) = R$. □

10.1.12 REMARK. Equation 10.1.9 is known as the *quantum Yang-Baxter equation* (QYBE) in statistical mechanics. The quantum enveloping algebras $U_q(s\ell(n))$ were introduced to construct matrix solutions to the QYBE.

Also, consider R acting on $H^{\otimes 2}$ by left multiplication and let $B = \tau\circ R$. Then 10.1.9 becomes the *braid relation*

$$B^{12}B^{23}B^{12} = B^{23}B^{12}B^{23}.$$

The connection between braids and quantum groups is discussed in [Bi].

For quasitriangular Hopf algebras, a much stronger property of the antipode holds. The following result appears in [Dr 89a], where it is credited to Lyubashenko.

10.1.13 THEOREM. *If (H,R) is quasitriangular, then for all $h\in H$,*

$$S^4h = ghg^{-1},$$

where $g = u(Su)^{-1}$ is a group-like element of H.

PROOF. We first prove that, for $u = \sum(Sb_i)a_i$ as in 10.1.4,

(*) $$\Delta u = (R^{21}R)^{-1}(u\otimes u) = (u\otimes u)(R^{21}R)^{-1}.$$

Using 10.1.1, $R^{21}R(\Delta u) = (\Delta u)R^{21}R$, and thus it suffices to show that $(\Delta u)R^{21}R = u \otimes u$. Write $\Delta' = \tau \circ \Delta$. Then

$$\Delta u = \sum_i \Delta(Sb_i)\Delta a_i = \sum_i [(S \otimes S)(\Delta' b_i)]\Delta a_i$$

and so by 10.1.1

$$(\Delta u)R^{21}R = \sum_i [(S \otimes S)(\Delta' b_i)]R^{21}R(\Delta a_i).$$

Now make $H \otimes H = H^{\otimes 2}$ into a right $H^{\otimes 4}$-module by defining

$$(h \otimes k) \bullet (x \otimes y \otimes z \otimes w) = (Sz)hx \otimes (Sw)ky.$$

Then

$$(\Delta u)R^{21}R = \sum (Sb_{i_2})b_k a_j a_{i_1} \otimes (Sb_{i_1})a_k b_j a_{i_2} = R^{21} \bullet [R^{12}(\Delta \otimes \Delta')(R)].$$

Using 10.1.6, 10.1.7, and 10.1.9,

$$R^{12}(\Delta \otimes \Delta')(R) = R^{12}R^{13}R^{23}R^{14}R^{24} = R^{23}R^{13}R^{12}R^{14}R^{24}.$$

Also

$$\begin{aligned} R^{21} \bullet R^{23} &= \sum_{i,j} (Sb_j)b_i \otimes a_i a_j \\ &= (S \otimes id)[(\sum_i S^{-1}b_i \otimes a_i)(\sum_j b_j \otimes a_j)] = 1, \end{aligned}$$

since $R^{-1} = (id \otimes S^{-1})(R) = \sum_i a_i \otimes S^{-1}b_i$. Thus

$$R^{21} \bullet (R^{23}R^{13}) = 1 \bullet R^{13} = u \otimes 1,$$

and

$$R^{21} \bullet (R^{23}R^{13}R^{12}R^{14}) = (u \otimes 1) \bullet R^{12}R^{14} = \sum_{i,j} ua_i a_j \otimes (Sb_j)b_i = u \otimes 1.$$

Finally, $(u \otimes 1) \bullet R^{24} = u \otimes u$, proving $(*)$. Now

$$\begin{aligned} \Delta(Su) &= (S \otimes S)\Delta' u = (S \otimes S)[\tau(R^{21}R)^{-1}(u \otimes u)] \quad \text{by } (*) \\ &= (S \otimes S)(u \otimes u)(S \otimes S)(RR^{21})^{-1} \\ &= (Su \otimes Su)(R^{21}R)^{-1} \quad \text{by 10.1.10.} \end{aligned}$$

Thus

$$\Delta g = \Delta u \Delta(Su)^{-1} = (u \otimes u)(R^{21}R)^{-1}R^{21}R(Su \otimes Su)^{-1} = g \otimes g. \qquad \square$$

We have already seen that for any finite-dimensional Hopf algebra H, S^4 has a special form; recall from §2.5 Radford's formula $S^{-4} = ad\,a \circ (ad\,\alpha)^*$, where $\alpha \in H^*$ is the distinguished group-like element of H^* and $a \in H$ is the distinguished group-like element of H. A connection between this formula and 10.1.13 is the following:

10.1.14 PROPOSITION [Dr 89a]. *Let (H, R) be a finite-dimensional quasitriangular Hopf algebra, $a \in H$ and $\alpha \in H^*$ the distinguished group-like elements, and $g = u(Su)^{-1} = (Su)^{-1}u$ as in 10.1.13. Set $\tilde{\alpha} = (\alpha \otimes id)R \in H$. Then*

$$g = a^{-1}\tilde{\alpha} = \tilde{\alpha}a^{-1}.$$

See also [R 92].

It is possible for a Hopf algebra to be quasitriangular for more than one R. Thus we make the following definition:

10.1.15 DEFINITION. 1) If (H, R) and (H', R') are quasitriangular, then they are *isomorphic as quasitriangular Hopf algebras* if there exists a Hopf algebra isomorphism $f : H \to H'$ such that $R' = (f \otimes f)(R)$.
2) Two quasitriangular structures R, R' on a Hopf algebra H are *equivalent* if $(H, R) \cong (H, R')$ as quasitriangular Hopf algebras.

10.1.16 EXAMPLE. Again, any cocommutative Hopf algebra H is quasitriangular by setting $R = 1 \otimes 1$. However H may also have non-trivial quasitriangular structures. For example if $H = \mathrm{k}\mathbf{Z}_2$ and char $\mathrm{k} \neq 2$, then a non-trivial R is given by

$$R = \tfrac{1}{2}(1 \otimes 1 + 1 \otimes g + g \otimes 1 - g \otimes g),$$

where we have written $\mathbf{Z}_2$ multiplicatively as $\{1, g\}$. We will see in 10.2.7 that this R is unique.

10.1.17 EXAMPLE. Let $H = H_4$, with relations as in 1.5.6, and assume char $\mathrm{k} \neq 2$. Then H has a one-parameter family of quasitriangular structures, as shown in [R 93a]: for each $\alpha \in \mathbf{k}$, define

$$R_\alpha = \frac{1}{2}(1 \otimes 1 + 1 \otimes g + g \otimes 1 - g \otimes g) + \frac{\alpha}{2}(x \otimes x - x \otimes gx + gx \otimes x + gx \otimes gx)$$

(the reader is warned that the notation in [R 93a] is slightly different; the x there is our xg). If $\alpha \neq 0$, then Radford shows that (H, R_α) is not equivalent to (H, R_0); however if $\mathbf{k}$ is algebraically closed, $(H, R_\alpha) \cong (H, R_1)$ for all $\alpha \neq 0$. If $\mathbf{k}$ is not algebraically closed, the (H, R_α) may fall into infinitely many isomorphism classes; in fact this happens when $\mathbf{k} = \mathbf{Q}$.

10.1.18 REMARK. As noted in 10.1.3, $H = (\mathbf{k}G)^*$ for G finite and non-abelian can not be quasitriangular. However we will see in 10.3.7 that any finite-dimensional H can be imbedded in a quasitriangular Hopf algebra, namely its Drinfeld double $D(H)$.

10.1.19 EXAMPLE. Let $H = U_q(\mathbf{g})$, for $\mathbf{g}$ a finite-dimensional semisimple Lie algebra over $\mathbf{k} = \mathbf{C}$; this is in fact the motivating example for quasitriangular Hopf algebras, see [Dr 86]. However H is not QT in the strict sense of 10.1.5, since R lies in a completion of $H \otimes H$ rather than in $H \otimes H$ itself; [Dr 86, p. 817] gives an expression for R as a power series in tensor products of the generators of $U_q(\mathbf{g})$. Nevertheless, for any finite-dimensional representation $\rho : U_q(\mathbf{g}) \to \operatorname{End}_{\mathbf{k}} V$, the image of R, that is $R_\rho = (\rho \otimes \rho)R \in M_n(\mathbf{k}) \otimes M_n(\mathbf{k})$, will provide a matrix solution to the QYBE. From an algebraic point of view, these Hopf algebras might be called "topologically" quasitriangular.

§10.2 Coquasitriangular and almost commutative Hopf algebras

We now consider the formal duals of the Hopf algebras studied in §10.1. The advantage of the dual point of view is that the topological difficulties mentioned above do not arise.

10.2.1 DEFINITION. A Hopf algebra H is called *coquasitriangular* (CQT) if there exists a bilinear form $\langle\ |\ \rangle : H \otimes H \to \mathbf{k}$, which is convolution invertible in $\operatorname{Hom}_{\mathbf{k}}(H \otimes H, \mathbf{k})$, such that for all $h, k, \ell \in H$,

$$\sum \langle h_1 \mid k_1 \rangle k_2 h_2 = \sum h_1 k_1 \langle h_2 \mid k_2 \rangle \tag{10.2.2}$$

$$\langle h \mid k\ell \rangle = \sum \langle h_1 \mid k \rangle \langle h_2 \mid \ell \rangle \tag{10.2.3}$$

$(10.2.4)$ $\langle hk \mid \ell \rangle = \sum \langle h \mid \ell_2 \rangle \langle k \mid \ell_1 \rangle.$

H is called *almost commutative* if only (10.2.2) is satisfied.

When H is finite-dimensional, one may check that H is CQT $\Leftrightarrow H^*$ is QT. For, given a form $\langle \mid \rangle$ on H as in 10.2.1, define $R \in H^* \otimes H^* \cong (H \otimes H)^*$ by $R(h \otimes k) = \langle k \mid h \rangle$; conversely, given $R = \sum_i a_i \otimes b_i \in H^* \otimes H^*$, define $\langle \mid \rangle : H \otimes H \to \mathbf{k}$ via $\langle h \mid k \rangle = \sum_i a_i(k) b_i(h)$. Note that there is a twist in these definitions, so that the form $\langle \mid \rangle$ is really the dual of $B = \tau \circ R$, as in 10.1.12.

CQT Hopf algebras have been studied by a number of people. They were suggested in an announcement by [Ly] on "vector symmetries", and mentioned by Majid in [Mj 91] where they are called dual quasitriangular. They are studied in detail by Larson and Towber in [LT], where they are called braided Hopf algebras, and by Schauenburg in [Sb], where they are called coquasitriangular; this is also the name used by Hayashi [Ha]. Coquasitriangular seems to us the most appropriate, as it is consistent with other Hopf terminology.

10.2.5 EXAMPLE. Let H be any commutative Hopf algebra. Then H is CQT by setting $\langle h \mid k \rangle = \varepsilon(h)\varepsilon(k)$, the trivial braiding.

10.2.6 EXAMPLE. Let $H = \mathbf{k}G$ be a group algebra. Since H is cocommutative and $\langle \mid \rangle$ is invertible, the three conditions in 10.2.1 become:

$(10.2.2)'$ $gh = hg$

$(10.2.3)'$ $\langle h \mid g\ell \rangle = \langle h \mid g \rangle \langle h \mid \ell \rangle$

$(10.2.4)'$ $\langle hg \mid \ell \rangle = \langle h \mid \ell \rangle \langle g \mid \ell \rangle$

for all $h, g, \ell \in G$. That is, G is abelian and the form $\langle \mid \rangle$ is a bicharacter on G.

In general the bicharacters on G correspond to homomorphisms of G to its character group $\hat{G}$, for if $\langle \mid \rangle$ is a bicharacter then $\phi_{\langle \mid \rangle} : G \to \hat{G}$ given by $\phi_{\langle \mid \rangle}(g)(h) = \langle h \mid g \rangle$ is a group morphism, and conversely. Thus the bicharacters are essentially known.

10.2.7 EXAMPLE. When G is a finite abelian group and $\mathbf{k}$ contains a primitive n^{th} root of unity, for $n = |G|$, we know $G \cong \hat{G}$ and thus $(\mathbf{k}G)^* = \mathbf{k}\hat{G}$. We may therefore determine the quasitriangular structures on $\mathbf{k}G$ by instead determining the bicharacters on $\mathbf{k}\hat{G}$. For example, consider $\mathbf{k}\mathbf{Z}_2$ as in 10.1.15, and write $\hat{G} = \{\varepsilon, \gamma\}$, where ε is the trivial character and $\gamma(1) = 1, \gamma(g) = -1$. Then there is only one non-trivial bicharacter on $\hat{G}$, namely $\langle \varepsilon \mid \varepsilon \rangle = \langle \varepsilon \mid \gamma \rangle = \langle \gamma \mid \varepsilon \rangle = 1$ and $\langle \gamma \mid \gamma \rangle = -1$. Write $R = \alpha_1(1 \otimes 1) + \alpha_2(1 \otimes g) + \alpha_3(g \otimes 1) + \alpha_4(g \otimes g) \in \mathbf{k}G \otimes \mathbf{k}G$. Evaluating as in the discussion after 10.2.4, we see that

$$\begin{aligned}
R(\varepsilon \otimes \varepsilon) &= \alpha_1 + \alpha_2 + \alpha_3 + \alpha_4 = 1 \\
R(\varepsilon \otimes \gamma) &= \alpha_1 - \alpha_2 + \alpha_3 - \alpha_4 = 1 \\
R(\gamma \otimes \varepsilon) &= \alpha_1 + \alpha_2 - \alpha_3 - \alpha_4 = 1 \\
R(\gamma \otimes \gamma) &= \alpha_1 - \alpha_2 - \alpha_3 + \alpha_4 = -1.
\end{aligned}$$

Solving these equations for the α_i, we see that $\alpha_1 = \alpha_2 = \alpha_3 = \frac{1}{2}$ and that $\alpha_4 = -\frac{1}{2}$; thus R is exactly as described in 10.1.15.

10.2.8 EXAMPLE. If $H = \mathbf{k}G$, for G a non-abelian group, then H can not be almost commutative, and so can not be coquasitriangular, as noted in 10.2.6. However we will see in §10.3 that any finite-dimensional Hopf algebra H is a quotient of a coquasitriangular Hopf algebra, namely the dual of its Drinfeld double.

10.2.9 EXAMPLE. H_4 is CQT since it is QT and self-dual. Its CQT structures are dual to those given in 10.1.16.

10.2.10 EXAMPLE. Probably the most important examples of CQT Hopf algebras are the quantum groups $\mathcal{O}_q(GL_n(\mathbf{k}))$ and $\mathcal{O}_q(SL_n(\mathbf{k}))$, and more generally the algebras $A(R)$ as constructed in [FRT]. Larson and Towber define the bilinear form $\langle \mid \rangle$ explicitly on these algebras. For, we have $A(R) = \mathcal{O}_R(M_n(\mathbf{k})) = \mathbf{k}\langle x_{ij} \mid 1 \leq i,j \leq n \rangle / I_R$, where the ideal of relations I_R is defined as in A.6; we note that here R is the "R-matrix" $[R^{k\ell}_{ij}] \in M_{n^2}(\mathbf{k})$ and *not* the general R in 10.1.1. Then $\langle \mid \rangle : A(R) \otimes A(R) \to \mathbf{k}$ may be defined by $\langle x_{ij} \mid x_{k\ell} \rangle = R^{\ell j}_{ik}$ [LT, 5.3].

We record a few general properties of CQT Hopf algebras. First, just as we are interested in the modules for QT Hopf algebras, it is the comodules which are important in the dual situation.

10.2.11 LEMMA. *Let H be an almost commutative Hopf algebra and let V, W be right H-comodules. Then $V \otimes W \cong W \otimes V$ as right comodules.*

PROOF. Although this is the formal dual of 10.1.2, we write it out for practice in dealing with the form $\langle\ |\ \rangle$. Thus, we define $\phi : V \otimes W \to W \otimes V$ via $v \otimes w \mapsto \sum \langle w_1 \mid v_1 \rangle w_0 \otimes v_0$. One may check that

$$\rho_{W\otimes V} \circ \phi(v \otimes w) = \sum \langle w_2 \mid v_2 \rangle w_0 \otimes v_0 \otimes w_1 v_1$$

and that

$$(\phi \otimes id) \circ \rho_{V\otimes W}(v \otimes w) = \sum \langle w_1 \mid v_1 \rangle w_0 \otimes v_0 \otimes v_2 w_2.$$

Applying 10.2.2 we see that $\rho_{W\otimes V} \circ \phi = (\phi \otimes id) \circ \rho_{V\otimes W}$, and thus that ϕ is a right comodule map. It is an isomorphism since $\langle\ |\ \rangle$ is invertible. □

The dual analog of 10.1.4 is also true, though it does not require that S be bijective, but rather obtains it as a consequence.

10.2.12 PROPOSITION [Sb 92, 3.3.2]. *Let $(H, \langle\ |\ \rangle)$ be a coquasitriangular Hopf algebra. Then S^2 is a coinner automorphism.*

More precisely, set $u(h) = \sum \langle h_2 \mid S h_1 \rangle$. Then u is $*$-invertible in H^*, and $S^2 = u * id * u^{-1}$.

We leave the proof to the reader. We note that the same fact was also shown somewhat later by Doi [D 93].

§10.3 The Drinfeld double

The basic construction is due to Drinfeld [Dr 86]; however we follow the treatment of Majid [Mj 90a] as modified by Radford [R 93a]. Thus let H be any finite-dimensional Hopf algebra over k with antipode S, and let S^* denote the antipode of H^*. The composition inverses of S and S^* are denoted by $\bar{S}$ and $\overline{S^*}$, respectively. We need to define some new actions of H on H^*:

10.3.1 DEFINITION. Let H be any Hopf algebra with bijective antipode, and let $h \in H, f \in H^*$.

1) The *left coadjoint action* of H on H^* is given by

$$h \rightharpoonup f = \sum h_1 \rightharpoonup f \leftharpoonup \bar{S}h_2 \tag{10.3.2}$$

2) The *right coadjoint action* of H on H^* is given by

$$f \leftharpoonup h = \sum \bar{S}h_1 \rightharpoonup f \leftharpoonup h_2 \tag{10.3.3}$$

These actions are called coadjoint actions since they are the transposes of respectively, the left and right adjoint actions of H on itself, as in 3.4.2, although note we have replaced h by $\bar{S}h$ so the action is on the appropriate side. That is, one may check that

$$\langle h \rightharpoonup f, k\rangle = \langle f, (ad_\ell \bar{S}h)(k)\rangle \text{ and } \langle f \leftharpoonup h, k\rangle = \langle f, (ad_r \bar{S}h)(k)\rangle.$$

10.3.4 EXAMPLE. Let $H = \mathrm{k}G$ for G finite, so that $H^* = (\mathrm{k}G)^*$. As we have seen in §9.5, $y \rightharpoonup p_x = p_{xy^{-1}}$ and $p_x \leftharpoonup y = p_{y^{-1}x}$, for $y \in G$ and $\{p_x\}$ the dual basis in H^*. It follows that

$$y \rightharpoonup p_x = y \rightharpoonup p_x \leftharpoonup y^{-1} = p_{yxy^{-1}}.$$

Now when H is finite-dimensional, we consider the left coadjoint action of H on H^* and the right coadjoint action of H^* on H. These actions make H^{*cop} into a left H-module coalgebra and H into a right H^{*cop}-module coalgebra. That is,

$$\Delta'(h \rightharpoonup f) = \sum (h_1 \rightharpoonup f_2) \otimes (h_2 \rightharpoonup f_1)$$

and

$$\Delta(h \leftharpoonup f) = \sum (h_1 \leftharpoonup f_2) \otimes (h_2 \leftharpoonup f_1).$$

10.3.5 DEFINITION. The *Drinfeld double* $D(H) = H^{*cop} \bowtie H$ has $H^{*cop} \otimes H$ as its underlying vector space. Multiplication is given by

$$(f \bowtie h)(f' \bowtie h') = \sum f(h_1 \rightharpoonup f_2') \bowtie (h_2 \leftharpoonup f_1')h'$$

for all $f \in H^*, h \in H$, with identity element $1_{H^*} \bowtie 1_H = \varepsilon_H \bowtie 1_H$. $D(H)$ has the coalgebra structure of the usual tensor product of coalgebras; that is,

$$\Delta_{D(H)}(f \bowtie h) = \sum (f_2 \bowtie h_1) \otimes (f_1 \bowtie h_2) \text{ and } \varepsilon_{D(H)} = \varepsilon_{H^{*cop}} \otimes \varepsilon_H = 1_H \otimes \varepsilon_H.$$

The antipode of $D(H)$ is given by

$$\begin{aligned} S(f \bowtie h) &= \sum (Sh_2 \rightharpoonup Sf_1) \bowtie (f_2 \rightharpoonup Sh_1) \\ &= \sum (Sf_2 \leftharpoonup h_1) \bowtie (Sh_2 \leftharpoonup Sf_1) \end{aligned}$$

10.3.6 THEOREM [Dr 86]. *Let H be any finite-dimensional Hopf algebra. Then $D(H)$ is a quasi-triangular Hopf algebra.*

More precisely, let $\{h_i\}$ be a basis of H and $\{h_i^*\}$ the corresponding dual basis of H^*. Then we may set

$$R = \sum_i (\varepsilon_H \bowtie h_i) \otimes (h_i^* \bowtie 1_H) \in D(H) \otimes D(H).$$

The proof of 10.3.6 is sketched in [Dr 86]; see also [Mj 91]. In fact the element R is independent of the choice of basis. For, consider the identification $H \otimes H^* \cong \mathrm{End}_{\mathbf{k}}(H)$ via $(h \otimes f)(k) = \langle f, k \rangle h$; this is essentially the map λ' in the proof of 9.4.2. Let $C \in H \otimes H^*$ be the element corresponding to id_H. Then $C = \sum_i h_i \otimes h_i^*$, for any choice of h_i, and thus $R = \varepsilon \otimes C \otimes 1$.

Note that $H \cong \varepsilon_H \bowtie H$ and that $H^{*cop} \cong H^{*cop} \bowtie 1_H$. Thus we have:

10.3.7 COROLLARY. *Let H be finite-dimensional. Then H is a subHopf-algebra of a quasitriangular Hopf algebra, and a Hopf algebra quotient of a coquasitriangular Hopf algebra.*

PROOF. The first statement is obvious. For the second, we have $H^* \hookrightarrow D(H^*)$, and so $D(H^*)^* \to H^{**} \cong H$ is surjective. But $D(H^*)^*$ is coquasitriangular. □

10.3.8 EXAMPLE. Drinfeld [Dr 86] uses this construction to show that $U_h(\mathbf{g})$ is "topologically" quasitriangular, for $\mathbf{g}$ a simple Lie algebra of classical type.

Here U_h means that we are working over power series $\mathbf{k}[[h]]$, and not just over $\mathbf{k}$. He proceeds as follows: let $\mathbf{h}$ be the Cartan subalgebra of $\mathbf{g}$ and let $\mathbf{b}_\pm$ be Borel subalgebras. Set $H = U_h(\mathbf{b}_+)$; then $H^* \cong U_h(\mathbf{b}_-)$ using power series duals. Since $\mathbf{h} = \mathbf{b}_+ \cap \mathbf{b}_-$, it follows that $D(H) \cong U_h(\mathbf{g}) \otimes U_h(\mathbf{h})$, and so is quasitriangular via R_D (which actually lies in a closure of $D(H)$ in this case). We then have a projection

$$D(H) \otimes D(H) \to U_h(\mathbf{g}) \otimes U_h(\mathbf{g}).$$

Denote by R the image of R_D in the closure of $U_h(\mathbf{g}) \otimes U_h(\mathbf{g})$. Then this R makes $U_h(\mathbf{g})$ quasitriangular in the sense of 10.1.19.

When H is cocommutative, $D(H)$ takes a particularly simple form; it is just a smash product. To see this we first give another form of 10.3.2 and 10.3.3.

10.3.9 LEMMA. *Let H be any Hopf algebra with bijective antipode. Then for all $h \in H, f \in H^\circ$,*

1) $h \rightharpoonup f = \sum \langle f_3(\overline{S^*}f_1), h \rangle f_2$
2) $h \leftharpoonup f = \sum \langle f, (\bar{S}h_3)h_1 \rangle h_2$

PROOF. We show 2); 1) is similar. Now by 10.3.2, 1.6.4, and 1.6.5,

$$h \leftharpoonup f \ = \ \sum \langle \overline{S^*}f_1, h_2 \rangle h_1 \leftharpoonup f_2$$

$$= \ \sum \langle \overline{S^*}f_1, h_3 \rangle \langle f_2, h_1 \rangle h_2$$

$$= \ \sum \langle f_1, \bar{S}h_3 \rangle \langle f_2, h_1 \rangle h_2$$

$$= \ \sum \langle f, (\bar{S}h_3)h_1 \rangle h_2.$$

$\square$

An alternate proof of 10.3.9 could be given using 3.4.5. For, since $\leftharpoonup$ is the transpose of the right adjoint action of H on itself, followed by replacing h by $\bar{S}h$, it can also be obtained by dualizing the right adjoint coaction of H. To see this, apply 1.6.4 to the coaction in 3.4.5, 2); this gives a left action of H^* on H, via

$$f \cdot h = \sum \langle f, (Sh_1)h_3 \rangle h_2$$

for all $h \in H, f \in H^*$. Replacing f by $\overline{S^*}f$ gives a right action

$$h \bullet f = \sum \langle \overline{S^*}f, (Sh_1)h_3 \rangle h_2 = \sum \langle f, (\bar{S}h_3)h_1 \rangle h_2,$$

which is the desired formula for $h \leftharpoonup f$.

The next fact is noted in [CW], with a somewhat different proof.

10.3.10 COROLLARY. *Let H be finite-dimensional and cocommutative. Then H^* is a left H-module algebra via $\rightharpoonup$ and $D(H) \cong H^* \# H$ as algebras.*

PROOF. By 10.3.9, 2) $h \leftharpoonup f = \langle f, 1\rangle h = \varepsilon_{H^*}(f)h$. Thus H^* acts trivially on H, and it then follows that the multiplication in $D(H)$ is just $(f \bowtie h)(f' \bowtie h') = \sum f(h_1 \rightharpoonup f') \bowtie h_2 h'$. We now use 10.3.9, 1) to check that H^* is an H-module algebra:

$$\begin{aligned} h \rightharpoonup (fg) &= \sum \langle f_3 g_3 \overline{S^*} g_1)(\overline{S^*} f_1), h\rangle f_2 g_2 \\ &= \sum \langle f_3, h_1\rangle\langle g_3, h_2\rangle\langle \overline{S^*} g_1, h_3\rangle\langle \overline{S^*} f_1, h_4\rangle f_2 g_2 \\ &= \sum \langle f_3, h_1\rangle\langle \overline{S^*} f_1, h_2\rangle f_2 \langle g_3, h_2\rangle\langle \overline{S^*} g_1, h_4\rangle g_2 \\ &= \sum (h_1 \rightharpoonup f)(h_2 \rightharpoonup g). \end{aligned}$$

□

We note that although H^* is an H-module algebra in this situation the coadjoint action $\rightharpoonup$ gives a very different result from the standard dual action $\rightharpoonup$. For, under $\rightharpoonup$, $H^*\#H = \mathcal{H}(H)$ is a simple algebra by 9.4.3, and thus cannot be a Hopf algebra since Ker ε would be an ideal of codimension one. However, under $\rightharpoonup$, $H^*\#H = D(H)$ is a Hopf algebra.

We will consider the general question as to when a smash product is a Hopf algebra in §10.6.

Returning to $D(H)$ for an arbitrary H, we prove it is unimodular, a result of Radford. We first record several alternate forms of the multiplication in 10.3.5, from [R 93a]:

10.3.11 LEMMA. *Let H be a finite-dimensional Hopf algebra. Then multiplication in $D(H)$ can be given as follows, for $f \in H^*, h \in H$:*

1) $(f \bowtie h)(f' \bowtie h') = \sum f(h_1 \rightharpoonup f' \leftharpoonup \bar{S}h_3) \bowtie h_2 h'$

2) $(f \bowtie h)(f' \bowtie h') = \sum f f'_2 \bowtie (\overline{S^*} f'_1 \rightharpoonup h \leftharpoonup f'_3)h'$

PROOF. We use 10.3.9, and show 1); the argument for 2) is similar. Now

$$\begin{aligned}(f \bowtie h)(f' \bowtie h') &= \sum f(h_1 \rightharpoonup f'_2) \bowtie (h_2 \leftharpoonup f'_1)h' \\ &= \sum f(h_1 \rightharpoonup f'_2) \bowtie \langle f'_1, (\bar{S}h_4)h_2\rangle h_3 h' \\ &= \sum f(h_1 \rightharpoonup \langle f'_1, (\bar{S}h_4)h_2\rangle f'_2) \bowtie h_3 h' \\ &= \sum f(h_1 \rightharpoonup (f' \leftharpoonup (\bar{S}h_4)h_2)) \bowtie h_3 h' \\ &= \sum f(h_1 \rightharpoonup f' \leftharpoonup \bar{S}h_3) \bowtie h_2 h'.\end{aligned}$$

□

10.3.12 THEOREM [R 93a]. *Let H be a finite-dimensional Hopf algebra, and choose $0 \neq t \in \int_H^r$ and $0 \neq T \in \int_{H^*}^\ell$. Then $T \bowtie t$ is a left and right integral for $D(H)$. In particular $D(H)$ is unimodular.*

PROOF. We show that $T \bowtie t$ is a right integral for $D(H)$. To do this, we first need an identity about integrals. Thus let $a \in H$ be the distinguished group-like element of H, as in 2.2.3 applied to H^*. Then we claim that

$$(*) \qquad \sum (\bar{S}t_3)a^{-1}t_1 \otimes t_2 = 1 \otimes t.$$

First, choose $0 \neq t' \in \int_H^\ell$ such that $St' = t$; this can be done by 2.1.3. Then it is known that $\sum t'_2 \otimes t'_1 = \sum t'_1 \otimes (S^2 t'_2)a$, by [R 93c, Theorem 3]. Applying $(\Delta \circ S) \otimes id$ to this equation, using $St' = t$ and 1.5.10, 2), we obtain

$$\sum t_1 \otimes t_2 \otimes \bar{S}t_3 = \sum t_2 \otimes t_3 \otimes (St_1)a.$$

Thus

$$\sum (\bar{S}t_3)a^{-1}t_1 \otimes t_2 = \sum (St_1)t_2 \otimes t_3 = 1 \otimes t,$$

proving the claim.

Now choose $f \in H^*, h \in H$. Then by 10.3.11, 1), we have

$$\begin{aligned}
(T \bowtie t)(f \bowtie h) &= \sum T(t_1 \rightharpoonup f \leftharpoonup \bar{S}t_3) \bowtie t_2 h \\
&= \sum T\langle t_1 \rightharpoonup f \leftharpoonup \bar{S}t_3, a^{-1}\rangle \bowtie t_2 h \\
&\qquad \text{since } Tf = \langle f, a^{-1}\rangle T, \text{ all } f \in H^* \\
&= \sum T\langle f, (\bar{S}t_3)a^{-1}t_1\rangle \bowtie t_2 h \\
&\qquad \text{using properties of } \langle\ ,\ \rangle, \rightharpoonup, \text{ and } \leftharpoonup \\
&= T\langle f, 1\rangle \bowtie th \quad \text{by } (*) \\
&= \langle f, 1\rangle \varepsilon(h)(T \bowtie t).
\end{aligned}$$

Thus $T \bowtie t$ is a right integral for $D(H)$. The proof that it is a left integral is similar. □

10.3.13 COROLLARY. *Suppose that H is a finite-dimensional Hopf algebra. Then the following are equivalent:*
1) $D(H)$ is semisimple
2) H and H^* are semisimple
3) H and H^* are cosemisimple
4) $D(H)$ is cosemisimple.

PROOF. This follows from Maschke's theorem 2.2.1, 10.3.12, and the fact that H is semisimple $\Leftrightarrow$ H^* is cosemisimple. □

We now consider $D(H)^*$; it is an interesting object for the study of co-quasitriangular Hopf algebras.

10.3.14 PROPOSITION. *Let H be a finite-dimensional Hopf algebra. Then $D(H)^*$ is a coquasitriangular Hopf algebra, with structure as follows:*
1) as an algebra, $D(H)^* = H^{op} \otimes H^*$, the usual tensor product of algebras;
2) set $R^* = \sum_s h_s^* \otimes h_s \in H^* \otimes H^{op}$ where $\{h_s\}, \{h_s^*\}$ are dual bases of H

and H^* as in 10.3.6; then the coproduct in $D(H)^*$ is given by

$$\begin{aligned}\Delta(k \otimes f) &= \sum k_1 \otimes R^*(f_1 \otimes k_2)(R^*)^{-1} \otimes f_2 \\ &= \sum_{s,t} k_1 \otimes h_s^* f_1 h_t^* \otimes (\bar{S}h_t)k_2 h_s \otimes f_2\end{aligned}$$

for all $k \in H^{op}, f \in H^*$

3) *the antipode of $D(H)^*$ is given by*

$$S(k \otimes f) = \tau(R^*)^{-1}(Sk \otimes S^* f)\tau(R^*).$$

4) *the coquasitriangular structure map $\langle \mid \rangle : D(H)^* \otimes D(H)^* \to \mathrm{k}$ is given by*

$$\langle h \otimes f \mid k \otimes g \rangle = \varepsilon(k) f(1) \langle g, h \rangle.$$

PROOF. 1) is clear, since the coalgebra structure of $D(H)$ is just the tensor product of coalgebras.

2) This expression for the coproduct shows that $D(H)^*$ is a "double crossed product" in the sense of [RS]; note that the second expression for Δ follows from the first using 10.1.10 and the multiplication in H^{op}. We follow [Lu] to show that this formula is in fact the formal dual of the multiplication in $D(H)$. Thus, choose $h, k, \ell \in H^{op}$ and $x, y, z \in H^*$, and consider $x \bowtie h, y \bowtie k$ in $D(H)$ and $\ell \otimes z$ in $D(H)^*$. Then

$$\begin{aligned}\langle (x \bowtie h)(y \bowtie k), \ell \otimes z \rangle &= \langle x \otimes h \otimes y \otimes k, \Delta(\ell \otimes z) \rangle \\ &= \sum_{s,t} \langle x, \ell_1 \rangle \langle h_s^* z_1 h_t^*, h \rangle \langle y, \bar{S}h_t \ell_2 h_s \rangle \langle z_2, k \rangle \\ &= \sum \langle x, \ell_1 \rangle \langle h_s^*, h_1 \rangle \langle z_1, h_2 \rangle \langle h_t^*, h_3 \rangle \langle y_1, \bar{S}h_t \rangle \\ &\qquad \langle y_2, \ell_2 \rangle \langle y_3, h_s \rangle \langle z_2, k \rangle \\ &= \sum \langle xy_2, \ell \rangle \langle y_3, h_1 \rangle \langle z, h_2 k \rangle \langle \overline{S^*} y_1, h_3 \rangle \\ &= \sum \langle y_1, \bar{S}h_3 \rangle \langle x(h_1 \rightharpoonup y_2), \ell \rangle \langle z, h_2 k \rangle \\ &= \langle \sum x(h_1 \rightharpoonup y_2 \leftharpoonup \bar{S}h_3) \otimes h_2 k, \ell \otimes z \rangle.\end{aligned}$$

The expression on the left is now the formula for $(x \bowtie h) \cdot (y \bowtie k)$ in $D(H)$ given in 10.3.11. This proves 2).

3) This can be verified directly; it is also the dual of the antipode in $D(H)$.

4) For R as in 10.3.6, we have $\langle\ |\ \rangle = R \circ \tau$. Thus

$$\langle h \otimes f \mid k \otimes g \rangle = R(k \otimes g \otimes h \otimes f) = \sum_i \varepsilon(k)\langle g, h_i\rangle\langle h_i^*, h\rangle f(1) = \varepsilon(k) f(1)\langle g, h\rangle.$$

□

An alternate formula for Δ is given in [R 93a], where some results are proved about the structure of $D(H)^*$.

Here we consider some recent work of Lu, in which it is shown that $D(H)^*$ has a close relationship to the "Heisenberg double" $\mathcal{H}(H) = H \# H^*$, in which H^* acts on H via $\rightharpoonup$. To see this we first need a lemma.

10.3.15 LEMMA [D 93, Lu]. *Let $H, \langle\ |\ \rangle$ be any coquasitriangular Hopf algebra. Then $\sigma = \langle\ |\ \rangle \circ \tau : H \otimes H \to \mathbf{k}$ is a Hopf cocycle in the sense of 7.1.4, where H acts trivially on* $\mathbf{k}$.

PROOF. Since H acts trivially on $\mathbf{k}$, we must show

$$\sum \sigma(k_1, m_1)\sigma(h, k_2 m_2) = \sum \sigma(h_1, k_1)\sigma(h_2 k_2, m),$$

or

$$\sum \langle m_1 \mid k_1\rangle\langle k_2 m_2 \mid h\rangle = \sum \langle k_1 \mid h_1\rangle\langle m \mid h_2 k_2\rangle,$$

for all $h, k, m \in H$. Now

$$\begin{aligned}
\sum \langle m_1 \mid k_1\rangle\langle k_2 m_2 \mid h\rangle &= \sum \langle\langle m_1 \mid k_1\rangle k_2 m_2 \mid h\rangle \\
&= \sum \langle\langle m_2 \mid k_2\rangle m_1 k_1 \mid h\rangle \quad \text{by 10.2.2} \\
&= \sum \langle m_2 \mid k_2\rangle\langle m_1 \mid h_2\rangle\langle k_1 \mid h_1\rangle \quad \text{by 10.2.4} \\
&= \sum \langle m \mid h_2 k_2\rangle\langle k_1 \mid h_1\rangle \text{ by 10.2.3.}
\end{aligned}$$

□

It follows from the lemma that we may form the twisted product ${}_\sigma H = \mathbf{k}_\sigma[H]$ as in 7.1.5 for any coquasitriangular H. In particular this applies to

$D(H)^*$. We can describe σ explicitly in this case, using 10.3.14. That is,

$$\sigma(h\otimes f, k\otimes g) = \varepsilon(h)\langle g, 1\rangle\langle f, k\rangle. \tag{10.3.16}$$

10.3.17 THEOREM [Lu]. *Let H be any finite-dimensional Hopf algebra and let $\sigma : D(H)^*\otimes D(H)^* \to \mathrm{k}$ be as above. Then as algebras,*

$$_\sigma D(H)^* \cong \mathcal{H}(H) = H\#H^*$$

PROOF. Let $k\otimes f$ and $\ell\otimes g$ be in $D(H)^*$, where $k,\ell\in H^{op}$ and $f,g\in H^*$. Using 10.3.14, 2) for the coproduct and 10.3.16 for the definition of σ, we have as in 7.1.5,

$$\begin{aligned}
\overline{(k\otimes f)}\cdot\overline{(\ell\otimes g)} &= \sum \sigma((k\otimes f)_1, (\ell\otimes g)_1)(k\otimes f)_2(\ell\otimes g)_2 \\
&= \textstyle\sum_{s,t,u,v} \sigma(k_1\otimes h_s^* f_1 h_t^*, \ell_1\otimes h_u^* g_1 h_v^*) \\
&\qquad ((\bar{S}h_t)k_2h_s\otimes f_2)((\bar{S}h_v)\ell_2h_u\otimes g_2) \\
&= \sum \varepsilon(k_1)\langle h_u^* g_1 h_v^*, 1\rangle\langle h_s^* f_1 h_t^*, \ell_1\rangle \\
&\qquad ((\bar{S}h_v)\ell_2h_u(\bar{S}h_t)k_2h_s\otimes f_2g_2) \\
&= \sum \langle h_s^* f_1 h_t^*, \ell_1\rangle \ell_2(\bar{S}h_t)kh_s\otimes f_2g \\
&= \sum \langle h_s^*, \ell_1\rangle\langle f_1, \ell_2\rangle\langle h_t^*, \ell_3\rangle \ell_4(\bar{S}h_t)kh_s\otimes f_2g \\
&= \sum \langle f_1, \ell_2\rangle \ell_4(\bar{S}\ell_3)k\ell_1\otimes f_2g \\
&= \sum k\langle f_1, \ell_2\rangle \ell_1\otimes f_2g \\
&= \sum k(f_1 \rightharpoonup \ell)\otimes f_2g.
\end{aligned}$$

This is exactly the product $(k\#f)(\ell\#g)$ in $\mathcal{H}(H) = H\#H^*$. □

There are a number of generalizations of these constructions, in which two Hopf algebras act or coact on each other. The most useful may be Majid's "bicrossedproduct" $K \bowtie H$ [Mj 90a]. This Hopf algebra generalizes $D(H)$, so that K may be any Hopf algebra and not just H^*; it also generalizes the "bismash product" of [Tk 81]. A slightly simpler form of the defining relations in [Mj 90a] is given in [R 93a]. See also the "double crossed product" of [RS].

§10.4 Braided monoidal categories

We have seen in Lemmas 10.1.2 and 10.2.11 that the category of modules of a quasitriangular Hopf algebra (respectively, the category of comodules of a coquasitriangular Hopf algebra) has the symmetry property $V \otimes W \cong W \otimes V$. In fact these categories have other very nice properties. A good reference for categories is [Mac].

First, a *monoidal category* is a category $\mathcal{C}$ together with a functor $\otimes : \mathcal{C} \times \mathcal{C} \to \mathcal{C}$, called tensor product, an object I of $\mathcal{C}$, called the unit object, and natural families of isomorphisms

$$a_{U,V,W} : (U \otimes V) \otimes W \to U \otimes (V \otimes W)$$

$$r_V : V \otimes I \to V, \qquad \ell_V : I \otimes V \to V$$

in $\mathcal{C}$ called the associativity, right unit, and left unit constraints, respectively, subject to the two conditions:

(MC1)

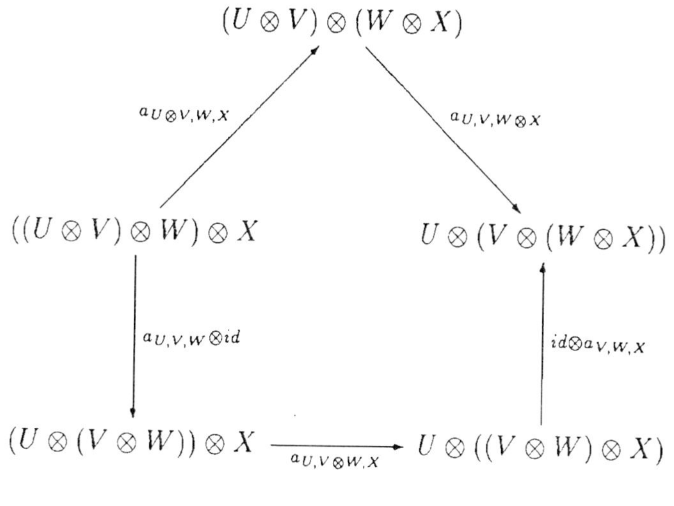

(MC2)

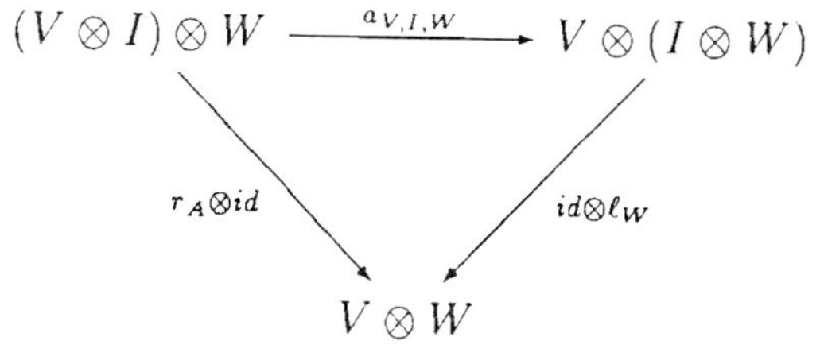

It follows from MacLane's coherence theorem that one may work as if the a, r, ℓ were all identities.

A *braiding* for a monoidal category $\mathcal{C}$ is a natural family of isomorphisms

$$t_{V,W} : V \otimes W \to W \otimes V$$

in $\mathcal{C}$ subject to the conditions

$(B1)$

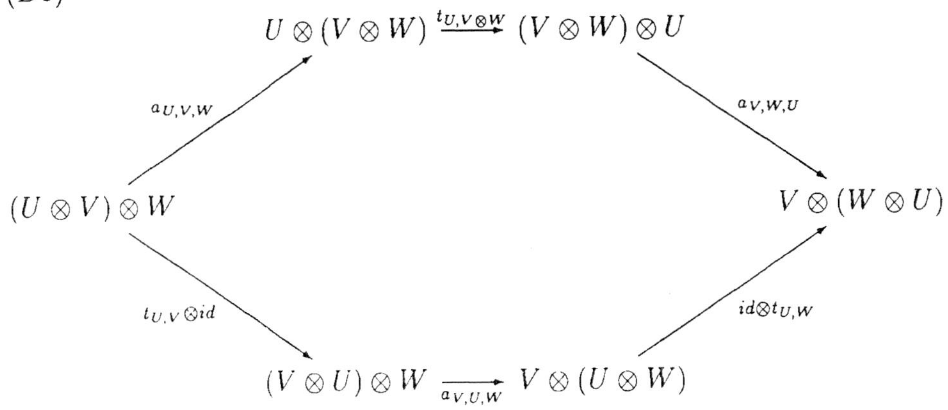

$(B2)$

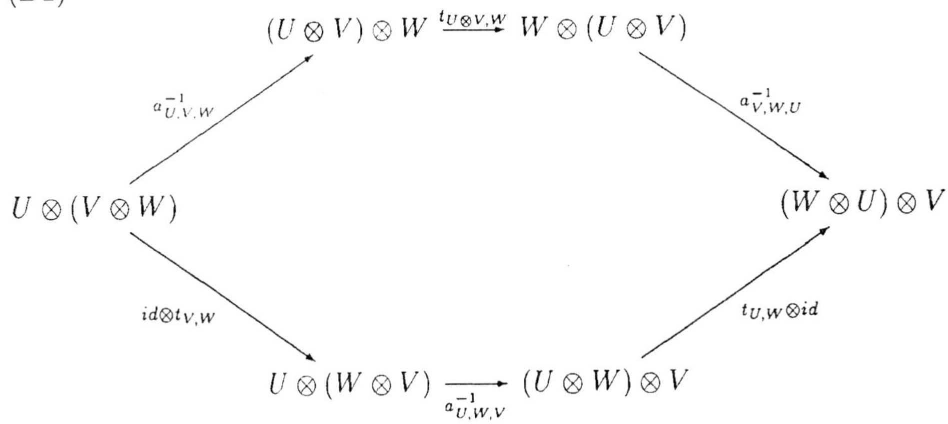

10.4.1 DEFINITION. A *braided monoidal category* is a monoidal category with a chosen braiding.

These categories were introduced by Joyal and Street in [JS 86], motivated by the theory of braids and links in topology. In that paper they are called *braided tensor categories*, which is probably a better name; however "monoidal" seems to be in wider use. The braiding is called a *quasisymmetry* in [Sb 92] and in [Mj 90b].

A *symmetric monoidal category* is a monoidal category with a symmetric braiding; that is, all $t_{r,w}$ satisfy the additional condition

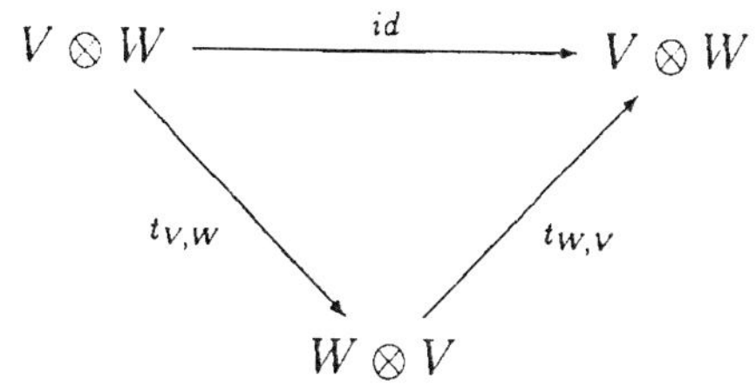

If $\mathcal{C}$ is either the category of modules for a cocommutative bialgebra, or the category of comodules for a commutative bialgebra, then $\mathcal{C}$ is a symmetric monoidal category by using $t_{V,W} = \tau_{V,W}$, the usual twist map; this has been known for a very long time. What is new is the following:

10.4.2 THEOREM. *Let B be a bialgebra. Then*
1) *B is quasitriangular $\Leftrightarrow$ the category $\mathcal{C} = {}_B\mathcal{M}$ of left B- modules is a braided monoidal category.*
2) *B is coquasitriangular $\Leftrightarrow$ the category $\mathcal{C} = \mathcal{M}^B$ of right B-comodules is a braided monoidal category*
Moreover B is triangular (resp. cotriangular) $\Leftrightarrow$ ${}_B\mathcal{M}$ (resp. $\mathcal{M}^B$) is symmetric monoidal.

PROOF (sketch). 1) First assume that there exists an element $Q \in B \otimes B$ such that for all $V, W \in {}_B\mathcal{M}$, each $t_{V,W}$ is given as in 10.1.2. That is, for all $v \in V, w \in W$

$$t_{V,W} : V \otimes W \to W \otimes V \text{ via } v \otimes w \mapsto Q(w \otimes v).$$

Now, $t_{V,W}$ is an isomorphism $\Leftrightarrow$ Q is invertible, and $t_{V,W}$ satisfies the braid diagrams (B1) and (B2) $\Leftrightarrow$

$$(\Delta \otimes id)(Q) = Q^{23}Q^{13} \quad \text{and} \quad (id \otimes \Delta)(Q) = Q^{12}Q^{13}.$$

These are exactly 10.1.6 and 10.1.7, if we use $Q = R^{-1}$ and take inverses.

Note that $Q = t_{B,B}(1 \otimes 1)$. Then since $t_{B,B}$ is a B-module map,

$$Q\tau(\Delta a) = t_{B,B}(\Delta a) = t_{B,B}(a \cdot (1 \otimes 1)) = a \cdot Q = (\Delta a)Q.$$

This gives almost cocommutativity as in 10.1.1.

It remains, however, to consider what happens if ${}_B\mathcal{M}$ is braided monoidal but we are not explicitly given the $t_{V,W}$ in terms of Q. In this case we *define* $Q = t_{B,B}(1 \otimes 1)$ and show that all the $t_{V,W}$ are of the desired form. To see this, consider any $V, W \in {}_B\mathcal{M}$, and choose $v \in V, w \in W$. Consider the left B-module maps $f_v : B \to V, f_w : B \to W$ given by $b \mapsto bv$ and $b \mapsto bw$, respectively. By the "naturality" of the maps $t_{V,W}$, the following diagram commutes:

$$\begin{array}{ccc} B \otimes B & \xrightarrow{f_v \otimes f_w} & V \otimes W \\ \downarrow t_{B,B} & & \downarrow t_{V,W} \\ B \otimes B & \xrightarrow[f_w \otimes f_v]{} & W \otimes V \end{array}$$

Thus

$$\begin{aligned} t_{V,W}(v \otimes w) &= t_{V,W} \circ (f_v \otimes f_w)(1 \otimes 1) = (f_w \otimes f_v) \circ t_{B,B}(1 \otimes 1) \\ &= (f_w \otimes f_v)(Q) = Q(w \otimes v). \end{aligned}$$

Thus all $t_{V,W}$ are of the desired form. The previous arguments show that B is quasitriangular.

2) As in 1), it is straightforward to check that if there exists a bilinear form $\langle \mid \rangle : B \otimes B \to \mathrm{k}$ such that for all $V, W \in \mathcal{M}^B$, each $t_{V,W}$ is given as in 10.2.11, that is

$$v \otimes w \mapsto \sum \langle w_1 \mid v_1 \rangle w_0 \otimes v_0,$$

then B is coquasitriangular $\Leftrightarrow \mathcal{M}^B$ is braided monoidal; this is done explicitly in [LT]. Thus as in 1) we have to consider the situation in which the $t_{V,W}$ are not given explicitly.

First, we define the bilinear form by $\langle \mid \rangle = (\varepsilon \otimes \varepsilon) \circ t_{B,B} \circ \tau$; this dualizes what we did in 1). Thus for all $a, b \in B$,

$$\langle a \mid b \rangle = (\varepsilon \otimes \varepsilon) t_{B,B}(b \otimes a).$$

One first checks that $t_{B,B}$ is given as in 10.2.11. For $f \in B^*$, consider the right comodule map $\rho_f : B \to B$ given by $\rho_f(b) = b \leftharpoonup f$, and similarly for $g \in B^*$ (recall that $\leftharpoonup$ is the transpose of left multiplication and compare with 1)). Then the naturality of $t_{B,B}$ gives

$$\begin{aligned}(f \otimes g)(t_{B,B}(a \otimes b)) &= (\varepsilon \otimes \varepsilon) \circ (\rho_f \otimes \rho_g)(t_{B,B}(a \otimes b)) \\ &= (\varepsilon \otimes \varepsilon) t_{B,B}((\rho_g \otimes \rho_f)(a \otimes b)) \\ &= \sum (\varepsilon \otimes \varepsilon) t_{B,B}(a_2 \otimes b_2) \langle f, b_1 \rangle \langle g, a_1 \rangle \\ &= \sum \langle b_2 \mid a_2 \rangle f(b_1) g(a_1) \\ &= (f \otimes g) \sum \langle b_2 \mid a_2 \rangle (b_1 \otimes a_1)\end{aligned}$$

for all $f, g \in B^*$. It follows that $t_{B,B}(a \otimes b) = \sum \langle b_2 \mid a_2 \rangle b_1 \otimes a_1$.

It then follows that $t_{V,W}$ is of the desired form if V and W are direct sums of copies of B, and consequently for arbitrary V and W since an arbitrary B-comodule can be imbedded in a direct sum of copies of B. By the above B must be coquasitriangular. □

10.4.3 REMARK. The difficulty with 1) above is that the basic examples $U_q(\mathbf{g})$ are only "topologically" quasi-triangular, as mentioned earlier; that is, $R \in B \hat{\otimes} B$, a completion of $B \otimes B$, rather than in B itself. Moreover one is then interested in the subcategory ${}_B\mathcal{M}$ consisting of the finite-dimensional representations of B. This set-up has been given a rigorous algebraic treatment in [LT, §7], where *completed-triangular* bialgebras are defined and a "completed" version of 10.4.2, 1) is proved. A somewhat different approach to circumvent this difficulty is proposed in [Mj 90b], using what are called *essentially quasitriangular* bialgebras.

It is a little difficult to assign credit for 10.4.2, since many people had these ideas at about the same time. The first part of 1) in which a given R determines the maps $t_{V,W}$ is essentially in [Dr 86], although without the formalism of braided monoidal categories. In the context of symmetric monoidal categories, the use of $Q = t(1 \otimes 1)$ in 1) and of $t(v \otimes w) = \sum \langle w_1 \mid v_1 \rangle w_0 \otimes v_0$ in 2) goes back to work of Pareigis [Pa 81] and Lyubashenko [Ly]. For braided monoidal categories, these maps appear in an equivalent form in [Mj 90b]. They also appear explicitly in [LT], [Sb 92] and in [JS 91].

A more general question is the following: given a bialgebra B (or a Hopf algebra H), when can B (or H) be "reconstructed" from its category of modules or comodules? This is in the spirit of the classical theorem of Tannaka, as mentioned in §9.3., and has been the object of a great deal of work over the last 20 years. To give any detailed statements of the results is beyond the scope of these notes. However we include a brief discussion. A good survey is the paper [JS 91]; see also the introduction in [Sb 92]. Classically, it was shown in [S-R] and [DM] that reconstruction could be done for a commutative Hopf algebra. That is, the bialgebra could be recovered from its category $\mathcal{C}$ of finite-dimensional comodules together with the forgetful functor to the category *Vec* of vector spaces provided the functor preserved suitable properties of $\otimes$. With somewhat more restrictive hypotheses, an antipode could also be constructed, and thus the Hopf algebra could be recovered. A more general set-up was considered independently in [Pa 79] in which *Vec* could be replaced by an arbitrary base category; see also [Pa 81]. In all of the above constructions, $\mathcal{C}$ was a symmetric monoidal category. This work was extended by Ulbrich in [U 90] who shows that the antipode can be reconstructed even if $\mathcal{C}$ is not symmetric, provided $\mathcal{C}$ is rigid (this means that for every object X in $\mathcal{C}, X^*$ is also in $\mathcal{C}$).

Lyubashenko sketches in [Ly] how to build a symmetric monoidal category $\mathcal{C}$ from a solution to the QYBE, and then how to construct a Hopf algebra H acting on all the vector spaces in the category. H has properties corresponding to the original solution of the QYBE, although quasitriangularity is not explicitly mentioned. His procedure enables one to construct a class of cotriangular Hopf algebras.

More recently, with the appearance of quantum groups, the categories of interest are braided monoidal. Majid in [Mj 90b] gives a reconstruction of a quasitriangular Hopf algebra whose finite-dimensional modules form a braided monoidal category; to do so he uses the category *vec* of finite-dimensional vector spaces. See also [Mj 91a]. There is also some related work on this topic by Yetter; see for example [Y 90, Y 92].

Not surprisingly, in view of our previous remarks, things work better in the coquasitriangular case. Schauenberg in [Sb 92] reconstructs a coquasitriangular Hopf algebra from a braided monoidal category, extending Ulbrich's constructions.

Majid proves a more general reconstruction theorem in [Mj 92]: he shows that certain monoidal categories are the comodule categories of coquasitriangular quasi-Hopf algebras; in this situation the functor to *vec* does not have to satisfy as strong a coherence condition as before. Quasi-Hopf algebras, introduced by Drinfeld [Dr 89b], are not actually Hopf algebras, since their comultiplication is only coassociative up to conjugation by an invertible element in $H \otimes H \otimes H$. These algebras are important in physics, as they provide solutions to the Knizhnik-Zamolodchikov (K-Z) equations in quantum field theory. A nice survey of quasiHopfalgebras is given in [St].

§10.5 Hopf algebras in categories; graded Hopf algebras

The idea of a Hopf algebra in a category is an old one, although without the formalism of category theory. For, the original graded Hopf algebras of Milnor and Moore are actually Hopf algebras in the category of $\mathbf{k}\mathbf{Z}$-comodules [MiMo], and enveloping algebras of Lie superalgebras live in the category of $\mathbf{k}\mathbf{Z}_2$-comodules. More generally a Lie color algebra is graded by an arbitrary abelian group G and a (symmetric) bicharacter on G; the enveloping algebras of such Lie algebras are Hopf algebras in the category of $\mathbf{k}G$-comodules [Sc]. These examples are not Hopf algebras in the usual sense, since their coproducts are only "graded multiplicative". We shall see what this condition means more generally.

In general an algebra, coalgebra, bialgebra, or Hopf algebra being in a category means that the structure maps are morphisms in the category. The categories of interest to us are ${}_H\mathcal{M}$ or $\mathcal{M}^H$, the category of modules or of comodules over a given Hopf algebra H. In particular an algebra A in ${}_H\mathcal{M}$ is simply a left H-module algebra, since as noted in §4.1, conditions 4.1.1, 2) and 3) say that m_A and u_A are morphisms in ${}_H\mathcal{M}$; similarly an algebra in $\mathcal{M}^H$ is a right H-comodule algebra, as in 4.1.2. A coalgebra C in ${}_H\mathcal{M}$ is a left H-module coalgebra, that is Δ_C and ε_C are H-module maps, and C is in $\mathcal{M}^H$ if it is a right H-comodule coalgebra, that is Δ_C and ε_C are H-comodule maps. See [A, pp 137-139].

To have a bialgebra B in a category $\mathcal{C}$, however, requires more than B being both an algebra and a coalgebra in the category; for we require that $\Delta_B : B \to B \otimes B$ be an algebra morphism *in the category*. That is, $\Delta_B(ab) = (\Delta_B a) \cdot (\Delta_B b)$, where the product $\cdot$ is a morphism in $\mathcal{C}$. In order to have this property, we have to understand how to form the tensor product of two algebras in the category. The usual tensor product of algebras A and

B is given by

$$(a \otimes b) \cdot (a' \otimes b') = (m_A \otimes m_B)(id \otimes \tau \otimes id)(a \otimes b \otimes a' \otimes b'), \tag{10.5.1}$$

for all $a, a' \in A, b, b' \in B$; that is, we have to twist $b \otimes a'$ before we use m_A and m_B. This suggests that the category $\mathcal{C}$ should have a way of twisting tensor products; in fact everything works if $\mathcal{C}$ is a braided monoidal category as in the previous section (we note this was done in [Pa 81] if $\mathcal{C}$ was symmetric monoidal and extended to braided monoidal categories in [Mj 91a]). Thus from now on we assume that $\mathcal{C} = {}_H\mathcal{M}$ or $\mathcal{C} = \mathcal{M}^H$ and that $\mathcal{C}$ is braided monoidal.

In 10.5.1 above, τ is replaced by the appropriate $t_{B,A}$ in $\mathcal{C}$. If $\mathcal{C} = {}_H\mathcal{M}$, we have $t_{V,W}(v \otimes w) = R^{-1}(w \otimes v)$ by 10.4.2, 1). Thus

$$(a \otimes b) \cdot (a' \otimes b') = (m_A \otimes m_B)(a \otimes R^{-1}(a' \otimes b) \otimes b'). \tag{10.5.2}$$

Similarly if $\mathcal{C} = \mathcal{M}^H$, we have $t_{V,W}(v \otimes w) = \sum \langle w_1 \mid v_1 \rangle w_0 \otimes v_0$. Thus

$$\begin{aligned} (a \otimes b) \cdot (a' \otimes b') &= (m_A \otimes m_B)(a \otimes \sum \langle a'_1 \mid b_1 \rangle a'_0 \otimes b_0 \otimes b') \\ &= \sum \langle a'_1 \mid b_1 \rangle a a'_0 \otimes b_0 b' \end{aligned} \tag{10.5.3}$$

where $a \mapsto \sum a_0 \otimes a_1, \quad b \mapsto \sum b_0 \otimes b_1$ is the usual comodule notation. These products are what must be used in the product $(\Delta_B a) \cdot (\Delta_B b)$ as mentioned above.

To give an explicit formula for this product, we introduce new summation notation for Δ_B. That is, for any $a \in B$, we write

$$\Delta_B a = \sum a^{(1)} \otimes a^{(2)} = \sum a^1 \otimes a^2. \tag{10.5.4}$$

The new notation is needed to distinguish Δ_B from Δ_H; otherwise ambiguity would result. For, when $B \in \mathcal{M}^H$, the comodule notation $a \mapsto \sum a_0 \otimes a_1$ means that $a_1 \in H$, and a further application of $id \otimes \Delta_H$ would give $\sum a_0 \otimes a_1 \otimes a_2 \in B \otimes H \otimes H$; this should not be confused with $\Delta_B a \in B \otimes B$. The new notation together with (10.5.3) then gives

$$\begin{aligned} \Delta_B(ab) &= \sum (a^1 \otimes a^2) \cdot (b^1 \otimes b^2) \\ &= \sum \langle (b^1)_1 \mid (a^2)_1 \rangle a^1 (b^1)_0 \otimes (a^2)_0 b^2 \end{aligned} \tag{10.5.5}$$

for all $a, b \in B$.

Similarly when $B \in {}_H\mathcal{M}$, the coproduct becomes

$$(10.5.6) \qquad \Delta_B(ab) = \sum (a^1 \otimes a^2) \cdot (b^1 \otimes b^2) = \sum a^1(c_j \cdot b^1) \otimes (d_j \cdot a^2)b^2$$

where we have written $R^{-1} = \sum c_j \otimes d_j$ as before.

A similar problem arises in discussing tensor products of modules in $\mathcal{C}$. For, if B is a bialgebra in $\mathcal{C}$, and V, W are left B-modules in $\mathcal{C}$, then in defining

$$b \cdot (v \otimes w) = (\sum b^1 \otimes b^2) \cdot (v \otimes w)$$

for $b \in B, \quad v \in V, \quad w \in W$, we need to twist the v past the b_2. Thus if $\phi_V : H \otimes V \to V$ and $\phi_W : H \otimes W \to W$ are the two given module actions, we have

$$(10.5.7) \; \phi_{V \otimes W} = (\phi_V \otimes \phi_W) \circ (id \otimes t_{H,V} \otimes id) \circ (\Delta \otimes id^2) : H \otimes V \otimes W \to V \otimes W.$$

Compare with the discussion after 1.8.1. Similarly if A is a B-module algebra in the category $\mathcal{C}$, and the B-action is given by ϕ_A, then we have

$$(10.5.8) \qquad b \cdot (ac) = m_A \circ \phi_{A \otimes A}(b \otimes a \otimes c),$$

for $b \in B, \quad a, c \in A$, where $\phi_{A \otimes A}$ is as in 10.5.7.

Finally, if B is a Hopf algebra in $\mathcal{C}$, the antipode S of B must also be a map in $\mathcal{C}$. It follows that S is a "twisted" antihomomorphism; that is,

$$(10.5.9) \qquad S(hk) = m_B \circ t_{B,B}(Sh \otimes Sk),$$

for all $h, k \in B$.

We now specialize to the concrete case of $\mathrm{k}G$-comodules, so that we may recover the classical examples. Thus, let $H = \mathrm{k}G$ for an abelian group G, and fix a bicharacter $\lambda : G \times G \to \mathrm{k}^*$; that is, λ is multiplicative in both entries. As noted in 10.2.6, this makes H into a coquasitriangular Hopf algebra, with $\langle g \mid h \rangle = \lambda(g, h)$, and thus $\mathcal{M}^H$ is a braided monoidal category as in 10.4.2. We also know that $\mathcal{M}^H$ consists of the G-graded modules, as in 1.6.7. Thus algebras in $\mathcal{M}^H$ are just G-graded algebras. If $A = \oplus_{x \in G} A_x$ and $B = \oplus_{y \in G} B_y$ are algebras in $\mathcal{M}^H$, then $A \otimes B$ is an algebra in $\mathcal{M}^H$, via

$$(10.5.10) \qquad (a_x \otimes b_y) \cdot (a'_z \otimes b'_w) = \lambda(y, z) a_x a'_z \otimes b_y b'_w$$

for all $a_x \in A_x, a'_z \in A_z, b_y \in B_y, b'_w \in B_w$, and $x, y, z, w \in G$; note that 10.5.10 is a special case of 10.5.3. With this we can give a general definition:

10.5.11 DEFINITION: Given an abelian group G and a bicharacter λ on G, B is a *G-graded Hopf algebra* (with respect to λ) if

1) B is a G-graded algebra
2) B is a G-graded coalgebra
3) Δ_B and ε_B are algebra maps, where $m_{B\otimes B}$ is given by 10.5.10
4) the antipode $S_B : B \to B$ is a G-graded map as in 10.5.9.

We specialize some of the earlier formulas to this case. For example if V and W are G-graded B-modules, then 10.5.7 gives that $V \otimes W$ is a G-graded B-module, as follows: if $b_x \in B_x$, then $\Delta b_x = \sum_{yz=x} \sum_i b_{y,i} \otimes b'_{z,i} \in \sum_{yz=x} B_y \otimes B_z$, since B is a G-graded coalgebra. Thus if $v_s \in V_s$ and $w_t \in W_t$, for $s, t \in G$, we have

$$b_x \cdot (v_s \otimes w_t) = \sum_{yz=x} \lambda(s, z) \sum_i b_{y,i} \cdot v_s \otimes b'_{z,i} \cdot w_t.$$

For the particular case of a homogeneous primitive element $b \in B_x$, that is $\Delta b = b \otimes 1 + 1 \otimes b$, it follows that

$$b \cdot (v_s \otimes w_t) = b \cdot v_s \otimes w_t + \lambda(s, x) v_s \otimes b \cdot w_t. \tag{10.5.12}$$

Thus, b acts as a graded derivation.

10.5.13 EXAMPLE: **Z**-graded Hopf algebras. These include the Hopf algebras of [MiMo]; in their case the bicharacter was given by $\lambda(p, q) = (-1)^{pq}$, for $p, q \in \mathbf{Z}$ (where $\mathbf{Z}$ is written under $+$ as usual). More generally, any bicharacter of $\mathbf{Z}$ is given by $\lambda(p, q) = c^{pq}$, for some $c \in \mathbf{k}^*$, and such a λ could be used to define a **Z**-graded Hopf algebra B.

10.5.14 EXAMPLE: Lie superalgebras and Lie color algebras. For $\mathbf{g} = \mathbf{g}_0 \oplus \mathbf{g}_1$, a Lie superalgebra, its universal enveloping algebra $U(\mathbf{g})$ is a $\mathbf{Z}_2$-graded Hopf algebra with $\lambda(\bar{p}, \bar{q}) = (-1)^{pq}$, for $\bar{p}, \bar{q} \in \mathbf{Z}_2$. We note that 10.5.12 gives the usual action for a graded derivation in this case: if $b \in \mathbf{g}$ is homogeneous, of degree $|b| \in \{\bar{0}, \bar{1}\}$ and b acts on a $\mathbf{Z}_2$-graded module $V = V_0 \oplus V_1$, then

$$b \cdot (v \otimes w) = (b \cdot v) \otimes w + (-1)^{|b|\,|v|} v \otimes b \cdot w,$$

where also $v \in V$ is homogeneous of degree $|v|$.

More generally, for any abelian G and *symmetric* bicharacter $\lambda : G \times G \to \mathbf{k}^*$ (that is $\lambda(x,y) = \lambda(y,x)^{-1}$, all $x, y \in G$), a *(G,λ)-Lie color algebra* $\mathfrak{g} = \oplus_{x \in G} \mathfrak{g}_x$ is a G-graded module with a product $[\ ,\ ] : \mathfrak{g} \otimes \mathfrak{g} \to \mathfrak{g}$ satisfying

1) $[a,b] = -\lambda(y,x)[b,a]$
2) $\lambda(x,z)[a,[b,c]] + \lambda(z,y)[c,[a,b]] + \lambda(y,x)[b,[c,a]] = 0$

for all $a \in \mathfrak{g}_x, b \in \mathfrak{g}_y, c \in \mathfrak{g}_z$.

For example, given any G-graded algebra A and a symmetric λ, A^- becomes a (G,λ)-Lie color algebra via

$$[a,b] = ab - \lambda(x,y)ba,$$

for all $a \in A_x, b \in A_y$. Universal enveloping algebras of Lie color algebras exist and become G-graded Hopf algebras; see [Sc]. We note that the symmetry condition on λ is needed for 1) above to be consistent; thus $\mathcal{M}^{\mathbf{k}G}$ is a symmetric monoidal category in this case. In fact $\mathfrak{g}$ may be thought of as a Lie algebra in the category $\mathcal{M}^{\mathbf{k}G}$; more generally, Gurevich has considered Lie algebras in arbitrary symmetric monoidal categories [Gu]. It is an interesting question as to whether the notion of a Lie algebra in a category makes sense for more general categories.

§10.6 Biproducts and Yetter-Drinfeld modules

In this last section we consider when a smash product is a Hopf algebra, giving criteria due to Radford, and apply this to see that Hopf algebras in the categories ${}_H\mathcal{M}$ or $\mathcal{M}^H$ as in the last section can be enlarged to ordinary Hopf algebras. We then consider Yetter-Drinfeld modules; these modules, related to $D(H)$, give a more general setting in which a smash product can become a Hopf algebra.

We first describe the coalgebra structure of the smash product. This is essentially Molnar's smash coproduct [Mo], although we use H^{cop} instead of H since we will be using right and not left comodule algebras.

10.6.1 DEFINITION. Let H be a bialgebra and C a coalgebra in $\mathcal{M}^H$. The *smash coproduct* $C\natural H^{cop}$ is defined to be $C \otimes H^{cop}$ as a vector space, with comultiplication given by

$$\Delta(c\natural h) = \sum c^1\natural(c^2)_1 h_2 \otimes (c^2)_0\natural h_1$$

and counit

$$\varepsilon(c\natural h) = \varepsilon_C(c)\varepsilon_H(h),$$

for all $c \in C, h \in H^{cop}$.

Here we have used the notation $\Delta_C c = \sum c^1 \otimes c^2$ as in 10.5.4 and the usual notation for Δ_H and for the H-comodule structure of C. Then the fact that Δ_C is an H-comodule map says that

$$\sum (c_0)^1 \otimes (c_0)^2 \otimes c_1 = \sum (c^1)_0 \otimes (c^2)_0 \otimes (c^1)_1 (c^2)_1 \tag{10.6.2}$$

for all $c \in C$, using 1.8.2.

10.6.3 LEMMA. *$C\natural H^{cop}$ is a coalgebra.*

PROOF. We check that Δ is coassociative. Now for $c \in C, h \in H^{cop}$,

$$\begin{aligned}
(id \otimes \Delta)\Delta(c\natural h) &= (id \otimes \Delta)\sum (c^1\natural (c^2)_1 h_2 \otimes (c^2)_0 \natural h_1) \\
&= \sum c^1 \natural (c^2)_1 h_3 \otimes ((c^2)_0)^1 \natural (((c^2)_0)^2)_1 h_2 \otimes (((c^2)_0)^2)_0 \natural h_1 \\
&= \sum c^1 \natural (c^{21})_1 (c^{22})_1 h_3 \otimes (c^{21})_0 \natural (c^{22})_{01} h_2 \otimes (c^{22})_{00} \natural h_1 \\
&\qquad \text{by 10.6.2 applied to } c^2 \\
&= \sum c^1 \natural (c^2)_1 (c^3)_2 h_3 \otimes (c^2)_0 \natural (c^3)_1 h_2 \otimes (c^3)_0 \natural h_1 \\
&\qquad \text{since } C \text{ is an } H\text{-comodule} \\
&= (\Delta \otimes id)\sum (c^1 \natural (c^2)_1 h_2 \otimes (c^2)_0 \natural h_1) \\
&= (\Delta \otimes id)\Delta(c\natural h).
\end{aligned}$$

A simpler computation shows that ε is a counit. □

With the smash coproduct in hand, we can define the biproduct.

10.6.4 DEFINITION. Let H be a bialgebra, and let B be an algebra in ${}_{H^{cop}}\mathcal{M}$ and a coalgebra in $\mathcal{M}^H$. The *(coopposite) biproduct $B \star H^{cop}$ of B and H^{cop}* is defined to be $B\# H^{cop}$ as an algebra and $B\natural H^{cop}$ as a coalgebra.

A theorem of Radford gives necessary and sufficient conditions for $B\star H^{cop}$ to be a bialgebra or a Hopf algebra. We give the coopposite version of his result.

10.6.5 THEOREM[R 85]. *Let H be a bialgebra, and let B be an algebra in ${}_{H^{cop}}\mathcal{M}$ and a coalgebra in $\mathcal{M}^H$. Then $B\star H^{cop}$ is a bialgebra $\Leftrightarrow$ B is a coalgebra in ${}_{H^{cop}}\mathcal{M}$ and an algebra in $\mathcal{M}^H$, ε_B is an algebra map, $\Delta 1_B = 1_B\otimes 1_B$ and the identities*

1) $\Delta_B(ab) = \sum a^1((a^2)_1\cdot b^1)\otimes(a^2)_0 b^2$

2) $\sum h_1\cdot b_0\otimes h_2 b_1 = \sum (h_2\cdot b)_0\otimes(h_2\cdot b)_1 h_1$

hold for all $a, b\in B$ and $h\in H$.

In addition if H^{cop} is a Hopf algebra with antipode $\bar S_H$ and B has an antipode S_B then $B\star H^{cop}$ is a Hopf algebra with antipode given by

$$S(b\star h) = \sum (1\star\bar S_H(b_1 h))(S_B b_0\star 1).$$

PROOF(sketch). We will show the sufficiency of the hypotheses for $B\star H^{cop}$ to be a bialgebra. First, $B\star H^{cop}$ is an algebra from 4.1.3, and it is a coalgebra by 10.6.2. Thus we need to show that Δ and ε are multiplicative. We will check this for Δ; we follow Radford's proof [R 85, pp. 327-329]. First note that for $a, b\in B$ and $h, k\in H^{cop}$,

$$\begin{aligned}\Delta(a\star h)(b\star k) &= \Delta(\sum a(h_2\cdot b)\star h_1 k)\\ &= \sum (a(h_2\cdot b))^1\star((a(h_2\cdot b))^2)_1(h_1k)_2\\ &\qquad\qquad\otimes((a(h_2\cdot b))^2)_0\star(h_1k)_1\\ &= \sum (a(h_3\cdot b))^1\star((a(h_3\cdot b))^2)_1 h_2k_2\otimes((a(h_3\cdot b))^2)_0\star h_1k_1\end{aligned}$$

and that

$$\Delta(a\star h)\Delta(b\star k) = \sum (a^1\star(a^2)_1h_2\otimes(a^2)_0\star h_1)(b^1\star(b^2)_1k_2\otimes(b^2)_0\star k_1)$$

$$= \sum a^1((a^2)_1h_2)_2\cdot b^1\star((a^2)_1h_2)_1(b^2)_1k_2\otimes(a^2)_0h_{12}\cdot(b^2)_0\star h_{11}k_1$$

To see that these two expessions are equal, it suffices to show this when $k=1$; moreover we can then ignore the last h's. That is, it suffices to show

$$(*)\qquad \sum (a(h_2\cdot b))^1\star((a(h_2\cdot b))^2)_1h_1\otimes((a(h_2\cdot b))^2)_0$$

$$= \sum a^1((a^2)_1 h_2)_2 \cdot b^1 \star ((a^2)_1 h_2)_1 (b^2)_1 \otimes (a^2)_0 h_1 \cdot (b^2)_0.$$

Applying condition 1), the left side of $(*)$ becomes

$$\sum a^1((a^2)_1 \cdot (h_2 \cdot b)^1) \star ((a^2)_0 (h_2 \cdot b)^2)_1 h_1 \otimes ((a^2)_0 (h_2 \cdot b)^2)_0$$

and the fact that B is an H^{cop}-module coalgebra gives

$$\sum a^1((a^2)_1 \cdot (h_3 \cdot b^1)) \star ((a^2)_0 (h_2 \cdot b^2))_1 h_1 \otimes ((a^2)_0 (h_2 \cdot b^2))_0.$$

Now use the fact that B is an H-comodule algebra and apply condition 2) to b^2 to get

$$\sum a^1((a^2)_1 \cdot (h_3 \cdot b^1)) \star ((a^2)_{01} h_2 (b^2)_1) \otimes (a^2)_{00} h_1 \cdot (b^2)_0.$$

This is equal to the right side of $(*)$, using that Δ_H is multiplicative and that B is an H-comodule. Thus Δ is multiplicative. We leave it to the reader to check that ε is also multiplicative. Thus $B \star H^{cop}$ is a bialgebra.

The fact that the given S makes $B \star H^{cop}$ into a Hopf algebra is a coopposite version of [R 85, Prop. 2, p.332].

The necessity of the various conditions can be seen as in [R 85]; most follow by appropriately specializing $(*)$. □

10.6.6 REMARK. It is worth noting the original formulation of 10.6.1 and 10.6.4, in which B was an algebra in ${}_H\mathcal{M}$ and a coalgebra in ${}^H\mathcal{M}$. Recall that when considering left comodules, the summation notation for $\rho : B \to H \otimes B$ becomes $b \mapsto \sum b_{-1} \otimes b_0$. Then Molnar's smash coproduct $B\natural H$ has comultiplication

$$\Delta(b\natural h) = \sum b^1 \natural (b^2)_{-1} h_1 \otimes (b^2)_0 \natural h_2.$$

Using this coproduct, the biproduct $B \star H$ becomes a bialgebra if B is a coalgebra in ${}_H\mathcal{M}$, an algebra in ${}^H\mathcal{M}$, ε_B is an algebra map, $\Delta 1_B = 1_B \otimes 1_B$, and the following identities hold:

1)' $\Delta_B(ab) = \sum a^1((a^2)_{-1} \cdot b^1) \otimes (a^2)_0 b^2$

2)' $\sum h_1 b_{-1} \otimes h_2 \cdot b_0 = \sum (h_1 \cdot b)_{-1} h_2 \otimes (h_1 \cdot b)_0$

If also B has an antipode S_B and H is a Hopf algebra with antipode S_H, then $B \star H$ is a Hopf algebra with antipode $S(b \star h) = \sum (1 \star S_H(b_{-1} h))(S_B b_0 \star 1)$.

Although this formulation may be more natural, and has the advantage of using H and not H^{cop}, the computations are more tiresome because of the notation.

We now apply the biproduct construction to bialgebras (or Hopf algebras) in categories. The first result is due to Majid; constructing a Hopf algebra in this way is a process he calls "bosonization".

10.6.7 PROPOSITION[Mj 93]. *Let (H, R) be a quasitriangular bialgebra, and let B be a bialgebra in ${}_H\mathcal{M}$. Then B is also a left H-comodule algebra by defining*

$$\rho : B \to H \otimes B \text{ via } b \mapsto R^{-1}(1 \otimes b) = \sum_j c_j \otimes d_j \cdot b = \sum b_{-1} \otimes b_0$$

for all $b \in B$. Using this coaction, 10.6.6 applies and the biproduct $B \star H$ is a bialgebra. If also B is a Hopf algebra in ${}_H\mathcal{M}$, then the biproduct is a Hopf algebra. Finally if (H, R) is triangular and B is quasitriangular in ${}_H\mathcal{M}$, then $B \star H$ is quasitriangular.

PROOF. We check that 1)′ and 2)′ of 10.6.6 hold. First, since B is a bialgebra in ${}_H\mathcal{M}$, its coproduct satisfies 10.5.6. Using the H-coaction defined above on a^2, we have $\sum (a^2)_{-1} \otimes (a^2)_0 = \sum_j c_j \otimes d_j \cdot a^2$; substituting, we see that 10.5.6 is exactly condition 1)′.

Again using the new coaction, condition 2)′ becomes

$$\sum h_1 c_j \otimes h_2 \cdot (d_j \cdot b) = \sum c_j h_2 \otimes d_j \cdot (h_1 \cdot b),$$

for all $b \in B, h \in H$; thus it suffices to show

$$\sum_j h_1 c_j \otimes h_2 d_j = \sum_j c_j h_2 \otimes d_j h_1.$$

However this is exactly $(\Delta h)R^{-1} = R^{-1}\tau(\Delta(h))$, which holds since H is quasitriangular. Thus 2)′ holds. The reader may check that B is also an algebra and a coalgebra in ${}^H\mathcal{M}$. Now by 10.6.6 $B \star H$ becomes a bialgebra.

Now assume that (H, R) is triangular and that $(B, \tilde{R})$ is quasitriangular in ${}_H\mathcal{M}$. Then $B \star H$ is quasitriangular via

$$\mathcal{R} = R_{24} R_{32} \cdot \tilde{R}_{12}$$

as is shown in [Mj 93, 4.1]. □

We note that Majid's braiding $t_{V,W}$ and so his coaction used $\tau(R)$ instead of R^{-1}; however when H is triangular $R^{-1} = \tau(R)$, and thus our expression for $\mathcal{R}$ agrees with his.

The dual case is similar; it appears in [FiM] for B in $\mathcal{M}^H$.

10.6.8 PROPOSITION. *Let $(H, \langle\ |\ \rangle)$ be a coquasitriangular bialgebra and let B be a bialgebra in $\mathcal{M}^H$. Then B is also a bialgebra in ${}_{H^{cop}}\mathcal{M}$ by defining*

$$h \cdot b = \sum \langle b_1 \mid h \rangle b_0$$

for all $h \in H^{cop}$ and $b \in B$. Using this action, 10.6.5 applies and the biproduct $B \star H^{cop}$ is a bialgebra. If also B is a Hopf algebra in $\mathcal{M}^H$, then $B \star H^{cop}$ is a Hopf algebra.

PROOF. It is not difficult see that the action gives B the structure of an algebra and a coalgebra in ${}_{H^{cop}}\mathcal{M}$. Since B is a bialgebra in $\mathcal{M}^H$, its coproduct satisfies 10.5.5. From the above action, $(a^2)_1 \cdot b^1 = \sum \langle (b^1)_1 \mid (a^2)_1 \rangle (b^1)_0$; substituting, we se that 10.5.5 is exactly condition 1).

Analogously to the proof of 10.6.7, condition 2) follows from the coquasitriangular hypothesis. For,

$$\sum h_1 \cdot b_0 \otimes h_2 b_1 = \sum \langle b_{01} \mid h_1 \rangle b_{00} \otimes h_2 b_1 = \sum \langle b_1 \mid h_1 \rangle b_0 \otimes h_2 b_2$$

and

$$\sum (h_2 \cdot b)_0 \otimes (h_2 \cdot b)_1 h_1 = \sum \langle b_1 \mid h_2 \rangle b_{00} \otimes b_{01} h_1 = \sum \langle b_2 \mid h_2 \rangle b_0 \otimes b_1 h_1.$$

Thus condition 2) holds $\Leftrightarrow \sum \langle b_1 \mid h_1 \rangle h_2 b_2 = \sum \langle b_2 \mid h_2 \rangle b_1 h_1$. But this is exactly 10.2.2. Thus 2) holds, and 10.6.5 implies that $B \star H^{cop}$ is a bialgebra. □

10.6.9 EXAMPLE. We consider the case when $H = \mathrm{k}G$, for G an abelian group and $\lambda : G \times G \to \mathrm{k}^*$ a given bicharacter on G. Thus as in 10.5.11, B is a G-graded bialgebra (or Hopf algebra). Now $H^{cop} = H$ since H is cocommutative, and thus we have an action of H on B given by

$$x \cdot b_y = \lambda(y, x) b_y$$

for all $x, y \in G$ and $b_y \in B_y$. We remark that this action corresponds to the standard action of the character group $\hat{G}$ of G on the graded algebra B; that is, if $\mu \in \hat{G}$, then $\mu \cdot b_y = \mu(y)y$. In our situation, λ induces a group morphism $G \to \hat{G}$ by $x \mapsto \hat{x} = \lambda(-, x)$ as in 10.2.6. Then $\hat{x} \cdot b_y = \hat{x}(y)b_y = x \cdot b_y$.

Now 10.6.8 applies and tells us that $B \star H$ is a bialgebra (or Hopf algebra). As an algebra $B \star H = B\#\mathrm{k}G$, the skew group ring; thus the skew group ring over any G-graded Hopf algebra becomes an ordinary Hopf algebra.

Looking again at 10.6.7 and 10.6.8, we see that in both cases condition 1) held because B was a bialgebra in the category ${}_H\mathcal{M}$ or $\mathcal{M}^H$ and condittion 2) held because H was quasi- or coquasi-triangular. This suggests that a more general result is true; in fact such a result can be stated in the context of "Yetter-Drinfeld" modules. These were introduced in [Y 90], where they were called crossed bimodules. It was then shown in [Mj 91c] that these modules can be identified with modules over the Drinfeld double, when H is finite-dimensional; see below. The present name comes from [RT].

10.6.10 DEFINITION. For any bialgebra H, a *left Yetter-Drinfeld module* is a k-space M which is both a left H-module and a left H-comodule and satisfies the compatibility condition

$$(10.6.11) \qquad \sum h_1 m_{-1} \otimes h_2 \cdot m_0 = \sum (h_1 \cdot m)_{-1} h_2 \otimes (h_1 \cdot m)_0$$

for all $h \in H, m \in M$. The category of left Yetter-Drinfeld modules is denoted by ${}_H^H\mathcal{YD}$.

We shall also be interested in the Yetter-Drinfeld category ${}_H\mathcal{YD}^H$; an object M in this category is both a left H-module and a right H-comodule and satisfies the condition

$$(10.6.12) \qquad \sum h_1 \cdot m_0 \otimes h_2 m_1 = \sum (h_2 \cdot m)_0 \otimes (h_2 \cdot m)_1 h_1$$

for all $h \in H, m \in M$.

10.6.13 EXAMPLE. For H a Hopf algebra, H itself is a Yetter-Drinfeld module by considering H as a left H -comodule via Δ and as a left H-module via the left adjoint action.

10.6.14 EXAMPLE. If H is quasitriangular, then we have ${}_H^H\mathcal{YD} \supseteq {}_H\mathcal{M}$ by using the coaction defined in 10.6.7, since the Yetter-Drinfeld condition 10.6.11 is just condition 2)$'$ in 10.6.6. Similarly if H is coquasitriangular then

${}_H\mathcal{YD}^H \supseteq \mathcal{M}^H$ by using the action defined in 10.6.8, since the Yetter-Drinfeld condition 10.6.12 is exactly condition 2) in 10.6.5.

Now ${}^H_H\mathcal{YD}$ has a "pre-braiding" on objects M, N, as follows:

$$t_{M,N} : M \otimes N \to N \otimes M \quad \text{via} \quad m \otimes n \mapsto \sum (m_{-1}) \cdot n \otimes m_0$$

for $m \in M, n \in N$. This map is a braiding on ${}^H_H\mathcal{YD}$ precisely when H^{cop} is a Hopf algebra [Y 90].

10.6.15. COROLLARY. *Given a bialgebra H such that H^{cop} is a Hopf algebra, let B be a bialgebra in the Yetter-Drinfeld category ${}^H_H\mathcal{YD}$. Then the biproduct $B \star H$ is a bialgebra. If also H is a Hopf algebra and B a Hopf algebra in ${}^H_H\mathcal{YD}$, then $B \star H$ is a Hopf algebra.*

PROOF. Since B is a bialgebra in ${}^H_H\mathcal{YD}$, it is certainly a coalgebra and an algebra in ${}_H\mathcal{M}$ and in ${}^H\mathcal{M}$, and also ε is an algebra map, $\Delta_B 1_B = 1_B \otimes 1_B$, and condition 1)′ of 10.6.6 holds (since 1)′ just says that Δ_B is multiplicative in the category, using $t_{B,B}$ above). Condition 2)′ is precisely the compatibility condition for ${}^H_H\mathcal{YD}$, as noted above. □

The Corollary is essentially in [Mj 91c], without the categorical formulation. We now give the connection to $D(H)$-modules mentioned above.

10.6.16 PROPOSITION[Mj 91c]. *Let H be a finite-dimensional Hopf algebra. Then the Yetter-Drinfeld category ${}_H\mathcal{YD}^H$ can be identified with the category ${}_{D(H)}\mathcal{M}$ of left modules over the Drinfeld double $D(H)$.*

PROOF. Our formulation is slightly different than Majid's because of our definition 10.3.5 of $D(H)$. First note that a k-space M will be a $D(H)$-module $\Leftrightarrow$ M is a left H-module, a left H^*-module, and

$$(*) \qquad h \cdot (f \cdot m) = \sum (h_1 \rightharpoonup f_2) \cdot ((h_2 \leftharpoonup f_1) \cdot m)$$

for all $h \in H, f \in H^*$, and $m \in M$; this follows from the definition of $(1 \bowtie h)(f \bowtie 1)$ in 10.3.5.

As in 1.6.4, M is a left H^*-module $\Leftrightarrow$ it is a right H-comodule, and if $m \mapsto \sum m_0 \otimes m_1$ is the comodule map, then $f \cdot m = \sum \langle f, m_1 \rangle m_0$ for any $f \in H^*$. Using this relationship along with 10.3.9 and 10.3.2, we see that $(*)$

is equivalent to

$$\begin{aligned}\sum\langle f, m_1\rangle h\cdot m_0 &= \sum\langle f_1, (\bar{S}h_{23})h_{21}\rangle(h_1 \rightharpoonup f_2)\cdot(h_{22}\cdot m)\\ &= \sum\langle f_1, (\bar{S}h_{23})h_{21}\rangle\langle f_2, (ad_\ell\bar{S}h_1)(h_{22}\cdot m)_1\rangle(h_{22}\cdot m)_0\\ &= \sum\langle f, (\bar{S}h_{23})h_{21}(\bar{S}h_1)_1(h_{22}\cdot m)_1 S(\bar{S}h_1)_2\rangle(h_{22}\cdot m)_0\\ &= \sum\langle f, (\bar{S}h_5)h_3\bar{S}h_2(h_4\cdot m)_1 h_1\rangle(h_4\cdot m)_0\\ &= \sum\langle f, \bar{S}h_3(h_2\cdot m)_1 h_1\rangle(h_2\cdot m)_0.\end{aligned}$$

Since this equality holds for all $f \in H^*$, it is equivalent to

$$\sum h\cdot m_0 \otimes m_1 = \sum(h_2\cdot m)_0 \otimes \bar{S}h_3(h_2\cdot m)_1 h_1.$$

Equivalently, for all $k \in H$,

$$\sum h\cdot m_0 \otimes km_1 = \sum(h_2\cdot m)_0 \otimes k(\bar{S}h_3)(h_2\cdot m)_1 h_1.$$

Replacing $h\otimes k$ by Δh and simplifying, we obtain

$$\sum h_1\cdot m_0 \otimes h_2 m_1 = \sum(h_2\cdot m)_0 \otimes (h_2\cdot m)h_1.$$

This is precisely 10.6.12, the Yetter-Drinfeld condition in ${}_H\mathcal{YD}^H$. Thus $D(H)$-modules are Yetter-Drinfeld modules, and conversely. □

10.6.17 REMARK. The reader should compare the Yetter-Drinfeld condition with the somewhat different compatibility condition in the category ${}^H_H\mathcal{M}$ of left H-Hopf modules, as in 1.9.1: the left version of that condition says that

$$\sum h_1 m_{-1} \otimes h_2\cdot m_0 = \sum(h\cdot m)_{-1} \otimes (h\cdot m)_0.$$

Recall that in that case, H is a left H-Hopf module using Δ and left multiplication of H on itself. The fact that H becomes a Yetter-Drinfeld module by using the adjoint action rather than the regular representation is analogous to what we saw in constructing the Drinfeld double in §10.3: namely the coadjoint action $\rightharpoonup$ was used rather than $\rightharpoonup$, the transpose of the regular representation, which was used in constructing the Heisenberg double $\mathcal{H}(H) = H\#H^*$. The fact that ${}_H\mathcal{YD}^H$ can be identified with ${}_{D(H)}\mathcal{M}$ as in 10.6.16, is analogous to the fact, mentioned in 8.5.2, that ${}_H\mathcal{M}^H$ can be identified with ${}_{H\#H^*}\mathcal{M}$.

More recently, a connection has been made between Hopf modules and Yetter-Drinfeld modules in [Sb 93] by considering "two-sided" Hopf modules.

It is interesting to note that the Drinfeld-Yetter condition arose first in abstract Hopf algebras, and later independently in the context of braids and quantum groups.

Appendix

Some quantum groups

We give here the generators and relations of some important quantum groups. Some basic references for quantum groups are [Dr 86], [Mn], and [FRT]. A very readable survey for algebraists is the paper [Sm], and we follow the notation given there.

Assume that $0 \neq q \in \mathbf{k}$ (and that q is not a root of 1, where appropriate).

A.1 $U_q(s\ell(2))$

We first recall the relations for the classical Lie algebra $\mathbf{g} = s\ell(2)$: here $s\ell(2)$ has $\mathbf{k}$-basis $\{e, f, h\}$ with relations $[ef] = h, [he] = 2e, [hf] = -2f$. The universal enveloping algebra $U(s\ell(2))$ becomes a Hopf algebra by setting, for each $x \in \mathbf{g}$,

$$\Delta x = x \otimes 1 + 1 \otimes x, \quad Sx = -x, \text{ and } \varepsilon(x) = 0.$$

The q-analogue of the enveloping algebra of the Lie algebra $s\ell(2)$ is defined as follows:

$U_q(s\ell(2)) = \mathbf{k}\langle E, F, K, K^{-1}\rangle$ with relations

$$KE = q^2 EK, \quad KF = q^{-2}FK, \quad EF - FE = \frac{K^2 - K^{-2}}{q^2 - q^{-2}}.$$

There is a Hopf algebra structure on $U_q(s\ell(2))$ defined by

$$\Delta(E) = E \otimes K^{-1} + K \otimes E, \quad S(E) = -q^{-2}E, \quad \varepsilon(E) = 0$$

$$\Delta(F) = F \otimes K^{-1} + K \otimes F, \quad S(F) = -q^{2}F, \quad \varepsilon(F) = 0$$

$$\Delta(K) = K \otimes K \qquad S(K) = K^{-1} \qquad \varepsilon(K) = 1.$$

Both Δ and ε extend to algebra homomorphisms, and S extends to an algebra anti-homomorphism.

To understand the relationship of $U(s\ell(2))$ to $U_q(s\ell(2))$, one should think of $K = \exp(H)$ for some H; this can be made precise if we work over power series instead of the original base field. See [Sm, pp 143-144].

A.2 $U_q(\mathfrak{g})$

Let $\mathfrak{g}$ be a complex semisimple Lie algebra of rank n, with Cartan matrix $A = [a_{ij}]$. There are integers $d_1, \ldots, d_n \in \{1, 2, 3\}$ such that $d_i a_{ij} = d_j a_{ji}$; i.e. $[d_i a_{ij}]$ is a symmetric matrix.

Define $U_q(\mathfrak{g}) = \mathbf{k}\langle E_i, K_i, K_i^{-1}, F_i \mid 1 \leq i \leq n\rangle$ with relations

$$K_i K_i^{-1} = K_i^{-1} K_i = 1 \quad K_i K_j - K_j K_i = 0$$

$$E_i F_j - F_j E_i = \delta_{ij} \left(\frac{K_i^2 - K_i^{-2}}{q^{2d_i} - q^{-2d_i}} \right)$$

$$K_i E_j K_i^{-1} = q^{d_i a_{ij}} E_j \quad K_i F_j K_i^{-1} = q^{-d_i a_{ij}} F_j$$

$$\sum_{\nu=0}^{1-a_{ij}} (-1)^\nu \begin{bmatrix} 1 - a_{ij} \\ \nu \end{bmatrix}_{q^{2d_i}} E_i^{1-a_{ij}-\nu} E_j E_i^\nu = 0 \qquad \text{for all } i \neq j$$

$$\sum_{\nu=0}^{1-a_{ij}} (-1)^\nu \begin{bmatrix} 1 - a_{ij} \\ \nu \end{bmatrix}_{q^{2d_i}} F_i^{1-a_{ij}-\nu} F_j F_i^\nu = 0 \qquad \text{for all } i \neq j.$$

These last two formulas, known as the *Serre relations*, involve the t-binomial coefficient $\begin{bmatrix} m \\ n \end{bmatrix}_t$ for $t = q^{2d_i}$. For t an indeterminate, it is given as follows:

$$\begin{bmatrix} m \\ n \end{bmatrix}_t = \frac{(t^m - t^{-m})(t^{(m-1)} - t^{-(m-1)}) \cdots (t^{(m-n+1)} - t^{-(m-n+1)})}{(t - t^{-1})(t^2 - t^{-2}) \cdots (t^n - t^{-n})}$$

The Hopf algebra structure on $U_q(\mathfrak{g})$ is defined by

$$\Delta(E_i) = E_i \otimes K_i^{-1} + K_i \otimes E_i, \quad S(E_i) = -q^{-2d_i} E_i, \quad \varepsilon(E_i) = 0,$$

$$\Delta(F_i) = F_i \otimes K_i^{-1} + K_i \otimes F_i, \quad S(F_i) = -q^{2d_i} F_i, \quad \varepsilon(F_i) = 0,$$

$$\Delta(K_i) = K_i \otimes K_i \qquad S(K_i) = K_i^{-1}, \qquad \varepsilon(K_i) = 1.$$

Both Δ and ε extend to algebra homomorphisms, and S extends to an algebra anti-homomorphism.

A.3 A quantum analog of the Heisenberg Lie algebra

Just as the classical Heisenberg algebra is the Lie subalgebra $\mathcal{H}$ of $s\ell(3)$ generated by $\{e_1, e_2\}$, we might try to define $U_q(\mathcal{H})$ to be the subalgebra of

$U_q(s\ell(3))$ generated by $\{E_1, E_2\}$. However to get a quantum group we also need to add some group-like elements. Thus we define

$$U_q(\mathcal{H}) = \mathrm{k}\langle E_1, E_2, K_1, K_2, K_1^{-1}, K_2^{-1}\rangle,$$

where the E_i and K_i satisfy the same relations as in $U_q(s\ell(3))$. Note that in this case, since $s\ell(3)$ has Cartan matrix $A = \begin{bmatrix} 2 & -1 \\ -1 & 2 \end{bmatrix}$ the Serre relations become

$$E_i^2 E_j + E_j E_i^2 = (q^2 + q^{-2}) E_i E_j E_i$$

for $i, j \in \{1, 2\}, i \neq j$. Moreover $U_q(\mathcal{H})$ is a subHopfalgebra of $U_q(s\ell(3))$.

This construction seems to be the right one from the point of view of physics; see [KSm] for a discussion (though the relations given there for $U_q(\mathcal{H})$ look a bit different from those above, in fact they are equivalent).

A.4 The coordinate ring of quantum 2×2 matrices

First recall the definition of $B = \mathcal{O}(M_n(\mathrm{k}))$ from 1.3.8. The coordinate ring of quantum $M_2(\mathbf{k})$ is defined to be:

$\mathcal{O}_q(M_2(\mathbf{k})) = \mathrm{k}\langle a, b, c, d\rangle$ subject to the relations

$$ba = q^{-2}ab \qquad ca = q^{-2}ac \qquad bc = cb$$

$$db = q^{-2}bd \qquad dc = q^{-2}cd \qquad ad - da = (q^2 - q^{-2})bc$$

Define a comultiplication and co-unit on $\mathcal{O}_q(M_2(\mathrm{k}))$ by

$$\Delta(a) = a \otimes a + b \otimes c, \quad \Delta(b) = a \otimes b + b \otimes d,$$

$$\Delta(c) = c \otimes a + d \otimes c, \quad \Delta(d) = c \otimes b + d \otimes d$$

$$\varepsilon(a) = \varepsilon(d) = 1, \quad \varepsilon(b) = \varepsilon(c) = 0.$$

Then $\mathcal{O}_q(M_2(\mathrm{k}))$ becomes a bialgebra. If we write

$$X = \begin{bmatrix} a & b \\ c & d \end{bmatrix} = \begin{bmatrix} X_{11} & X_{12} \\ X_{21} & X_{22} \end{bmatrix},$$

the coproduct and co-unit are given by $\Delta(X_{ij}) = \sum_k X_{ik} \otimes X_{kj}$, and $\varepsilon(X_{ij}) = \delta_{ij}$, as for $\mathcal{O}(M_2(\mathbf{k}))$. Define $\det_q X = ad - q^2 bc$. Then one may check that

$\det_q X$ is a group-like element which is not invertible in $\mathcal{O}_q(M_2(\mathbf{k}))$; thus $\mathcal{O}_q(M_2(\mathbf{k}))$ is not a Hopf algebra.

Observe that when $q = 1$, we obtain $\mathcal{O}(M_2(\mathbf{k}))$.

A.5 Quantum matrices, quantum $SL_n(\mathbf{k})$ and quantum $GL_n(\mathbf{k})$

Define $\mathcal{O}_q(M_n(\mathbf{k})) = \mathbf{k}[X_{ij} \mid 1 \le i, j \le n]$ subject to the relations that if $i < j$ and $k < m$, the map $\mathcal{O}_q(M_2(\mathbf{k})) \to \mathbf{k}\langle X_{ik}, X_{im}, X_{jk}, X_{jm}\rangle$ given by

$$\begin{bmatrix} a & b \\ c & d \end{bmatrix} \longrightarrow \begin{bmatrix} X_{ik} & X_{im} \\ X_{jk} & X_{jm} \end{bmatrix}$$

is a ring isomorphism, where a, b, c, d are as in A.4.

Write S_n for the symmetric group on n letters, and write $\ell(\sigma)$ for the Coxeter length of $\sigma \in S_n$; thus $\ell(\sigma)$ is the minimal number of terms required to express σ as a product of the simple transpositions $(i, i+1)$. The *quantum determinant* is the element

$$\det_q X = \sum_{\sigma \in S_n} (-q^2)^{\ell(\sigma)} X_{1,\sigma 1} X_{2,\sigma 2} \dots X_{n,\sigma n}.$$

where $X = [X_{ij}]$; when $n = 2$ this gives $\det_q X$ as in A.4.

Now $\det_q X$ is a central element of $\mathcal{O}_q(M_n(\mathbf{k}))$, and in fact the center of $\mathcal{O}_q(M_n(\mathbf{k}))$ is precisely $\mathbf{k}[\det_q X]$ if q is not a root of 1. It can be shown that $\Delta(\det_q X) = (\det_q X) \otimes (\det_q X)$ (see [Mn] for a nice treatment of this) and thus as for $n = 2, \mathcal{O}_q(M_n(\mathbf{k}))$ is not a Hopf algebra. It is a bialgebra as before, with $\Delta X_{ij} = \sum_k X_{ik} \otimes X_{kj}$ amd $\varepsilon(X_{ij}) = \delta_{ij}$.

We can then define, as in 1.5.7:

a) $\mathcal{O}_q(SL_n(\mathbf{k})) = \mathcal{O}_q(M_n(\mathbf{k}))/(\det_q X - 1)$, and

b) $\mathcal{O}_q(GL_n(\mathbf{k})) = \mathcal{O}_q(M_n(\mathbf{k}))[(\det_q X)^{-1}]$.

For $q = 1$, the constructions give $\mathcal{O}(SL_n(\mathbf{k}))$ and $\mathcal{O}(GL_n(\mathbf{k}))$, respectively.

$\mathcal{O}_q(SL_n(\mathbf{k}))$ and $\mathcal{O}_q(GL_n(\mathbf{k}))$ are bialgebras, with the bialgebra structure inherited from $\mathcal{O}_q(M_n(\mathbf{k}))$. To see that they are also Hopf algebras, we need to describe the antipode S. It is uniquely determined by the requirement that $X(SX) = (SX)X = I_n$; to describe it explicitly we define the quantum analog of the adjoint matrix. Thus if $X = [X_{k\ell}]$ as above, let Y denote the

$(n-1)\times(n-1)$ generic matrix obtained by deleting the ith row and j^{th} column of X, and set

$$A_{ji} = \det_q Y \in \mathcal{O}_q(M_{n-1}(\mathrm{k})).$$

It follows that

$$SX_{ij} = (\det_q X)^{-1}(-q^2)^{i-j}A_{ji}.$$

When $n=2$ the antipode takes a fairly simple form. That is, in the notation of A.4,

$$S\begin{bmatrix} a & b \\ c & d \end{bmatrix} = (ad-q^2bc)^{-1}\begin{bmatrix} d & -q^{-2}b \\ -q^2c & a \end{bmatrix}$$

That is, $Sa = (ad-q^2bc)^{-1}d$, etc.

A.6 The R-matrix approach to coordinate rings of quantum matrices

This approach is due to [FRT], and enables one to construct other quantum groups, such as $\mathcal{O}_q(SO_n(\mathrm{k}))$ and $\mathcal{O}_q(SP_{2n}(\mathrm{k}))$, in addition to $\mathcal{O}_q(SL_n(\mathrm{k}))$ and $\mathcal{O}_q(GL_n(\mathrm{k}))$ which we have already seen.

Let $\mathrm{k}\langle t_{ij} \mid 1 \le i,j \le n\rangle$ be the free algebra on the indeterminates t_{ij} and let $T=[t_{ij}]$ be the corresponding "generic matrix". Let $R \in M_n(\mathrm{k})\otimes M_n(\mathrm{k})$; we use R to define an ideal I_R of relations in $\mathrm{k}\langle t_{ij}\rangle$, as follows: let $T_1 = T\otimes I_n$ and $T_2 = I_n \otimes T$ be in $M_n(\mathrm{k}\langle t_{ij}\rangle)^{\otimes 2}$. Then I_R is generated by the entries in the $n^2\times n^2$ matrix

$$RT_1T_2 - T_2T_1R.$$

Define the coordinate ring of the quantum matrix algebra associated to R to be

$$\mathcal{O}_R(M_n(\mathrm{k})) = \mathrm{k}\langle t_{ij}\rangle / I_R.$$

There is a bialgebra structure on $\mathcal{O}_R(M_n(\mathrm{k}))$ determined by setting $\Delta t_{ij} = \sum_k t_{ik}\otimes t_{kj}$ and $\varepsilon(t_{ij}) = \delta_{ij}$.

To give an explicit example, we need to fix some notation. If A and B are $n\times n$ matrices, we view $A\otimes B$ as an $n^2\times n^2$ matrix as follows: consider an $n^2\times n^2$ matrix as an $n\times n$ matrix of $n\times n$ blocks. Then $A\otimes B$ is the $n^2\times n^2$ matrix with $a_{ij}b_{k\ell}$ as the $k\ell^{th}$ entry in the ij^{th} block. Thus T_1 has as its $k\ell^{th}$ block the diagonal matrix $t_{k\ell}I_n$, and T_2 has a copy of T in its kk^{th} block and zero in the off-diagonal blocks.

If $\{e_1, \cdots, e_n\}$ is a basis of $\mathbf{k}^{(n)}$, we write $R_{ij}^{k\ell}$ for the coefficient of $e_i \otimes e_j$ in $R(e_k \otimes e_\ell)$. If we order the basis of $\mathbf{k}^{(n)} \otimes \mathbf{k}^{(n)}$ by

$$\{e_1 \otimes e_1, e_1 \otimes e_2, \ldots, e_1 \otimes e_n, e_2 \otimes e_1, \ldots, e_n \otimes e_{n-1}, e_n \otimes e_n\}$$

then $R_{ij}^{k\ell}$ is the entry in the $j\ell^{th}$ position of the ik^{th} block. Thus the defining relations in I_R are

$$\sum_{k,\ell} R_{ij}^{k\ell} t_{km} t_{\ell p} = \sum_{k,\ell} t_{j\ell} t_{ik} R_{k\ell}^{mp}$$

for all i, j, m, p.

For example, if $T = \begin{bmatrix} a & b \\ c & d \end{bmatrix}$ and $R = \begin{bmatrix} q^2 & 0 & 0 & 0 \\ 0 & 1 & 0 & 0 \\ 0 & q^2 - q^{-2} & 1 & 0 \\ 0 & 0 & 0 & q^2 \end{bmatrix}$ we recover the six defining relations in A.4. Thus in this case

$$\mathcal{O}_R(M_2(\mathbf{k})) \cong \mathcal{O}_q(M_2(\mathbf{k})).$$

References

[A] E. Abe, Hopf Algebras, Cambridge University Press, Cambridge, 1980 (original Japanese version published by Iwanami Shoten, Tokyo, 1977).

[Am] S.A. Amitsur, Rings of quotients and Morita contexts, J. Algebra 17 (1971), 273-298.

[AST] M. Artin, W. Schelter, and J. Tate, Quantum deformations of GL_n, Comm. Pure and Appl. Math 44 (1991), 879-895.

[Ba 88] M. Beattie, A generalization of the smash product of a graded ring, J. Pure and Applied Algebra 52 (1988), 219-226.

[Ba 92a] M. Beattie, On the Blattner-Montgomery duality theorem for Hopf algebras, AMS Contemporary Math 124 (1992), 23-28.

[Ba 92b] M. Beattie, Strongly inner actions, coactions, and duality theorems, Tsukuba J. Math 16 (1992), 279-295.

[BaU] M. Beattie and K.H. Ulbrich, A Skolem-Noether theorem for Hopf algebra actions, Comm. Algebra 18 (1990), 3713-3724.

[BeC] J. Bergen and M. Cohen, Actions of commutative Hopf algebras, Bull. LMS 18 (1986), 159-164.

[BeM 86] J. Bergen and S. Montgomery, Smash products and outer derivations, Israel J. Math 53 (1986), 321-345.

[BeM 92] J. Bergen and S. Montgomery, Ideals and quotients in crossed products of Hopf algebras, J. Algebra 151 (1992), 374–396.

[BeM 93] J. Bergen and S. Montgomery, eds, Advances in Hopf Algebras, Marcel Dekker, to appear.

[Bi] J. Birman, New points of view in knot theory, Bull. AMS 28 (1993), 253-287.

[B] R.J. Blattner, UCLA Lecture Notes, 1987 (unpublished).

[BCM] R.J. Blattner, M. Cohen, and S. Montgomery, Crossed products and inner actions of Hopf algebras, Trans. AMS 298 (1986), 671-711.

[BM 85] R.J. Blattner and S. Montgomery, A duality theorem for Hopf module algebras, J. Algebra 95 (1985), 153-172.

[BM 89] R.J. Blattner and S. Montgomery, Crossed products and Galois extensions of Hopf algebras, Pacific J. Math 137 (1989), 37-54.

[BGR] W. Borho, P. Gabriel, and R. Rentschler, Primideale in Einhüllenden auflösbarer Lie-Algebren, Lecture Notes in Math 357, Springer, Berlin, 1974.

[Ca] P. Cartier, Groupes algébriques et groupes formels, in Coll. sur la théorie des groupes algébriques, Bruxelles, CBRM, 1962.

[CHR] S.U. Chase, D.K. Harrison, and Alex Rosenberg, Galois theory and cohomology of commutative rings, AMS Memoirs No. 52, 1965.

[CS] S.U. Chase and M.E. Sweedler, Hopf Algebras and Galois Theory, Lecture Notes in Math 97, Springer, Berlin, 1969.

[CnN] C-Y. Chen and W. Nichols, A duality theorem for Hopf module algebras over Dedekind rings, Comm. in Algebra 18 (1990), 3209-3221.

[Che] C. Chevalley, Theory of Lie Groups, Princeton University Press, Princeton, 1946.

[Cl] L. Childs, On the Hopf Galois theory for separable field extensions, Comm. Alg. 17 (1989), 809-825.

[Ch 87] W. Chin, Prime ideals in differential operator rings and crossed products of infinite groups, J. Algebra 106 (1987), 78-104.

[Ch 92] W. Chin, Crossed products of semisimple cocommutative Hopf algebras, Proc. AMS 116 (1992), 321-327.

[ChM] W. Chin and I. Musson, Hopf algebra duality, injective modules, and quantum groups, preprint.

[ChQ] W. Chin and D. Quinn, Rings graded by polycyclic-by-finite groups, Proc. AMS 102 (1988), 235-241.

[CPS] E. Cline, B. Parshall, and L. Scott, Induced modules and affine quotients, Math. Ann. 230 (1977), 1-14.

[C 86] M. Cohen, Smash products, inner actions, and quotient rings, Pacific J. Math 125 (1986), 46-65.

[CF 86] M. Cohen and D. Fischman, Hopf algebra actions, J. Algebra 100 (1986), 363-379.

[CF 92] M. Cohen and D. Fischman, Semisimple extensions and elements of trace 1, J. Algebra 149 (1992), 419-437.

[CFM] M. Cohen, D. Fischman, and S. Montgomery, Hopf Galois extensions, smash products, and Morita equivalence, J. Algebra 133 (1990), 351-372.

[CM] M. Cohen and S. Montgomery, Group-graded rings, smash products, and group actions, Trans. AMS 282 (1984), 237-258.

[CR] M. Cohen and L. Rowen, Group graded rings, Comm. in Algebra 11 (1983), 1253-1270.

[CW] M. Cohen and S. Westreich, From supersymmetry to quantum commutativity, J. Algebra, to appear.

[CuR] C.W. Curtis and I. Reiner, Representation Theory of Finite Groups and Associative Algebras, Interscience, New York, 1962.

[Da] E.C. Dade, Group-graded rings and modules, Math Z. 174 (1980), 241-262.

[DNRvO] S. Dascalescu, C. Nastasescu, S. Raianu, F. van Oystaeyen, Graded coalgebras and Morita-Takeuchi contexts, to appear.

[DM] P. Deligne and J.S. Milne, Tannakien Categories, Lecture Notes in Math 900, Springer, Berlin, 1982.

[DGa] M. Demazure and P. Gabriel, Groupes Algébriques I, North Holland, Amsterdam, 1970.

[DG] M. Demazure, A. Grothendieck, et al, SGA 3: Schémas en Groupes, 1962/64, Lecture Notes in Math 151, 152, 153, Springer, Berlin, 1970.

[Di] J. Dieudonné, Introduction to the Theory of Formal Groups, Marcel Dekker, New York, 1973.

[D 83] Y. Doi, On the structure of relative Hopf modules, Comm. Algebra 11 (1983), 243-255.

[D 85] Y. Doi, Algebras with total integrals, Comm. Alg. 13 (1985), 2137-2159.

[D 89] Y. Doi, Equivalent crossed products for a Hopf algebra, Comm. Alg. 17 (1989), 3053-3085.

[D 90] Y. Doi, Hopf extensions of algebras and Maschke-type Theorems, Israel J. Math 72 (1990), 99-108.

[D 93] Y. Doi, Braided bialgebras and quadratic bialgebras, Comm. Alg. 21 (1993), 1731-1750.

[DT 86] Y. Doi and M. Takeuchi, Cleft comodule algebras for a bialgebra, Comm. Alg. 14 (1986), 801-818.

[DT 89] Y. Doi and M. Takeuchi, Hopf-Galois extensions of algebras, the Miyashita-Ulbrich action, and Azumaya algebras, J. Algebra 121 (1989), 488-516.

[Dr 86] V.G. Drinfeld, Quantum groups, Proc. Int. Cong. Math, Berkeley, 1 (1986), 789-820.

[Dr 89a] V.G. Drinfeld, On almost cocommutative Hopf algebras, Leningrad Math J. 1 (1990), 321-342 (Russian original in Algebra and Analysis, 1989).

[Dr 89b] V.G. Drinfeld, Quasi-Hopf algebras, Leningrad Math J. 1 (1990), 1419-1457 (Russian original in Algebra and Analysis, 1989).

[FRT] L.D. Faddeev, N. Yu. Reshetikhin, L.A. Takhtadzhyan, Quantization of Lie groups and Lie algebras, Leningrad Math J. 1 (1990), 193-225 (Russian orinial in Algebra and Analysis, 1989; see also LOMI preprint E 14-87).

[F-S] W. Ferrer-Santos, Finite generation of the invariants of finite-dimensional Hopf algebras, J. Algebra, to appear.

[Fi] D. Fischman, Schur's double centralizer theorem: a Hopf algebra approach, J. Algebra, to appear.

[FiM] D. Fischman and S. Montgomery, A Schur double centralizer theorem for cotriangular Hopf algebras and generalized Lie algebras, J. Algebra to appear.

[FM] J.W. Fisher and S. Montgomery, Semiprime skew group rings, J. Algebra 52 (1978), 241-247.

[Ga] P. Gabriel, Groupes Formels, Exp. VII B, SGA 3, Schémas en Groupes, Lecture Notes in Math 151, Springer, Berlin, 1970.

[Gr] J.A. Green, Locally finite representations, J. Algebra 41 (1976), 137-171.

[GPa] C. Greither and B. Pareigis, Hopf Galois theory for separable field extensions, J. Algebra 106 (1987), 239-258.

[Gu] D.I. Gurevich, Generalized translation operators in Lie groups, Soviet J. Contemp. Math. Anal. 18 (1983), 57-70.

[Ha] T. Hayashi, Quantum groups and quantum determinants, J. Algebra 152 (1992), 146-165.

[H 54] G. Hochschild, Representations of restricted Lie Algebras of characteristic p, Proc. AMS 5 (1954), 603-605.

[H 65] G. Hochschild, The Structure of Lie Groups, Holden-Day, San Francisco, 1965.

[H 81] G. Hochschild, Basic Theory of Algebraic Groups and Lie Algebras, Springer-Verlag, Berlin, 1981.

[HR] R.G. Heyneman and D.E. Radford, Reflexivity and coalgebras of finite type, J. Algebra (1974), 215-246.

[HS] R.G. Heyneman and M.E. Sweedler, Affine Hopf algebras I, J. Algebra 13 (1969), 192-241.

[J] N. Jacobson, Lie Algebras, Interscience, New York, 1962.

[Ji 85] M. Jimbo, A q-difference analog of $U(\mathrm{g})$ and the Yang-Baxter equation, Lett. Mat. Phys. 10 (1985), 63-69.

[Ji 86] M. Jimbo, A q-analog of $U(g\ell(n+1))$, Hecke algebra and the Yang-Baxter equation, Lett. Mat. Phys. 11 (1986), 247-252.

[JR] S.A. Joni and G.C. Rota, Coalgebras and bialgebras in combinatorics, AMS Contemporary Math 6 (1982), 1-47.

[JL] A. Joseph amd G. Letzter, Local finiteness of the adjoint action for quantized enveloping algebras, J. Algebra 153 (1992), 289-318.

[JS 86] A. Joyal and R. Street, Braided monoidal categories, Macquarie Univ. Reports 860081, 1986.

[JS 91] A. Joyal and R. Street, An introduction to Tannaka duality and quantum groups, in Category Theory, Lecture Notes in Math. 1488, Springer, 1991, pp. 413-492.

[JS 93] A. Joyal and R. Street, Braided tensor categories, Advances in Math, to appear.

[K] I. Kaplansky, Bialgebras, University of Chicago Lecture Notes, 1975.

[KiS] E. Kirkman and L.W. Small, A q-analog of the harmonic oscillator, Israel J. Math., to appear.

[Ko 91] M. Koppinen, A Skolem-Noether theorem for coalgebra measurings, Arch. Math 57 (1991), 34-40.

[Ko 92] M. Koppinen, A duality theorem for crossed products of Hopf algebras, J. Algebra 146 (1992), 153-174.

[Ko 93] M. Koppinen, Coideal subalgebras in Hopf algebras: freeness, integrals, smash products, Comm. in Algebra 21 (1993), 427-444.

[KT] H.F. Kreimer and M. Takeuchi, Hopf algebras and Galois extensions of an algebra, Indiana Univ. Math J. 30 (1981), 675-692.

[L 67] R.G. Larson, Cocommutative Hopf algebras, Canadian J. Math 19 (1967), 350-360.

[L 71] R.G. Larson, Characters of Hopf algebras, J. Algebra 17 (1971), 352-368.

[LR 87] R.G. Larson and D.E. Radford, Semisimple cosemisimple Hopf algebras, Amer. J. Math 109 (1987), 187-195.

[LR 88] R.G. Larson and D.E. Radford, Finite-dimensional cosemisimple Hopf algebras in characteristic 0 are semisimple, J. Algebra 117 (1988), 267-289.

[LS] R. G. Larson and M. Sweedler, An associative orthogonal bilinear form for Hopf algebras, Amer. J. Math 91 (1969), 75-93.

[LTa] R.G. Larson and E.J. Taft, The algebraic structure of linearly recursive sequences under Hadamard product, Israel J. Math. 72 (1990), 118-132.

[LT] R.G. Larson and J. Towber, Two dual classes of bialgebras related to the concepts of "quantum group" and "quantum Lie algebra", Comm. Algebra 19 (1991), 3295-3345.

[LoP] M. Lorenz and D.S. Passman, Two applications of Maschke's theorem, Comm. Alg. 8 (1980), 1853-1866.

[Lu] J-H Lu, On the Drinfeld double and the Heisenberg double of a Hopf algebra, to appear.

[Ly] V.V. Lyubashenko, Hopf algebras and vector symmetries, Russian Math Surveys 41 (1986), 153-154.

[Mac] S. MacLane, Categories for the Working Mathematician, Springer, New York, 1971.

[Mj 90a] S. Majid, Physics for algebraists: non-commutative and noncocommutative Hopf algebras by a bicrossproduct construction, J. Algebra 130 (1990), 17-64.

[Mj 90b] S. Majid, Quasitriangular Hopf algebras and Yang-Baxter equations, Int. J. Modern Physics A 5(1990), 1-91.

[Mj 90c] S. Majid, More examples of bicrossproduct and double cross product Hopf algebras, Israel J. Math. 72 (1990), 133-148.

[Mj 91a] S. Majid, Reconstruction theorems and rational conformal field theories, Int. J. Modern Physics A 6(1991), 4359-4374.

[Mj 91b] S. Majid, Quantum groups and quantum probability, in Quantum Probability and Related Topic VI (Proc. Trento 1989), L. Accordi et al, eds. World Science 1991.

[Mj 91c] S. Majid, Doubles of quasitriangular Hopf algebras, Comm. Alg. 19 (1991), 3061-3073.

[Mj 92] S. Majid, Tannaka-Krein theorem for quasi-Hopf algebras and other results, AMS Cont. Math 134 (1992), 219-232.

[Mj 93] S. Majid, Crossed products by braided groups and bosonization, J. Algebra, to appear.

[Mn] Yu. I. Manin, Quantum Groups and Non-commutative Geometry, Université de Montréal, Publ. CRM, 1988.

[Ml] T. Marlowe, The diagonal of a pointed coalgebra and incidence-like structure, J. Pure Appl. Algebra 35(1985), 157-169.

[MMN] T. Masuda, K. Mimachi, Y. Nakagami, M. Noumi, and K. Ueno, Representations of quantum groups and a q-analogue of orthogonal polynomials, C.R. Acad. Sci. Paris 307 (1988), 559-564.

[Ma 90a] A. Masuoka, Coalgebra actions on Azumaya algebras, Tsukuba J. Math 14 (1990), 107-112.

[Ma 90b] A. Masuoka, Existence of a unique maximal subcoalgebra whose action is inner, Israel J. Math 72 (1990), 149-157.

[Ma 91] A. Masuoka, On Hopf algebras with commutative coradical, J. Algebra 144 (1991), 451-466.

[Ma 92] A. Masuoka, Freeness of Hopf algebras over coideal subalgebras, Comm. in Algebra 20 (1992), 1353-1373.

[MaD] A. Masuoka and Y. Doi, Generalization of cleft comodule algebras, Comm. in Algebra 20 (1992), 3703-3721.

[MaW] A. Masuoka and D. Wigner, Faithful flatness of Hopf algebras, J. Algebra, to appear.

[Mc] J.C. McConnell, Representations of solvable Lie algebras and the Gelfand-Kirillov conjecture, Proc. LMS 29 (1974), 453-484.

[McR] J.C. McConnell and J.C. Robson Noncommutative Noetherian rings, Wiley-Interscience, New York, 1987.

[McS] J.C. McConnell and M.E. Sweedler, Simplicity of smash products, Proc. London Math. Soc. 23 (1971), 251-256.

[Mh] W. Michaelis, Properness of Lie algebras and enveloping algebras, Proc. AMS 101 (1987), 17-23.

[Mi] A. Milinski, Skolem-Noether theorems and coalgebra actions, to appear.

[MiMo] J.W. Milnor and J.C. Moore, On the structure of Hopf algebras, Annals of Math 81 (1965), 211-264.

[Mo] R.K. Molnar, Semi-direct products of Hopf algebras, J. Algebra 47 (1977), 29-51.

[M 80] S. Montgomery, Fixed Rings of Finite Automorphism Groups of Associative Rings, Lecture Notes in Math 818, Springer, Berlin, 1980.

[M 85] S. Montgomery, Duality for actions and coactions of groups, AMS Contempory Math 43 (1985), 191-207.

[M 88] S. Montgomery, Crossed products of Hopf algebras and enveloping algebras, in Perspectives in Ring Theory, Kluwer 1988, 253-268.

[M 92] S. Montgomery, Hopf Galois extensions, AMS Contemporary Math 124 (1992), 129-140.

[M 93a] S. Montgomery, Biinvertible actions of Hopf algebras, Israel J. Math 83 (1993), 45-72.

[M 93b] S. Montgomery, Some remarks on filtrations of Hopf algebras, Comm. Alg. 21 (1993), 999-1007.

[MSch] S. Montgomery and H.-J. Schneider, in preparation.

[MS 81] S. Montgomery and L.W. Small, Fixed rings of Noetherian rings, Bull. LMS 13 (1981), 33-38.

[MS 86] S. Montgomery and L.W. Small, Some remarks on affine rings, Proc. AMS 98 (1986), 537-544.

[MSm] S. Montgomery and S.P. Smith, Skew derivations and $U_q(sl(2))$, Israel J. Math 72 (1990), 158-166.

[Mr] R. Morris (editor), Umbral Calculus and Hopf algebras, Contemporary Math 6, AMS, Providence, 1982.

[MuF] D. Mumford and J. Fogarty, Geometric Invariant Theory, 2nd edition, Springer-Verlag, Berlin, 1982.

[Na] M. Nagata, Complete reducibility of rational representations of a matric group, J. Math Kyoto Univ. 1 (1961), 89-99.

[NaT] Y. Nakagami and M. Takesaki, Duality for crossed products of von Neumann algebras, Lecture Notes in Math 731, Springer, Berlin, 1979.

[NvO] C. Nastasescu and F. van Oystaeyen, Graded Ring Theory, North-Holland, Amsterdam, 1982.

[Ne] K. Newman, A correspondence between bi-ideals and subHopfalgebras in cocommutative Hopf algebras, J. Algebra 36 (1975), 1-5.

[NZ 89] W. D. Nichols and M. B. Zoeller, A Hopf algebra freeness theorem, Amer. J. Math 111 (1989), 381-385.

[NZ 92] W. D. Nichols and M. B. Richmond (Zoeller), Freeness of infinite dimensional Hopf algebras, Comm. Algebra 20 (1992), 1489-1492.

[O] U. Oberst, Affine Quotienten schemata nach affinen algebraischen Gruppen und induzierte Darstellungen, J. Algebra 44 (1977), 502-538.

[OSch 73] U. Oberst and H.J. Schneider, Über Untergruppen endlicher algebraischer Gruppen, Manuscripta Math. 8 (1973), 217-241.

[OSch 74] U. Oberst and H. J. Schneider, Untergruppen formeller Gruppen von endlichen Index, J. Algebra 31 (1974), 10-44.

[OQ] J. Osterburg and D. Quinn, A Noether-Skolem theorem for group-graded rings, J. Algebra 113 (1988), 483-490; addendum, J. Algebra 120 (1989), 414-415.

[Pa 77] B. Pareigis, Non-additive ring and module theory I: General theory of monoids, Publ. Math. Debrecen 24(1977), 189-204; II: $\mathcal{C}$-categories, $\mathcal{C}$-functors, and $\mathcal{C}$-morphisms, ibid, 351-361.

[Pa 81] B. Pareigis, A non-commutative, non-cocommutative Hopf algebra in "nature", J. Algebra 70 (1981), 356-374.

[Pa 90] B. Pareigis, Forms of Hopf algebras and Galois theory, Topics in Algebra, Vol. 26, Part 1, pp. 75-93, Banach Center Publications, Warsaw, 1990.

[Ps] I.B.S. Passi, Group rings and their augmentation ideals, Lecture Notes in Math 715, Springer, Berlin, 1979.

[P] D.S. Passman, Infinite Crossed Products, Academic Press, New York, 1989.

[PQ] D.S. Passman and D. Quinn, Burnside's theorem for Hopf algebras, Proc. AMS, to appear.

[PT] B. Peterson and E. Taft, The Hopf algebra of linearly recursive sequences, Aequationes Math. 20 (1980), 1-17.

[Q 85] D. Quinn, Group-graded rings and duality, Trans. AMS 292 (1985), 155-167.

[Q 89] D. Quinn, Integrality over fixed rings, J. London Math. Soc. 40 (1989), 206-214.

[Q 91] D. Quinn, Integral extensions of non-commutative rings, Israel J. Math 73 (1991), 113-121.

[R 76] D.E. Radford, The order of the antipode of a finite-dimensional Hopf algebra is finite, Amer. J. Math 98 (1976), 333-355.

[R 77] D. E. Radford, Freeness (projectivity) criteria for Hopf algebras over Hopf subalgebras, J. Pure Appl. Algebra 11 (1977), 15-28.

[R 78] D.E. Radford, On the structure of commutative pointed Hopf algebras, J. Algebra 50 (1978), 284-296.

[R 82] D.E. Radford, On the structure of pointed coalgebras, J. Algebra 77 (1982), 1-14.

[R 85] D.E. Radford, The structure of Hopf algebras with a projection, J. Algebra 92 (1985), 322-347.

[R 91] D.E. Radford, On the quasi-triangular structures of semisimple Hopf algebras, J. Algebra 141 (1991), 354-358.

[R 92] D.E. Radford, On the antipode of a quasi-triangular Hopf algebra, J. Algebra 151 (1992), 1-11.

[R 93a] D.E. Radford, Minimal quasi-triangular Hopf algebras, J. Algebra 157 (1993), 285-315.

[R 93b] D.E. Radford, Irreducible representations of $U_q(\mathbf{g})$ arising from $Mod^{\bullet}_{C^{\frac{1}{2}}}$, to appear.

[R 93c] D.E. Radford, The trace function and Hopf algebras, J. Algebra, to appear.

[RTW] D.E. Radford, E.J. Taft, and R.L. Wilson, Forms of certain Hopf algebras, Manuscripta Math. 17 (1975), 333-338.

[RT] D.E. Radford and J. Towber, Yetter-Drinfeld categories associated to an arbitrary bialgebra, J. Pure Appl. Alg. 1993, to appear.

[ReS] N. Reshetikin and M. Semenov-Tian-Shansky, Quantum R-matrices and factorization problems, J. Geom. Phys. 5 (1988), 533.

[S-R] N. Saavedra-Rivano, Catégories Tannakiennes, Lecture Notes in Amth 265, Springer, Berlin, 1972.

[Sb 92] P. Schauenburg, On Coquasitringular Hopf Algebras and the Quantum Yang-Baxter Equation, Algebra Berichte 67, Verlag Reinhard Fischer, 1992.

[Sb 93] P. Schauenburg, Hopf modules and Yetter-Drinfeld modules, J. Algebra, to appear.

[Sc] M. Scheunert, Generalized Lie algebras, J. Math. Phys. 20 (1979), 712-720.

[Sch 90a] H.J. Schneider, Principal homogeneous spaces for arbitrary Hopf algebras, Israel J. Math 72 (1990), 167-195.

[Sch 90b] H.J. Schneider, Representation theory of Hopf Galois extensions, Israel J. Math. 72 (1990), 196-231.

[Sch 92] H. J. Schneider, Normal basis and transitivity of crossed products for Hopf algebras, J. Algebra 152 (1992), 289-312.

[Sch 93a] H.J. Schneider, On inner actions of Hopf algebras and stabilizers of representations, J. Algebra (1993), to appear.

[Sch 93b] H.J. Schneider, Some remarks on exact sequences of quantum groups, to appear

[Sm] S.P. Smith, Quantum groups: an introduction and survey for ring theorists, in Noncommutative Rings, MSRI Publ. 24, Springer, 1992, pp.131-178.

[St] J. Stasheff, Drinfeld's quasi-Hopf algebras and beyond, AMS Cont. Math 134 (1992), 297-307.

[S 67] M.E. Sweedler, Hopf algebras with one group-like element, Trans. AMS 127 (1967), 515-526; correction, Trans. AMS 154 (1971), 427-428.

[S 68] M.E. Sweedler, Cohomology of algebras over Hopf algebras, Trans. AMS 127 (1968), 205-239.

[S] M.E. Sweedler, Hopf Algebras, Benjamin, New York, 1969.

[S 71] M.E. Sweedler, Connected fully reducible affine group schemes in positive characteristic are abelian, J. Math. Kyoto Univ. II (1971), 51-70.

[T 71] E. J. Taft, The order of the antipode of a finite-dimensional Hopf algebra, Proc. Nat. Acad. Sci. USA 68 (1971), 2631-2633.

[TW] E.J. Taft and R.L. Wilson, On antipodes in pointed Hopf algebras, J. Algebra 29 (1974), 27-32.

[Tk 71] M. Takeuchi, Free Hopf algebras generated by coalgebras, J. Math Soc. Japan 23 (1971), 561-582.

[Tk 72] M. Takeuchi, A correspondence between Hopf ideals and subHopfalgebras, Manuscripta Math 7 (1972), 251-270.

[Tk 79] M. Takeuchi, Relative Hopf modules - equivalence and freeness criteria, J. Algebra 60 (1979), 452-471.

[Tk 81] M. Takeuchi, Matched pairs of groups and bismash products of Hopf algebras, Comm. Alg. 9 (1981), 841-882.

[Tk 92a] M. Takeuchi, Some topics on $GL_q(n)$, J. Algebra 147 (1992), 379-410.

[Tk 92b] M. Takeuchi, Hopf algebra techniques applied to the quantum group $U_q(sl(2))$, AMS Contemporary Math 134 (1992), 309-323.

[Tn] T. Tannaka, Über den Dualitätssatz der nichtkommutativen topologischen Gruppen, Tôhoku J. Math 45 (1939), 1-12.

[U 81] K.H. Ulbrich, Vollgraduierte Algebren, Abh. Math. Sem. Univ. Hamburg 51 (1981), 136-148.

[U 82] K.H. Ulbrich, Galois erweiterungen von nicht-kommutativen ringen, Comm. Algebra 10 (1982), 655-672.

[U 90] K.H. Ulbrich, On Hopf algebras and rigid monoidal categories, Israel J. Math 72 (1990), 252-256.

[vdB] M. van den Bergh, A duality theorem for Hopf algebras, in Methods in Ring Theory, NATO ASI Series vol 129, Reidel, Dordrecht, 1984, 517-522.

[vO] F. van Oystaeyen, On Clifford systems and generalized crossed products, J. Algebra 87 (1984), 396-415.

[W] W. C. Waterhouse, Introduction to Affine Group Schemes, Springer-Verlag, Berlin, 1979.

[Wo 87] S.L. Woronowicz, Twisted $SU(2)$ group. An example of a non-commutative differential calculus, Publ. RIMS, Kyoto Univ. 23 (1987), 117-181.

[Wo 88] S.L. Woronowicz, Tannaka-Krein duality for compact matrix pseudogroups, Invent. Math 93 (1988), 35-76.

[Y 90] D.N. Yetter, Quantum groups and representations of monoidal categories, Math. Proc. Cambridge Phil. Soc. 108 (1990), 261-290.

[Y 92] D.N. Yetter, Framed tangles and a theorem of Deligne on braided deformations of Tannakian categories, AMS Cont. Math 134 (1992), 325-349.

[Z] J.J. Zhang, Twisted graded algebras and equivalences of graded categories, preprint.

[Zh] Y. Zhu, Quantum double construction of quasitriangular Hopf algebras and Kaplansky's conjecture, preprint.

Index

Other Titles in This Series

(*Continued from the front of this publication*)

(See the AMS catalog for earlier titles)